Reaction Mechanism and Properties of Cement-Based Materials

Reaction Mechanism and Properties of Cement-Based Materials

Weiting Xu

MDPI

Basel • Beijing • Wuhan • Barcelona • Belgrade • Novi Sad • Cluj • Manchester

Weiting Xu
School of Materials Science
and Engineering
South China University
of Technology
Guangzhou
China

Editorial Office
MDPI AG
Grosspeteranlage 5
4052 Basel, Switzerland

This is a reprint of articles from the Special Issue published online in the open access journal *Materials* (ISSN 1996-1944) (available at: www.mdpi.com/journal/materials/special_issues/UZT4YZQJYY).

For citation purposes, cite each article independently as indicated on the article page online and using the guide below:

Lastname, A.A.; Lastname, B.B. Article Title. *Journal Name* **Year**, *Volume Number*, Page Range.

ISBN 978-3-7258-2138-9 (Hbk)
ISBN 978-3-7258-2137-2 (PDF)
https://doi.org/10.3390/books978-3-7258-2137-2

Contents

About the Editor

Weiting Xu

Dr. Xu Weiting, editor of the Special Issue, received her PhD from the Department of Architecture and Civil Engineering at the City University of Hong Kong. She is currently an associate professor at the School of Materials Science and Engineering of South China University of Technology. Her research interests include high-performance cement-based materials, shrinkage reduction and toughening mechanism of concrete, prevention and control of concrete cracks, recycling of solid waste, organic–inorganic composite cementitious materials, and molecular dynamics simulation.

materials

Review

Chloride Penetration of Surface-Coated Concrete: Review and Outlook

Jing Liao [1], Yuchi Wang [2,*], Xiping Sun [2,*] and Yuanzhan Wang [1]

1 Simulation State Key Laboratory of Hydraulic Engineering Intelligent Construction and Operation, Tianjin University, 135 Yaguan Road, Jinnan District, Tianjin 300072, China; liaojing@tju.edu.cn (J.L.); yzwang@tju.edu.cn (Y.W.)
2 Tianjin Research Institute for Water Transport Engineering, M.O.T., 2618 Xingang 2nd Road, Binhai New District, Tianjin 300000, China
* Correspondence: yuchiwang@foxmail.com (Y.W.); xpsun@tju.edu.cn (X.S.)

Abstract: Concrete coatings show significant promise in shielding concrete substrates from corrosion by effectively resisting harmful ions and moisture. Thanks to their practicality, high efficiency, and cost-effectiveness, coatings are considered a potent technique for enhancing the chloride resistance of reinforced concrete structures. Over recent decades, extensive research has concentrated on employing coatings to bolster concrete's ability to withstand chloride penetration. This paper provides a holistic review of the current studies on chloride infiltration in concrete surfaces treated with coating materials, primarily focused on chloride resistance improvement efficiency and chloride transport modeling. Firstly, by comparing the functions of assorted coatings, four inherent protection mechanisms are summarized and elaborated thoroughly. Afterwards, the chloride resistance improvement efficiency of assorted coatings reported in current studies are reviewed and compared in great detail, with a specific focus on inorganic, organic, and organic–inorganic composite coatings. Furthermore, the theoretical research about methodologies for chloride transport behavior prediction is summarized. Finally, this paper outlines the potential research directions in this field and the theoretical, technical, and practical application challenges. This review not only identifies critical areas necessitating further investigation and problem-solving in this domain but also aids in selecting appropriate coating materials and refining corrosion management strategies.

Keywords: concrete coating; chloride resistance; improvement efficiency; silane coating; nanomaterial-modified coating; organic–inorganic composite coating

Citation: Liao, J.; Wang, Y.; Sun, X.; Wang, Y. Chloride Penetration of Surface-Coated Concrete: Review and Outlook. *Materials* **2024**, *17*, 4121. https://doi.org/10.3390/ma17164121

Academic Editor: Carlos Leiva

Received: 25 July 2024
Revised: 12 August 2024
Accepted: 15 August 2024
Published: 20 August 2024

1. Introduction

Concrete, a bulk construction material, plays an important role in modern infrastructure due to its prominent merits in terms of availability, moldability, compressive strength, durability, and cost [1]. Nevertheless, reinforced concrete (RC) structures in the marine environment may experience durability failure due to corrosion issues caused by chloride attack, which may trigger safety failure, demand repair or reestablishment, and bring economic losses and over-consumption of natural non-renewable resources [2]. For example, it is reported that the United Arab Emirates (UAE) spent about 5.2% GDP on repair or rehabilitation of RC structures during 2009–2011 [3]. This reminds stakeholders to deal with durability issues with discretion. There is an array of strategies available to prevent steel corrosion and enhance the longevity of RC structures, such as utilizing more durable cement material, incorporating corrosion inhibitors in concrete mixtures, applying protective coatings to the concrete surface, using stainless steel reinforcement or galvanized reinforcement, implementing cathodic protection systems, etc. Among these strategies, surface coating stands out due to its affordability, simplicity in application, and robust protective capabilities, and can be utilized during construction or service.

During the last few decades, the effect of concrete surface coating on chloride penetration has received increasing attention. For example, Saricimen et al. [4] analyzed the effect of a cementitious material coating, a mixture of Portland cement, fine silica, and an active chemical agent on the chloride resistance of in situ RC structures. The obtained chloride profiles of all components showed that chloride content decreased after the application of Portland cement coating. Moon et al. [5] investigated the chloride resistance of calcium–silicate compound-coated concrete using the steady-state chloride migration test. The results indicated the coated concrete had almost an 80% lower effective chloride diffusion coefficient compared with the control group. Medeiros and Helene [6] investigated and compared the chloride diffusion coefficients of concrete coated with a silane–siloxane mixture dispersed in water, silane–siloxane mixture dispersed in solvent, acrylic dispersed in solvent, and polyurethane. The results indicated that the aforementioned coating systems reduced the chloride diffusion coefficients by 9%, 17%, 20%, and 86%, respectively. This indicated the conventional organic coating polyurethane showed the most effective protection against chloride erosion. However, as is well known, conventional organic coatings represented by polyurethane have fatal defects of environmental pollution [7]. Kong et al. [8] successfully fabricated a superhydrophobic concrete coating (with a static contact angle of 159.1 degrees) into a concrete substrate. A non-toxic micro-diatomaceous earth, nano-Al_2O_3, and low-energy material were combined to fabricate a superhydrophobic coating with a micro–nanostructure. This study underscored the efficacy of protecting concrete against chloride erosion based on dynamic potential polarization curve measurement. Likewise, Liao et al. [9] put forward a hydrophobic nanocomposite coating fabricated by the combination of tannic acid, aminopropyl triethoxysilane, and a low-energy compound, hexadecyltrimethoxysilane ($C_{19}H_{41}O_3Si$). They investigated the effect of concentration mix ratio on coating performance and advised the best concentration combination. They also claimed superiority of this coating in economic performance by comparison with epoxy resin coating. Li et al. [10] investigated the effect of coating category on the instantaneous and long-term performance of coated concrete. They stated that although the coated concrete with organic coating showed superior improvement in instantaneous chloride resistance characterized by passing coulomb electric flux compared to an infiltrating coating (silane-based coating), the latter showed slower aging under similar climate exposure conditions and higher chloride resistance in the long run. This can be attributed to the difference in protection mechanisms of these two kinds of coating.

To conclude, with the efforts of the research community, many concrete surface-coating materials are presently available to apply. And studies to date have shown encouraging results on the improvement effect of the surface coatings on chloride resistance, a crucial indicator of coating efficacy. The coating category, protection mechanism, fabrication method, etc. affect the performance of coated concrete in chloride resistance (instantaneous or long term). And different coatings perform differently in economic and environmental respects. Therefore, a comprehensive review of current research on chloride penetration in surface-coated concrete is necessary for providing a comprehensive evaluation of these techniques and guidance for the design and development of new techniques.

This study aims to present a comprehensive review of chloride penetration in coated concrete, focusing on coating protection mechanisms, improvement efficacy, chloride transport modeling, etc. Economic and environmental performance, as well as the influence of construction technology, are also considered, but are not extensively elaborated upon. Following this introduction, improvement mechanisms for coating on chloride resistance of concrete are first discussed in Section 2, serving as the basis for the review of the studies on chloride resistance of coated concrete in Section 3, focusing on inorganic and organic coatings. Research into the modeling of chloride transport behavior in coated concrete is outlined in Section 4. Finally, Section 5 presents an outlook, outlining challenges and offering recommendations for future research directions, with the aim of promoting practical engineering applications.

2. Protection Mechanism of Surface Coating

A comprehensive understanding of the protection mechanism of surface coating against chloride erosion is crucial for evaluating the improvement efficacy of current techniques in chloride resistance and optimizing their structures and composition to achieve better performance and characteristics. In this section, studies into the chloride erosion mechanism in concrete and several protection mechanisms of different coating types are reviewed, based on which the measurement methods for chloride resistance of surface-coated concrete are summarized.

2.1. Chloride Erosion Mechanism in Concrete

Under normal circumstances, the pH value of the concrete pore solution is above 12 [11], which is in an alkaline state. And the surface of the embedded steel bar is in a passivation state in an alkaline state, which does not rust easily [12,13]. With the chloride erosion and the increase in the surrounding chloride concentration, the pH value decreases, which leads to the corrosion of the steel bar, and ultimately leads to damage to and performance decline of the concrete structure. There are two main chloride transport mechanisms considered in this study: (i) chloride ion diffusion in the pore solution driven by concentration gradient: the external chloride will diffuse into the concrete as ions along the pore solution under the concentration gradient during long-term service in high-salinity environments, such as marine environments or road salts with high chloride content used for icing prevention; and (ii) penetration of water carrying chloride ions: the external solution with high chloride concentration will penetrate into the pores of concrete, causing a chloride concentration change. In addition, the chloride transport process is also affected by electromigration, physiochemical adsorption, etc. which are not the focus of this review.

2.2. Improvement Mechanism of Surface Coating on Chloride Resistance of Concrete

The protective mechanism is intricately tied to the specific coating type, with varying coatings offering protection via distinct mechanisms.

According to [6,14], three fundamental protection mechanisms are identified for coating materials: (i) forming a continuous physical film on the original concrete surface (Figure 1a); (ii) reacting with certain cementitious material constituents to generate non-soluble fillers in the pores and cracks of concrete (Figure 1b); and (iii) forming a continuous waterproof layer on the concrete surface (Figure 1c). It is worth noting that a particular coating may correspond to a single protection mechanism or it may have multiple protection mechanisms acting simultaneously.

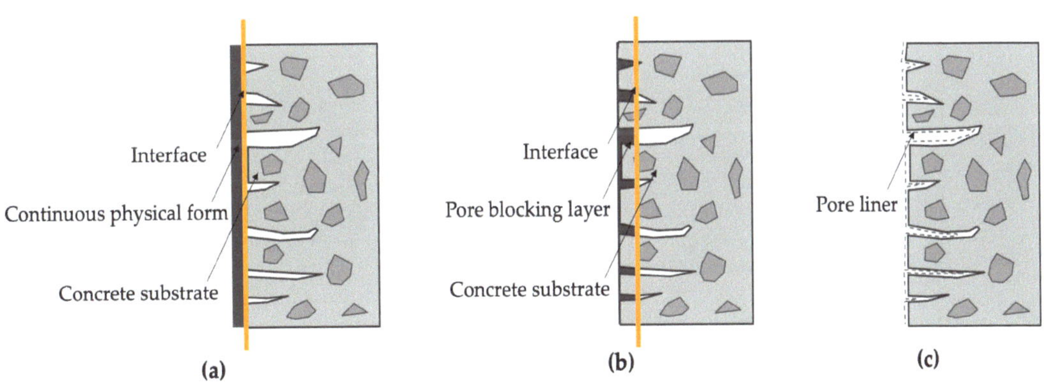

Figure 1. Types of surface coating-protection mechanisms: (**a**) film forming; (**b**) pore blocking; (**c**) pore liner.

When the film is denser than the original concrete structure, it can play a protective role. The most common film-forming coatings are polymer coatings, such as acrylic emulsion coatings, epoxy coatings, polyurethane coatings, and fluorocarbon coatings, etc., which form a uniform film after application through a chemical reaction or physical process that covers the surface of the substrate. In the study of Barbucci et al. [15], the water permeability coefficients of a conventional organic polymer coating coated concrete were decreased by 72–95% when the film thickness was 0.02–0.04 cm. It has been stated that the adhesion and cracking bridging capacity of polymer coating are excellent [16], although the aging problem is prominent.

Certain coatings can react chemically with the surface of the concrete and form substances that fill the pores of the concrete, such as silicate gels. It is reported that the ethyl silicate can react with water and calcium hydroxide to generate non-soluble fillers, which means they serve as a kind of penetrant [17]. In addition, ethyl silicate can also form a layer of silicate film to play a protective role when coated on the surface of the concrete. The coatings act as both film protection and penetrant can as sealant [17]. Similar coatings include alkali-activated cement coatings [18]. When the cement in concrete reacts with water, it releases basic ions, such as hydroxide ions (OH^-) and calcium hydroxide ($Ca(OH)_2$). These alkaline substances react with carbon dioxide in the air on the concrete surface to form carbonates. This process is called alkali excitation. The formation of carbonate helps to fill the pores of concrete, reduce the penetration of water and harmful substances, and improve the compactness and durability of concrete [18].

The typical pore liner coatings, i.e., silane-based water repellent agents, can form a continuous waterproof layer on the concrete surface. The silane-based water repellent agents penetrate into the micropores and capillary pores in the concrete and form hydrophilic silyl compounds. These compounds fill the microscopic pores of the concrete, reducing the ability of water and dissolved salts to enter the interior of the concrete, thereby reducing the seepage properties of the concrete [19]. Silane coatings have excellent weather resistance and can resist the influence of natural environmental factors such as ultraviolet radiation, high temperature, and humidity changes. These characteristics make the coating not susceptible to aging, discoloration or failure, and maintain a long-term stable protective effect. In [20], it was suggested that silane-based coating offered a residual water protection effect even after 20 years of service, with a confidence level of 97%.

With the continuous progress in the field of coatings, additives and materials science, as well as people's concern for environmental protection and sustainable development, scientists' research investment in waterproof materials and technologies is increasing, and continuous innovation in materials, processes and construction technologies has promoted the development and application of the new technology of hydrophobic composite concrete coating (Figure 2a). The new hydrophobic coating constructs a hydrophobic layer on the surface of the concrete, reducing the possibility of surface erosion by water. A common practice is to construct composite coatings using nanomaterials to modify the coating substrates composed of low-surface-energy substances [21]. Currently, the primary coating substrates include polymers, copolymers, and polymer blends [22]. Common nanomaterials are silica, titanium dioxide, graphene oxide and so on. Mechanical mixing or ultrasonic dispersion can be used to break the van der Waals forces between nanoparticles [23]. Silane coupling agents and surfactants can be used to graft functional groups on the surface of nanoparticles to achieve chemical dispersion [24]. This dispersion of nanoparticles allows the coated surface to reach or approach a superhydrophobic surface, with a water contact angle of about 150 degrees [25], as shown in Figure 2b.

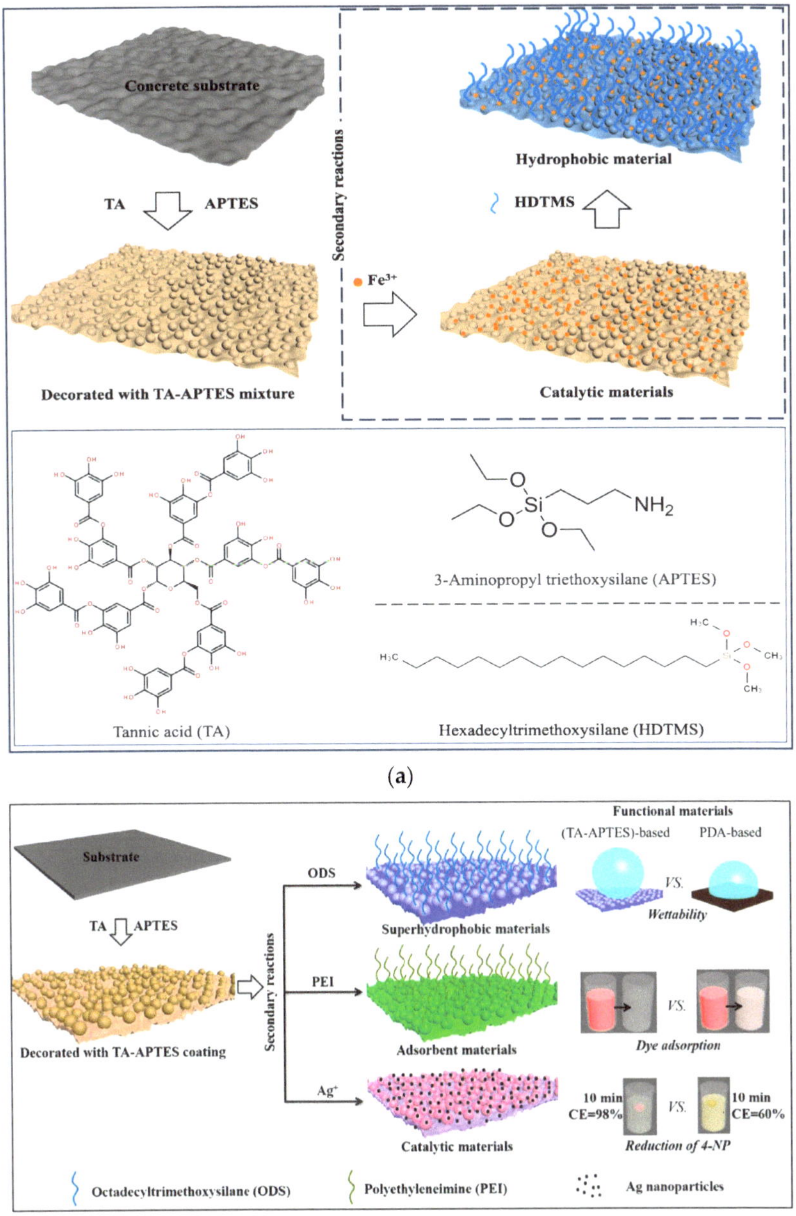

Figure 2. (**a**) Schematic illustration of the hydrophobicity coatings modified by hexadecyltrimethoxysilane [9]; (**b**) schematic illustration of the hydrophobicity coatings modified by octadecyltrimethoxysilane [25].

2.3. Measurement Methods for Chloride Resistance of Concrete

The methods for measuring the resistance of concrete to chlorine erosion mainly include the determination of chloride diffusion coefficient and chemical analysis. The former includes the standardized steady-state migration method according to NT Build 355 [26], the non-steady-state chloride migration test according to NT Build 492 [27] and the rapid chloride permeability test according to ASTM C1202/97 [28]. It worth noting that the rapid chloride permeability test can only directly obtain total passed charge in coulombs, which needs to be converted into the chloride diffusion coefficient by the method proposed by Luping and Nelson [29]. Chemical analysis of chloride concentration (profile) determination is usually combined with the bulk diffusion test introduced in NT Build 443 [30] or ASTM C1556 [31], which is also adopted for chloride resistance determination. In addition, the chloride concentration profile can be obtained by natural diffusion tests, offering a flexible simulation method that mimics the actual field environment closely [32–35]. Moreover, the chloride profiles obtained by in situ RC structures are also capable of chloride resistance determination [4,20,36–39].

3. Chloride Resistance of Surface-Coated Concrete

This section reviews the experimental and theoretical research on chloride ingress of surface-coated concrete in terms of various coating material categories, including inorganic, organic, and organic–inorganic composite coatings. By comparing the results of different studies, the improvement efficacy in chloride resistance is analyzed and compared emphatically.

It is worth mentioning that for coatings with the same category, the protection mechanisms may differ from each other.

3.1. Inorganic Coatings

Inorganic coatings typically exhibit superior weather resistance and chemical resistance, and can withstand ultraviolet light, temperature fluctuations, pH changes, and humidity, offering more stable performance in harsh environmental conditions. Furthermore, inorganic coatings demonstrate better stability and are less prone to decomposition, especially in the case of fire. In this section, we mainly review the studies on inorganic coatings. Table 1 lists chloride diffusion coefficients of concrete coated with various kinds of inorganic materials.

Silicate coating, a typical inorganic coating, has been studied by some researchers in terms of both mechanism and specific performance. The precise mechanism through which silicate enhances concrete performance remains uncertain. The silicates act as penetrants/pore blockers due to the generation of SiO_2 precipitate [40]. In the study of Higgins [41], it is stated that the reaction between concrete components, i.e., water, calcium hydroxide and the silicates, generate calcium silicate hydrate gel (C-S-H). Another theory suggests that the silicate gel generated can fill the pores in concrete, thereby improving its performance [40]. The specific mechanisms of application may vary from case to case. The study of Moon et al. [5] investigated the chloride resistance of mortar coated with a calcium–silicate compound by steady-state migration test. The results indicated the calcium–silicate compound decreased the steady-state chloride diffusion coefficient by 87% and 72% for high strength and low strength, respectively. The analysis results of mercury intrusion porosimetry (MIP) and scanning electron microscopy (SEM) demonstrated that the hydration of the main component in the coating, calcium–silicate, generated water-insoluble silicate compounds, which fill the pores as micro-fillers and form a very dense film on the surface of the mortar. The study of Thompson et al. [40] found that the aqueous sodium silicate coating decreased the electric flux by 45% at 38.6% liquid and 32% at 37.6% liquid. Their studies also indicated that the concentration of solution affected the improvement efficacy. Franzoni et al. [42] compared the chloride resistance of concrete coated with an aqueous sodium silicate solution and an ethyl silicate solution. Their research indicated that sodium silicate exhibited limited effectiveness in protecting against chloride penetration, regardless of whether it was combined with nano SiO_2. SEM-EDS (energy-dispersive X-ray spectroscopy) analysis found that the sodium silicate-coated concrete surface showed deposited layers, which were detached and exhibited some visible cracking. Conversely, the ethyl silicate coating was compact and less detached, which can be attributed to the consolidating effect of the generated amorphous silica gel in the concrete pores (Figure 3). The study of Pigino et al. [43] showed that ethyl silicate coating decreased the chloride diffusion coefficients of concrete by 98.5% (from 20.4×10^{-12} m^2/s to 0.3×10^{-12} m^2/s) when the w/c is 0.65 and by 98.6% (from 7.6×10^{-12} m^2/s to 0.1×10^{-12} m^2/s) when the w/c is 0.45. Sandrolini et al. [44] stated that the superior chloride resistance of ethyl silicate coating can be attributed to (i) the excellent impregnation ability (3–5 mm depth) due to its viscosity being relatively low and its monomer size relatively small; and (ii) the pozzolanic reaction with calcium hydroxide. In addition, ethyl silicate coatings have broad application potential due to their emission of only volatile and non-destructive by-products during the hardening process [42].

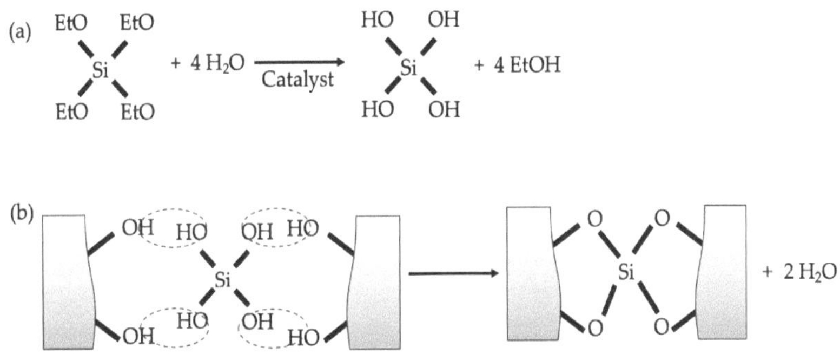

Figure 3. The reaction of ethyl silicate coating with concrete: (**a**) hydrolysis; (**b**) condensation.

Silane coatings refer to coatings prepared based on silane compounds, i.e., alkoxy and alkyl silanes, a typical molecular structures of which is shown in Figure 4. The alkoxy groups can react with surface hydration products to generate a silicon oxide layer. The alkylic groups possess organic properties, usually highly hydrophilic, making water molecules tend to form droplets on surfaces and slide off quickly, rather than adsorbing into the surface. This property allows the silicon oxide layer to prevent moisture penetration and surface moisture to a certain extent, helping to protect the concrete under the coating from water erosion and moisture erosion. Medeiros and Helene [6] studied the effectiveness of silane coating on improving chloride resistance. The results indicated that the silane dispersed in the water and solvent decreased the chloride diffusivity of saturated concrete by 9% and 17%, respectively. A similar minor reduction magnitude was reported in another study, which was 11% and 17% [45]. The study of Nanukuttan et al. [36] investigated the chloride penetration of silane-coated concrete subjected to a natural environment for up to 7 years. The diffusion coefficients obtained by Fick's second law fitting were similar for coated and uncoated conditions. However, in the study of Sakr and Bassuoni [46], the silane effectively improved the chloride resistance of concrete. For example, the chloride diffusion coefficient of coated concreted decreased from 97.28×10^{-12} m^2/s to 11.50×10^{-12} m^2/s (88% reduction) when the w/c ratio was 0.6. Nevertheless, results from their studies indicated that in terms of improvement in resistance to physical salt attack or frost attack, silane coating was ineffective. The incorporation of nanomaterials, i.e., nano-silica or nano-clay, significantly improved the performance of silane-based composite coating. For example, the chloride diffusion coefficients further decreased by 58%, 58%, 38% and 2% compared with those of single silane-coated concrete when incorporating 5% and 10% dosages of nano-clay and nano-silica, respectively [46]. The improvement efficacy of nano-clay was superior to that of nano-silica, which may be due to the fact that the smaller nano-silica particles are more likely to aggregate and disperse less effectively. Similarly, Ibrahim et al. [47] reported the chloride resistance improvement efficacy of nano-silica and nano-calcium carbonate-modified silane coating. The results indicated that the passing charges decreased by 32% for single silane coating compared to a control group.

Compared with single silane coating, the incorporation of a 2.5% dosage of nano-silica showed an 80% decreased passing charge value and 76% for nano-calcium carbonate with the same dosage. In addition, Gu et al. [48] proposed a superhydrophobic coatings based on nano-silica-modified isobutyltriethoxysilane. The superhydrophobic cellular coating was sprayed as the top layer of the coating. The results from rapid chloride migration tests demonstrated that this coating showed superior chloride resistance. The study of Schueremans et al. [49] demonstrated the superior long-term effectiveness of silane coatings. Their findings from on-site experiments concluded that the protection system was still effective even after 12 years.

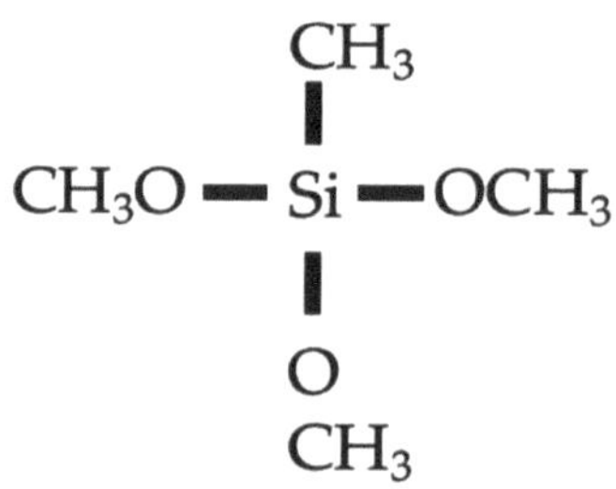

Figure 4. Molecular structure of typical silane.

Geopolymer is an inorganic polymer material characterized by the formation of a three-dimensional network structure through the biochemical reaction of alkaline activators and silica–aluminum skeleton raw materials (such as metallurgical residues, volcanic ash, etc.) at low temperatures. Compared to traditional cement-based materials, geopolymers do not require the use of cement as a cementing material, thus significantly reducing carbon emissions. Geopolymer has excellent mechanical properties, fire resistance, chemical stability and corrosion resistance, so it is used as a new inorganic coating material. The study of Zhuang et al. [50] fabricated a compound geopolymer coating composed of granulated blast furnace slag and metakaolin. Their studies found that this coating showed optimal properties when the content of granulated blast furnace slag was 10% at a liquid/solid ratio of 0.60 mL/g. In another published article [51], they investigated the microstructure and anticorrosion mechanism of this geopolymer coating. The reaction products between the geopolymer slurry and the concrete components were compact and the average open pore size was less than 15 nm. Another typical geopolymer is alkali-activated binder. The study of Phiangphimai et al. [18] fabricated an alkali-activated cement powder composed of geopolymer powder, Portland cement and silica fume. They found the chloride diffusion coefficient of coated concrete decreased by 98% when the ratio of sodium silicate to sodium hydroxide was 1.0 and 97% when the ratio of sodium silicate to sodium hydroxide was 2.0.

Currently, graphene oxide, a promising nanomaterial, has increasingly garnered attention in research circles. At present, most practices are to add graphene oxide to cement slurry to improve its performance. Although this approach can improve the mechanical performance [52–54] and durability [55] of cement pastes, the efficacy is far from satisfactory and the fresh mixing performance is adversely affected [56,57]. Antolín-Rodríguez et al. [58] investigated the effectiveness of graphene oxide as a surface treatment technique on improving the chloride penetration of coated concrete. Their results showed the chloride diffusion coefficient determined by rapid chloride migration tests decreased by 4%, 18.9%, 36.6%, 40.3% and 51% when the graphene oxide contents were 26.2, 52.4, 78.6, 104.8, and 131.1 μg/cm^2. In the study of Antolín-Rodríguez [59], the SEM analysis demonstrated that the microstructure of the surface was densified due to the facilitation effect of graphene on the hydration process. In another study of Antolín-Rodríguez [60], the amount of graphene oxide coating significantly showed a direct correlation with the protection efficacy. However, it is essential to determine an optimal level of graphene oxide content that strikes a balance between efficacy and economy in the treatment process.

Table 1. Chloride diffusivity of concrete coated with inorganic coating materials.

Ref.	Coating Type	w/c	Chloride Diffusion Coefficient (D = 10^{-12} m^2/s)/Electric Flux (C)	Test Method	Control Group	Coating Process
[43]	Calcium–silicate compound M1	0.485	0.3D	SMM	2.3D	Spraying
[43]	Calcium–silicate compound-M2.45	0.485	1.8D	SMM	6.5D	Spraying
[40]	Aqueous sodium silicate—N	0.48	8993C	RCPT	13219C	Brushing
[40]	Aqueous sodium silicate—OW	0.48	7238C	RCPT		Brushing
[44]	Ethyl silicate	0.65	0.3D	RCMT	20.4D	Brushing
[44]	Ethyl silicate	0.45	0.1D	RCMT	7.6D	Brushing
[18]	Alkali-activated/cement powder -S/H ratio of 1.0	0.5	0.081D	RCMT	6.212D	-
[18]	Alkali-activated/cement powder -S/H ratio of 2.0	0.5	0.192D	RCMT	6.212D	-
[61]	Graphene oxide (26.2 μg/cm^2)	0.5	1.52D 5.84D	SSM RCMT	1.59D 6.08D	Spraying
[61]	Graphene oxide (52.4 μg/cm^2)	0.5	1.36D 5.65D	SSM RCMT	1.59D 6.08D	Spraying
[61]	Graphene oxide (78.6 μg/cm^2)	0.5	1.03D 3.85D	SSM RCMT	1.59D 6.08D	Spraying
[61]	Graphene oxide (104.8 μg/cm^2)	0.5	0.55D 3.63D	SSM RCMT	1.59D 6.08D	Spraying
[61]	Graphene oxide (131.1 μg/cm^2)	0.5	0.39D 2.97D	SSM RCMT	1.59D 6.08D	Spraying
[49]	Silane	0.6	11.5D 1788C	RCPT	97.28D OVF	Brushing
[49]	silane/nano-clay (5%)	0.6	4.88D 832C	RCPT	97.28D OVF	Brushing
[49]	silane/nano-clay (10%)	0.6	4.86D 785C	RCPT	97.28D OVF	Brushing
[49]	silane/nano-silica (5%)	0.6	7.14D 1012C	RCPT	97.28D OVF	Brushing
[49]	silane/nano-silica (10%)	0.6	11.30D 1530C	RCPT	97.28D OVF	Brushing

Table 1. *Cont.*

Ref.	Coating Type	*w/c*	Chloride Diffusion Coefficient ($D = 10^{-12}$ m^2/s)/Electric Flux (C)	Test Method	Control Group	Coating Process
[49]	Silane	0.4	5.33D 963C	RCPT	20.93D 4015C	Brushing
[49]	silane/nano-clay (5%)	0.4	3.13D 425C	RCPT	20.93D 4015C	Brushing
[49]	silane/nano-clay (10%)	0.4	3.64D 488C	RCPT	20.93D 4015C	Brushing
[49]	silane/nano-silica (5%)	0.4	3.29D 466C	RCPT	20.93D 4015C	Brushing
[49]	silane/nano-silica (10%)	0.4	4.44D 609C	RCPT	20.93D 4015C	Brushing
[6]	Silane/siloxane dispersed in water	0.52	3.13D	RCPT	3.45D	Soaking
[6]	Silane/siloxane dispersed in solvent	0.52	2.85D	RCPT	3.45D	Soaking
[50]	Silane	0.6	268C	RCPT	OVF	Brushing
[50]	silane/nano-clay (2.5%)	0.6	641C	RCPT	OVF	Brushing
[50]	silane/nano-clay (5%)	0.6	750C	RCPT	OVF	Brushing
[50]	silane/nano-CaCO3 (2.5%)	0.6	488C	RCPT	OVF	Brushing
[50]	silane/nano-CaCO3 (5%)	0.6	726C	RCPT	OVF	Brushing
[50]	Silane	0.4	211C	RCPT	2326C	Brushing
[50]	silane/nano-clay (2.5%)	0.4	307C	RCPT	2326C	Brushing
[50]	silane/nano-clay (5%)	0.4	336C	RCPT	2326C	Brushing
[50]	silane/nano-CaCO3 (2.5%)	0.4	378C	RCPT	2326C	Brushing
[50]	silane/nano-CaCO3 (5%)	0.4	596C	RCPT	2326C	Brushing
[10]	silane	0.6	229C	RCPT	2745C	Brushing
[62]	Alkyl alkoxysilane	0.45	8.7D	FT	9.0D	/

Note: (i) M1 denotes the ratio of sand to cement is 1, M2.45 denotes the ratio of sand to cement is 2.45; (ii) N denotes the aqueous sodium silicate had 37.6% solids, OW denotes the aqueous sodium silicate had 38.6% solids; (iii) S/H ratio denotes the ratio of sodium silicate to sodium hydroxide. (iv) SMM denotes the steady-state migration method; RCPT denotes rapid chloride permeability test; RCMT denotes rapid chloride migration test. (v) OVF denotes overflow, which means the RCPT process is terminated before 6 h.

3.2. Organic Coatings

In this study, coatings made of organic chemicals are classified as organic coatings. Table 2 lists the chloride diffusion coefficients of concrete coated with various kinds of organic materials.

Table 2. Chloride diffusivity of concrete coated with organic coating materials.

Ref.	Coating Type	*w/c*	Chloride Diffusion Coefficient ($D = 10^{-12}$ m²/s)/Electric Flux (C)	Test Method	Control Group	Coating Process
[63]	Acrylic coating	0.6	6.53D	RCMT	14.4D	/
[63]	Epoxy coating	0.6	0D	RCMT	14.4D	/
[63]	Acrylic coating	0.4	3.86D	RCMT	6.9D	/
[63]	Epoxy coating	0.4	0D	RCMT	6.9D	/
[64]	Acrylic coating, AC1	0.45	2.08D	IT	19.18D	/
[64]	Acrylic coating, AC2	0.45	3.49D	IT	19.18D	/
[64]	Polymer emulsion coating, PE1	0.45	8.40D	IT	19.18D	/
[64]	Polymer emulsion coating, PE2	0.45	15.97D	IT	19.18D	/
[64]	Epoxy coating, EP1	0.45	7.67D	IT	19.18D	/
[64]	Epoxy coating, EP2	0.45	2.59D	IT	19.18D	/
[64]	Polyurethane coating, PU1	0.45	1.83D	IT	19.18D	/
[64]	Polyurethane coating, PU2	0.45	0.70D	IT	19.18D	/
[64]	Chlorinated rubber coating, CR1	0.45	9.56D	IT	19.18D	/
[64]	Chlorinated rubber coating, CR2	0.45	8.40D	IT	19.18D	/
[62]	Acrylic sealant	0.45	11.2D	FT	9.0D	Brushing
[62]	Polyurethane sealant	0.45	8.8D	FT	9.0D	Brushing
[62]	Acrylic coating	0.45	9.0D	FT	9.0D	Brushing
[62]	Polyurethane coating	0.45	4.7D	FT	9.0D	Brushing
[65]	Polyurethane–SiO_2 (0%)	0.6	2110C	RCPT	/	Spraying
[65]	Polyurethane–SiO_2 (1%)	0.6	1780C	RCPT	/	Spraying
[65]	Polyurethane–SiO_2 (2%)	0.6	1520C	RCPT	/	Spraying
[65]	Polyurethane–SiO_2 (3%)	0.6	1470C	RCPT	/	Spraying
[65]	Epoxy resin–SiO_2 (0%)	0.6	610C	RCPT	/	Spraying
[65]	Epoxy resin–SiO_2 (1%)	0.6	450C	RCPT	/	Spraying
[65]	Epoxy resin–SiO_2 (2%)	0.6	390C	RCPT	/	Spraying
[65]	Epoxy resin–SiO_2 (3%)	0.6	430C	RCPT	/	Spraying
[65]	Polyurethane–TiO_2 (0%)	0.6	2110C	RCPT	/	Spraying
[65]	Polyurethane–TiO_2 (1%)	0.6	1670C	RCPT	/	Spraying
[65]	Polyurethane–TiO_2 (2%)	0.6	1480C	RCPT	/	Spraying
[65]	Polyurethane–TiO_2 (3%)	0.6	1720C	RCPT	/	Spraying
[65]	Epoxy resin–TiO_2 (0%)	0.6	610C	RCPT	/	Spraying
[65]	Epoxy resin–TiO_2 (1%)	0.6	460C	RCPT	/	Spraying
[65]	Epoxy resin–TiO_2 (2%)	0.6	430C	RCPT	/	Spraying
[65]	Epoxy resin–TiO_2 (3%)	0.6	490C	RCPT	/	Spraying
[66]	Epoxy coating—submicron/nano carbon (0%)	0.45	102D	PT	134D	Soaking
[66]	Epoxy coating—submicron/nano carbon (0.25%)	0.45	98D	PT	134D	Soaking
[66]	Epoxy coating—submicron/nano carbon (0.5%)	0.45	92D	PT	134D	Soaking
[66]	Epoxy coating—submicron/nano carbon (0.75%)	0.45	78D	PT	134D	Soaking
[66]	Epoxy coating—submicron/nano carbon (1.0%)	0.45	70D	PT	134D	Soaking
[67]	Modified acrylic styrene emulsion	0.5	5.944D	RCMT	7.799D	/
[10]	epoxy glass flake paint	0.6	752C	RCPT	2745C	Brushing
[10]	Polyurethane paint	0.6	550C	RCPT	2745C	Brushing
[46]	Poly(methyl methacrylate) (PMMA)	0.6	17.5D 2981C	RCPT	97.28D OVF	Brushing
[46]	PMMA/nano-clay (5%)	0.6	4.88D 832C	RCPT	97.28D OVF	Brushing
[46]	PMMA/nano-clay (10%)	0.6	4.86D 785C	RCPT	97.28D OVF	Brushing
[46]	PMMA/nano-silica (5%)	0.6	7.14D 1012C	RCPT	97.28D OVF	Brushing
[46]	PMMA/nano-silica (10%)	0.6	11.30D 1530C	RCPT	97.28D OVF	Brushing
[46]	Poly (methyl methacrylate) (PMMA)	0.4	11.87D 1785C	RCPT	20.93D 4015C	Brushing
[46]	PMMA/nano-clay (5%)	0.4	4.00D 569C	RCPT	20.93D 4015C	Brushing
[46]	PMMA/nano-clay (10%)	0.4	5.78D 826C	RCPT	20.93D 4015C	Brushing

Table 2. *Cont.*

Ref.	Coating Type	*w/c*	Chloride Diffusion Coefficient ($D = 10^{-12}$ m^2/s)/Electric Flux (C)	Test Method	Control Group	Coating Process
[46]	PMMA/nano-silica (5%)	0.4	4.70D 761C	RCPT	20.93D 4015C	Brushing
[46]	PMMA/nano-silica (10%)	0.4	10.38D 1542C	RCPT	20.93D 4015C	Brushing
[47]	Vinyl ester	0.6	4485C	RCPT	OVF	Brushing
[47]	Vinyl ester/nano-clay (2.5%)	0.6	63C	RCPT	OVF	Brushing
[47]	Vinyl ester/nano-clay (5%)	0.6	153C	RCPT	OVF	Brushing
[47]	Vinyl ester/nano-$CaCO_3$ (2.5%)	0.6	74C	RCPT	OVF	Brushing
[47]	Vinyl ester/nano-$CaCO_3$ (5%)	0.6	44C	RCPT	OVF	Brushing
[47]	Vinyl ester	0.4	1576C	RCPT	2326C	Brushing
[47]	Vinyl ester/nano-clay (2.5%)	0.4	10C	RCPT	2326C	Brushing
[47]	Vinyl ester/nano-clay (5%)	0.4	66C	RCPT	2326C	Brushing
[47]	Vinyl ester/nano-$CaCO_3$ (2.5%)	0.4	34C	RCPT	2326C	Brushing
[47]	Vinyl ester/nano-$CaCO_3$ (5%)	0.4	23C	RCPT	2326C	Brushing
[68]	TA coating (Fe(III))—self-polymerization	0.5	4.97D 1163C	RCPT	5.2D 1617C	Soaking
[68]	TA coating (Fe(III))—one step method	0.5	4.18D 1114C	RCPT	5.2D 1617C	Soaking
[68]	TA coating (Fe(III))—multi-step method	0.5	4.13D 1249C	RCPT	5.2D 1617C	Soaking
[9]	TA-APTES-HDTMS (1%)	0.5	779C	RCPT	1657.81C	Soaking
[9]	TA-APTES-HDTMS (3%)	0.5	554C	RCPT	1657.81C	Soaking
[9]	TA-APTES-HDTMS (5%)	0.5	494C	RCPT	1657.81C	Soaking
[9]	TA-APTES-HDTMS (7%)	0.5	429C	RCPT	1657.81C	Soaking
[9]	TA-APTES-HDTMS (10%)	0.5	391C	RCPT	1657.81C	Soaking
[9]	TA-APTES-HDTMS (15%)	0.5	367C	RCPT	1657.81C	Soaking
[69]	D-CO2-monoethanolamine (MEA) (10%)	0.35	10.7D 296.09C	RCPT	13.3D 339.07C	/
[69]	D-CO_2-monoethanolamine (MEA) (30%)	0.35	9.1D 285.16C	RCPT	13.3D 339.07C	/
[69]	F-CO_2-monoethanolamine (MEA) (10%)	0.35	13.6D 336.30C	RCPT	13.3D 339.07C	/
[69]	F-CO_2-monoethanolamine (MEA) (30%)	0.35	13.5D 336.00C	RCPT	13.3D 339.07C	/
[69]	D-CO_2-monoethanolamine (MEA) (10%)	0.4	13.0D 339.10C	RCPT	18.1D 458.9C	/
[69]	D-CO_2-monoethanolamine (MEA) (30%)	0.4	12.4D 293.77C	RCPT	18.1D 458.9C	/
[69]	D-CO_2-monoethanolamine (MEA) (10%)	0.5	19.0D 440.86C	RCPT	20.4D 539.43C	/
[69]	D-CO_2-monoethanolamine (MEA) (30%)	0.5	18.2D	RCPT	20.4D 539.43D	/

Conventional organic coating refers to coatings made from organic polymers, including resins, solvents, and additives. Common organic polymers are polyester resin, polyacrylic resin, polyurethane resin, and so on. Aguiar et al. [63] evaluated the effect of acrylic and epoxy polymeric coatings on chloride penetration resistance of concrete. The results indicated that use of both acrylic and epoxy coatings contributed to a significantly decreased chloride diffusion coefficient. In the presence of epoxy resin coating, the diffusion coefficient of concrete even achieved null. Almusallam et al. [64] investigated five generic types of polymeric coatings. The acrylic, polymer emulsion, epoxy polyurethane, and chlorinated coating decreased the chloride diffusion coefficients by 89%/82%, 56%/17%, 60%/86%, 90%/96%, and 50%/56%, respectively. The values before and after the slash "/" represent that the same coating comes from different manufacturers. This suggests that the determination of coating efficacy should rely on preliminary tests rather on the types alone. Shi et al. [67] investigated the improvement effect of a modified styrene–acrylic emulsion on chloride diffusion of coated concrete. Compared with a control group, the penetration depth decreased from 18.5 mm to 13.25 mm, and the diffusion coefficient decreased by 24% with polymer thickness of 70 μm. Li et al. [10] investigated and compared the chloride resistance and service life prediction of epoxy glass-flake paint, polyurethane paint, cement-based permeable crystallization waterproof coating, and silane-based water-repellent coating. The results indicated that the corresponding coulomb electric flux decreases were 92%, 90%, 73% and 80%. Although the chloride resistance of polymer coatings was superior to silane coatings, the aging test results indicated that weather fastness of silane was more superior. The excellent weather resistance of silane has also been reported in the study of Christodoulou et al. [20], which stated that even after 20 years, the silane impregnation still exhibited a residual hydrophobic effect. Qu et al. [65] studied the improvement effect of nanomaterial-modified water-based polyurethane and epoxy resin coatings on chloride resistance of coated concrete. The addition of nano-SiO_2 and nano-TiO_2 improved the coulomb electric flux of neat resin coated concrete by 51–74% and 24–27%, respectively. In addition, the nanomaterials also improved the anti-aging capability of water-based organic coatings. The long-term anti-aging capability improvement effect of nano-TiO_2 was more significant than that of nano-SiO_2. Basha et al. [66] investigated the improvement effect of epoxy coating modified by submicron/nano-carbon obtained from the waste material generated from heavy fuel oil ash combustion. The results indicated the submicron/nano-carbon with dosage of 1% performed best in increasing the chloride resistance of coated concrete (with 66% reduction). In brief, nanomaterials show promise in enhancing the performance of conventional organic coatings by effectively filling micropores and the hydrophobic effect, as shown in Figure 5.

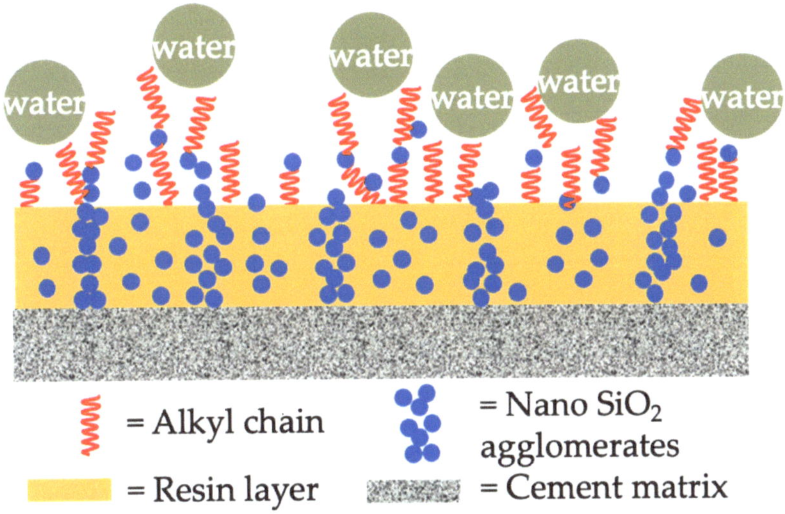

Figure 5. Microstructure model of organic coating modified by nanomaterials.

Although conventional organic coatings show predominance in chloride resistance improvement, their defects, i.e., bringing toxic by-products during the production process and releasing voltaic organic compounds during the coating process, are also troublesome. Thus, the research community have tried to fabricate novel environmentally friendly anti-corrosion coatings. Inspired by the robust adhesion between mussels and rock, a bionic coating for mussels based on polyphenol material tannic acid (TA) for concrete protection was proposed by Xiao et al. [68]. The results indicated the TA-trivalent iron cation (Fe(III)) coating showed comparable chloride resistance improvement effect to conventional organic coatings. On this basis, Liao et al. [9] performed hydrophobic modification of the TA coating, seeking to further improve the performance of the coating. They utilized the organosilicon compound hexadecyltrimethoxysilane (HDTMS) to modify a TA-aminopropyltriethoxysilane (APTES) coating and successfully fabricated a hydrophobic nanocomposite coating. They also investigated the effect of TA, APTES, and HDTMS concentration on coating performance and fabricated an optimal TA–APTES–HDTMS coating. The coulomb electric flux was decreased by more than 70% after coating. Additionally, methyl methacrylate and vinyl ester, which are not adequately effective to protect concrete from chloride erosion alone, can also be used in combination with nanomaterials for hydrophobic modification and performance improvement. In the study of Sakr and Bassuoni [46], the chloride resistance of sole and nano-clay- or nano-silica-modified methyl methacrylate-coated concrete was investigated. The results indicated that the incorporation of 5% nano-clay, 10% nano-clay, 5% nano-silica, and 10% nano-silica decreased the diffusion coefficients by 66%/72%, 51%/72%, 60%/59%, and 12%/35%, respectively. The values before and after the slash "/" represent substrate concretes with w/c of 0.4 and 0.6. In the study of Ibrahim et al. [47], they investigated the chloride resistance of sole and nano-clay- or nano-$CaCO_3$-modified vinyl ester-coated concrete. The results indicated that both 2.5% and 5% nano-clay or nano-$CaCO_3$ incorporated vinyl ester-coated concrete showed almost 100% decreased coulomb electric flux compared with neat vinyl ester-coated concrete. In addition, Kong et al. [70] synthesized a sustainable superhydrophobic concrete coating by micro-diatomaceous earth, nano-Al_2O_3, and stearic acid, which exhibited excellent anti-corrosion performance on electrochemical corrosion tests. Luo et al. [71] successfully fabricated a robust superhydrophobic coating with myristic acid and ZrO_2 particles. However, the chloride resistance of this novel hydrophobic coating has not been investigated.

In the study of Han et al. [69], the chloride resistance improvement effect of CO_2 dissolved monoethanolamine (MEA) on coated concrete was investigated. The HCO_3^- and CO_3^{2-} in CO_2-dissolved MEA solution can react with the hydration product, i.e., Ca $(OH)_2$, C_3S, and C_2S, to generate calcium carbonate precipitation, which can act as filler to densify the surface of the concrete substrate. The test results indicated that the performance of this coating was affected by various factors, i.e., MEA concentration, the quality of the substrate, coating time and pre-condition method. The treatment of this coating with a 10% dosage of MEA on concrete with 0.4 w/c ratio showed optimal chloride resistance (with 31.5% reduction).

3.3. Organic–Inorganic Composite Coatings

Organic–inorganic composite coating refers to the coating system that contains both organic and inorganic components. It is worth noting that the composite coatings made by combining an organic resin (such as a polymer) with an inorganic material (such as silica, zinc oxide, etc.) or inorganic particles (such as nanoparticles) belong to this type, which have been detailed in previous sections. In this section, the polymer-modified cementitious/geopolymer coatings are mainly focused on. Table 3 lists the chloride diffusion coefficients of concrete coated with various kinds of organic–inorganic composite coatings.

Zhang et al. [62] analyzed and compared the interfacial chloride concentration and pseudo-chloride diffusion coefficient of acrylic-modified cementitious coating and other pure polymer coatings. The results showed that the interfacial chloride concentration of composite coating-treated concrete was far lower than that of untreated concrete, but higher than that of sole acrylic- or polyurethane-treated concrete. The same was found for the pseudo-diffusion coefficients. Diamanti et al. [72] investigated the influence of polymer–cement ratio on the improvement efficiency of acrylic-modified ordinary Portland cement coating in chloride resistance of coated concrete. The results indicated that the coating porosity decreased with increasing polymer content, and thus the effectiveness of the barrier to chloride ingress was increased. Liang et al. [73] investigated the influence of fiber introduction on the protection effectiveness of silicone-modified polyacrylate modified calcium sulfoaluminate cement coating. The results indicated the polypropylene fiber showed obvious effects in improving the tensile strength, bonding strength, wear resistance, and anti-UV aging resistance, and the chloride penetration resistance was significantly increased (with 49% reduction rate compared to uncoated concrete). Zhang et al. [74] proposed a waterborne epoxy resin and silane coupling agent-modified metakaolin-based geopolymer coating, which exhibited excellent performance in shrinkage reduction, adhesion, anti-abrasion, and chloride corrosion resistance. However, the specific chloride resistance has not been investigated yet.

Table 3. Chloride diffusivity of concrete coated with organic–inorganic composite coating materials.

Ref.	Coating Type	*w/c*	Chloride Diffusion Coefficient (D = 10^{-12} m^2/s)/Electric Flux (C)	Test Method	Control Group	Coating Process
[10]	Cement-based permeable crystallization waterproof coating	0.6	279C	RCPT	2745C	Brushing
[62]	Two-component acrylic-modified cementitious coating	0.45	4.7D	FT	9.0D	Brushing
[72]	Acrylic–cementitious coating (PC ratio of 0.35)	0.65	0.50D	SSM	8.03D	/
[72]	Acrylic–cementitious coating (PC ratio of 0.55)	0.65	0.28D	SSM	8.03D	/
[72]	Acrylic–cementitious coating (PC ratio of 0.35)	0.5	0.13D	SSM	2.41D	/
[72]	Acrylic–cementitious coating (PC ratio of 0.55)	0.5	0.15D	SSM	2.41D	/
[73]	Silicone-modified polyacrylate-modified cement-based coating	/	2665.42C	RCPT	4054.00C	/
[73]	silicone-modified polyacrylate-modified cement-based coating incorporating polypropylene fiber	/	2070.64C	RCPT	4054.00C	/

4. Modeling Chloride Penetration into Surface-Coated Concrete

The goal of modeling chloride penetration into concrete is to predict chloride concentration profiles, which are crucial for determining the chloride concentration at the reinforcement depth and thus forecasting the service life of the concrete structure. Chloride transport in coated concrete is influenced by several factors, including both the properties of the concrete substrate and coating. Current studies on chloride transport modeling typically treat coated concrete as a bilayer material [75] consisting of the treated layer and the substrate layer, as illustrated in Figure 6. In the case of coatings with a film-forming mechanism, the treated layer represents the additional physical layer formed by the coating. For coatings that function through a pore-blocking mechanism, the treated layer corresponds to the concrete surface layer that has been impregnated with coating materials. However, for coatings that employ a pore-liner mechanism, there is currently a lack of specific chloride transport modeling approaches.

Compared to the chloride profiles in uncoated concrete, the chloride profiles in coated concrete often appear as a piecewise function, as illustrated in Figure 7. This is due to the differing diffusion coefficients between the treated layer and the substrate layer. Moradllo et al. [36,37] fitted the analytical solution of Fick's second law (Equation (1)) directly to the chloride profiles of untreated substrate concrete, without accounting for the chloride profiles of the surface-treated layer. The obtained surface chloride concentration is termed interfacial chloride concentration C_i; and the obtained diffusion coefficient D_a is termed pseudo-diffusion coefficient [62]. In the study of Moradllo et al. [36,37], the pseudo-diffusion coefficient was firstly smaller than the true diffusion coefficient of the substrate concrete D_0, and then increased to over D_0 with the degradation of coating materials. To put it simply, both C_i and pseudo-diffusion coefficients are time-dependent, which violates the premise for the existence of the simple closed-form solution (Equation (1)), that is, both C_i and pseudo-diffusion coefficients are assumed to be constant. For remedying this limitation, Petcherdchoo [17] proposed a time-dependent pseudo-coating model (Equation (2)). This model was validated by comparison with experimental data:

$$C(x,t) = C_i\left[1 - erf\left(\frac{x}{2\sqrt{D_a t}}\right)\right] \tag{1}$$

$$
\begin{aligned}
C(x,t) &= C_{0,t}\left[erfc\left(\frac{x}{2\sqrt{D_a t}}\right)\right] + k_l t\left[\left(1+\frac{x^2}{2D_a t}\right)erfc\left(\frac{x}{2\sqrt{D_a t}}\right) - \left(\frac{x}{\sqrt{\pi D_a t}}\right)e^{-\frac{x^2}{4D_a t}}\right] \\
D_a &= \frac{D_{28}}{1-m}\left[\left(1+\frac{28}{365t}\right)^{1-m} - \left(\frac{28}{365t}\right)^{1-m}\right]\cdot\left(\frac{28}{365t}\right)^{m}
\end{aligned}
\tag{2}
$$

where D_a denotes the obtained apparent diffusion coefficient; C_i denotes the interfacial chloride concentration; t denotes exposure time; D_{28} denotes the initial chloride diffusion at 28 days' curing; m denotes the aging coefficient; $C_{0,l}$ denotes the initial chloride content; and k_l is the fitting constant.

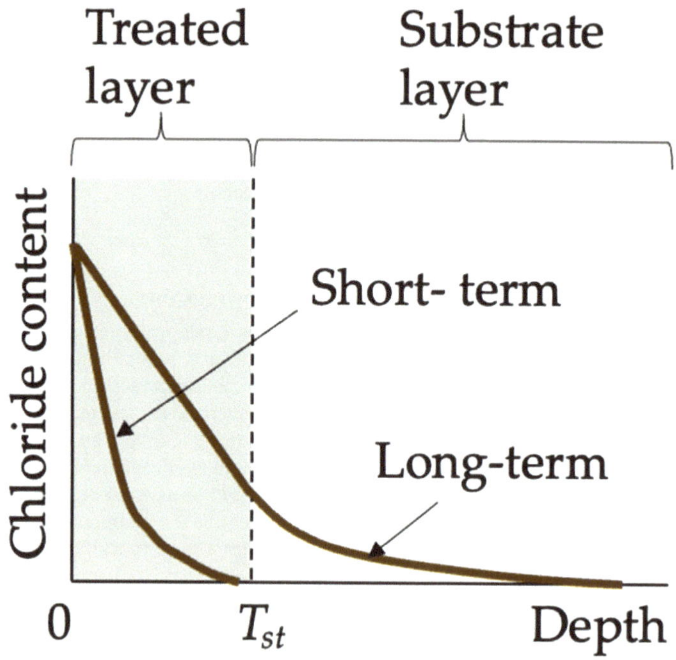

Figure 6. Chloride profiles in coated concrete.

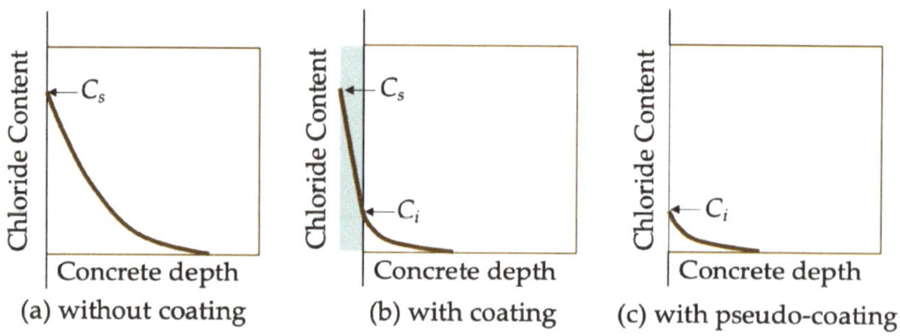

Figure 7. Schematic diagram of pseudo-diffusion model.

The pseudo-diffusion coefficient is affected by multiple factors and mechanisms, such as the diffusion properties of both treated and substrate layers and chloride binding. In some cases, integrating multiple factors into a one-parameter pseudo-diffusion coefficient may make decisions more efficient. At the same time, some information may be lost or blurred, because the uniqueness of specific factors may be consolidated into an average or composite value, making decisions based on this integrated value unreasonable and unscientific. To remedy this limitation, Zhang et al. [76] adopted the finite difference method to solve the chloride diffusion model in saturated coated concrete. In the diffusion model (Equations (3)–(6)), the properties of both treated layer and untreated substrate layer are separately considered. It is worth noting that the water-percolated porosity rather than the total porosity of the treated layer is used. It is necessary to use the water-percolated porosity, which contributes to chloride diffusion. For instance, the silane coating may not alter the total porosity, but obviously decreases the water-percolated porosity. In addition, Zhang et al. [76] analyzed the dependence of the pseudo-diffusion coefficient on the chloride diffusion coefficient of treated layer D_{st}, water-percolated porosity P_{st}, and the thickness of treated layer T_{st}. The results indicated that the pseudo-diffusion coefficient decreased with T_{st}, but increased with P_{st} and D_{st}.

$$\frac{\partial C}{\partial t} = \frac{\partial}{\partial x}\left(D\frac{\partial C}{\partial x}\right) \tag{3}$$

The initial condition:

$$C = 0,\ t = 0,\ x \geq 0 \tag{4}$$

Boundary conditions:

$$x = \infty,\ C = 0;\ x = 0,\ C = C_0,\ t > 0 \tag{5}$$

where C_0 denotes the external chloride content.

The continuity conditions in the interface of the treated layer and concrete substrate layer:

$$\begin{aligned} &P_{st}D_{st}\left.\frac{\partial C}{\partial x}\right|_{x=T_{st}} = P_{su}D_{su}\left.\frac{\partial C}{\partial x}\right|_{x=T_{st}} \\ &C_{st} = C_{su} \end{aligned} \tag{6}$$

where P_{st} denotes water-percolated porosity; P_{su} denotes the porosity of untreated concrete substrate; and D_{st} and D_{su} denote the chloride diffusion coefficient of treated layer and untreated concrete substrate, respectively. C_{st} and C_{su} denote the interfacial chloride content; and T_{st} denotes the thickness of treated layer.

Although the finite difference method can solve the chloride concentration profile, the calculation process is inefficient due to the low thickness of the treatment layer and the high requirement of space division. This limitation is especially prominent in the reliability calculation based on the Monte Carlo method. In response to this, Wang et al. [77] proposed an efficient mathematical method for predicting chloride transport in bilayer materials, that is, a rapid numerical approach. The diffusion equation is usually a partial differential equation with time and space variables. According to the separation-of-variables method, the function of the change in chloride ion concentration with respect to time and space can be written as the product of the part that depends only on space x and the part that depends only on time t, as in Equation (7). The analytical solution in Equation (7) is only valid when the diffusion coefficient is constant. By a step function, the chloride diffusion coefficient can be considered constant at each time step that is small enough. This approximation method allows the realization of rapid calculation, which significantly improves the calculation efficiency [75]:

$$C(x,T) = C_0 + \sum_{n=1}^{\infty} A_n f_n(x_n,\lambda_n)\exp\left(-\lambda_n^2 t\right) \tag{7}$$

where A_n is a coefficient that can be determined by known conditions; f_n is a function that can be determined by known conditions; λ_n and is the characteristic root.

5. Outlook

Effective coating materials for improving the chloride resistance of concrete have been widely developed and investigated, but more targeted studies are needed to quantify chloride transport behavior in coated concrete in the natural environment. This section presents an outlook for future research based on current studies regarding the chloride resistance of surface-coated concrete. Future research can focus on the following areas.

(i) There is significant disagreement in current research regarding the efficiency of silane-based coatings in enhancing concrete's resistance to chloride. Further detailed studies are needed, preferably considering the effects of coating processes, coating concentration, and the properties of concrete itself on improvement efficiency.

(ii) Although traditional coatings have demonstrated excellent performance in enhancing concrete's resistance to chloride, their environmental pollution limits their widespread use. Future work shall focus on developing green, environmentally friendly, safe, efficient, and low-cost multifunctional composite coatings.

(iii) Existing studies indicate that coating processes, sources of coating raw materials, and concentration ratios all influence the effectiveness of coatings in enhancing resistance to chloride. Further research should address this point specifically.

(iv) While current research has investigated the resistance of coatings to chloride, the focus has primarily been on short-term performance and laboratory studies. Long-term field tests should be conducted to gather data on the chloride ion transport behavior of coated concrete under natural climatic conditions.

(v) Current chloride transport modeling of coated concrete only considers diffusion mechanisms or modified parameters based on Fick's diffusion law to incorporate convective phenomena. The effect of coatings on moisture transport, bound and free chloride conversion, etc., behaviors also need to be considered. Moreover, the change in the treated layer properties with time will obviously affect the interfacial chloride concentration, thus affecting the chloride transport behavior. In future research, an effective model with interpretability, strong predictive ability, and practical application feasibility shall be developed, which can help to predict the service life.

(vi) The determination of the porosity, chloride profiles, and diffusion coefficient of the treated layer is still unclear, for which advanced techniques may be required to address this limitation.

(vii) The application and selection of coating shall consider the sustainability of coated concrete, including material selection, life cycle costs, and environmental impact assessments, to promote its widespread use in engineering practice.

Through the in-depth exploration of the above research directions, the performance and reliability of coated concrete in practical applications can be further improved, and more sustainable and durable solutions can be provided for engineering construction.

6. Conclusions

Concrete coatings have increasingly garnered attention for their promising potential in enhancing the chloride resistance of concrete structures, whether they are new or in service. A comprehensive review of reported studies on chloride penetration of coated concrete is presented in this study, including a specific elaboration of protection mechanisms, discussion of chloride resistance enhancement efficiency, summary of chloride transport modeling methodologies, and outlook for future research directions. For protection mechanisms, coatings can serve as continuous compact physical barriers, penetrants, waterproof films, or a combination of these three roles. Special emphasis shall be placed on identifying protective mechanisms, which may significantly influence the effectiveness, deterioration, and methodologies used to quantify chloride transport in the coatings. For chloride resistance enhancement efficiency, this study has thoroughly reviewed and compared the data reported in current studies. Although a great volume of research has been conducted in this field, few studies can provide data for quantitatively analyzing chloride transport behavior, especially for long-term in situ data, which are crucial for practical application. For modeling of chloride transport, the current methodologies can provide a rough quantification of chloride concentration profiles. More accurate physical models shall be developed in future research. Overall, while substantial progress has been made in developing effective coatings and studying the chloride resistance of coated concrete, further research and development in this critical area are necessary to enable broader applications of coatings in marine engineering.

Author Contributions: Conceptualization, J.L. and Y.W. (Yuanzhan Wang); methodology, J.L.; software, J.L.; validation, J.L., Y.W. (Yuanzhan Wang) and X.S.; formal analysis, J.L.; investigation, J.L.; resources, Y.W. (Yuchi Wang); writing—original draft preparation, J.L.; writing—review and editing, J.L. and Y.W. (Yuzhan Wang); visualization, J.L.; supervision, Yuchi Wang; project administration, Y.W. (Yuchi Wwang); funding acquisition, Y.W.(Yuchi Wang) and X.S. All authors have read and agreed to the published version of the manuscript.

Funding: This research was funded by the National Key Research and Development Program of China (2022YFB2603000), the National Natural Science Foundation of China (51979191), and the Fundamental Research Funds for the Central Research Institutes (TKS20200105).

Conflicts of Interest: The authors declare no conflicts of interest.

References

1. Wang, Y.; Liao, J.; Zhang, B. A Review of Chloride Penetration of Recycled Concrete with Enhancement Treatment and Service Life Prediction. *Materials* **2024**, *17*, 1349. [CrossRef] [PubMed]
2. Wang, Y.; Liao, J.; Liu, Z. Service Life Prediction and Environmental Impact Evaluation of Recycled Aggregate Concrete with Supplementary Cementitious Materials. *Constr. Build. Mater.* **2023**, *395*, 132270. [CrossRef]
3. Alexander, M.; Thomas, M. Service Life Prediction and Performance Testing—Current Developments and Practical Applications. *Cem. Concr. Res.* **2015**, *78*, 155–164. [CrossRef]
4. Saricimen, H.; Maslehuddin, M.; Iob, A.; Eid, O.A. Evaluation of a Surface Coating in Retarding Reinforcement Corrosion. *Constr. Build. Mater.* **1996**, *10*, 507–513. [CrossRef]
5. Moon, H.Y.; Shin, D.G.; Choi, D.S. Evaluation of the Durability of Mortar and Concrete Applied with Inorganic Coating Material and Surface Treatment System. *Constr. Build. Mater.* **2007**, *21*, 362–369. [CrossRef]
6. Medeiros, M.H.F.; Helene, P. Surface Treatment of Reinforced Concrete in Marine Environment: Influence on Chloride Diffusion Coefficient and Capillary Water Absorption. *Constr. Build. Mater.* **2009**, *23*, 1476–1484. [CrossRef]
7. Komada, M.; Asai, Y.; Morii, M.; Matsuki, M.; Sato, M.; Nagao, T. Maternal Bisphenol A Oral Dosing Relates to the Acceleration of Neurogenesis in the Developing Neocortex of Mouse Fetuses. *Toxicology* **2012**, *295*, 31–38. [CrossRef] [PubMed]
8. Kong, X.; Du, H.; Yin, S.; Zhang, W.; Shen, Y.; Fu, Y. Study on Preparation and Properties of Super-hydrophobic Cement-based Coating with High Stability. *Funct. Mater.* **2022**, *9*, 53.
9. Liao, J.; Wang, Y.; Wang, Y.; Lai, X. Preparation and Evaluation of TA/APTES-HDTMS Hydrophobic Nanocomposite Coating for Enhancing Corrosion Resistance of Concrete. *Constr. Build. Mater.* **2023**, *408*, 133792. [CrossRef]
10. Li, G.; Yang, B.; Guo, C.; Du, J.; Wu, X. Time Dependence and Service Life Prediction of Chloride Resistance of Concrete Coatings. *Constr. Build. Mater.* **2015**, *83*, 19–25. [CrossRef]
11. Wang, Y.; Liu, C.; Tan, Y.; Wang, Y.; Li, Q. Chloride Binding Capacity of Green Concrete Mixed with Fly Ash or Coal Gangue in the Marine Environment. *Constr. Build. Mater.* **2020**, *242*, 118006. [CrossRef]
12. Proske, T.; Hainer, S.; Rezvani, M.; Graubner, C.-A. Eco-Friendly Concretes with Reduced Water and Cement Contents—Mix Design Principles and Laboratory Tests. *Cem. Concr. Res.* **2013**, *51*, 38–46. [CrossRef]
13. Ryan, P.C.; O'Connor, A.J. Probabilistic Analysis of the Time to Chloride Induced Corrosion for Different Self-Compacting Concretes. *Constr. Build. Mater.* **2013**, *47*, 1106–1116. [CrossRef]
14. Arnon, B.; Diamond, S.; Berke, N. Corrosion control. In *Steel Corrosion in Concrete–Fundamentals and Civil Engineering Practice*; E and FN SPON: London, UK, 1997; pp. 94–145.
15. Barbucci, A.; Delucchi, M.; Cerisola, G. Organic Coatings for Concrete Protection: Liquid Water and Water Vapour Permeabilities. *Prog. Org. Coat.* **1997**, *30*, 293–297. [CrossRef]
16. Dulaijan, S.U.; Maslehuddin, M.; Al-Zahrani, M.M.; Sharif, A.M.; AlJuraifani, E.A.; Al-Idi, S.H. Performance evaluation of resin based surface coatings. In Proceedings of the 6th International Conference in Deterioration and Repair of Reinforced Concrete, Arabian Gulf, Bahrain, 20–22 November 2000; pp. 345–362.
17. Petcherdchoo, A. Pseudo-Coating Model for Predicting Chloride Diffusion into Surface-Coated Concrete in Tidal Zone: Time-Dependent Approach. *Cem. Concr. Compos.* **2016**, *74*, 88–99. [CrossRef]
18. Phiangphimai, C.; Joinok, G.; Phoo-ngernkham, T.; Damrongwiriyanupap, N.; Hanjitsuwan, S.; Suksiripattanapong, C.; Sukontasukkul, P.; Chindaprasirt, P. Durability Properties of Novel Coating Material Produced by Alkali-Activated/Cement Powder. *Constr. Build. Mater.* **2023**, *363*, 129837. [CrossRef]
19. Dai, J.-G.; Akira, Y.; Wittmann, F.H.; Yokota, H.; Zhang, P. Water Repellent Surface Impregnation for Extension of Service Life of Reinforced Concrete Structures in Marine Environments: The Role of Cracks. *Cem. Concr. Compos.* **2010**, *32*, 101–109. [CrossRef]
20. Christodoulou, C.; Goodier, C.I.; Austin, S.A.; Webb, J.; Glass, G.K. Long-Term Performance of Surface Impregnation of Reinforced Concrete Structures with Silane. *Constr. Build. Mater.* **2013**, *48*, 708–716. [CrossRef]

21. Gnanaraj, J.; Vasugi, K.A. Comprehensive Review of Hydrophobic Concrete: Surface and Bulk Modifications for Enhancing Corrosion Resistance. *Eng. Res. Express* **2024**, *6*, 032101. [CrossRef]
22. Yin, B.; Wu, C.; Hou, D.; Li, S.; Jin, Z.; Wang, M.; Wang, X. Research and Application Progress of Nano-Modified Coating in Improving the Durability of Cement-Based Materials. *Prog. Org. Coat.* **2021**, *161*, 106529. [CrossRef]
23. Li, G.; Cui, H.; Zhou, J.; Hu, W. Improvements of Nano-TiO_2 on the Long-Term Chloride Resistance of Concrete with Polymer Coatings. *Coatings* **2019**, *9*, 323. [CrossRef]
24. Li, G.; Yue, J.; Guo, C.; Ji, Y. Influences of Modified Nanoparticles on Hydrophobicity of Concrete with Organic Film Coating. *Constr. Build. Mater.* **2018**, *169*, 1–7. [CrossRef]
25. Wang, Z.; Han, M.; Zhang, J.; He, F.; Xu, Z.; Ji, S.; Peng, S.; Li, Y. Designing Preferable Functional Materials Based on the Secondary Reactions of the Hierarchical Tannic Acid (TA)-Aminopropyltriethoxysilane (APTES) Coating. *Chem. Eng. J.* **2019**, *360*, 299–312. [CrossRef]
26. *NT BUILD 355*; Concrete, Mortar and Cement-Based Materials: Chloride Diffuson Coefficient from Migration Cell Experiment. Nordtest: Taastrup, Denmark, 1997.
27. *NT BUILD 492*; Concrete, Mortar and Cement-Based Materials: Chloride Migration Coefficient from Non-Steady-State Migration Experiments. Nordtest: Taastrup, Denmark, 1999.
28. *ASTM C 1202–97*; Standard Test Method for Electrical Indication of Concrete's Ability to Resist Chloride Ion Penetration. American Society for Testing and Materials: West Conshohocken, PA, USA, 1997.
29. Tang, L.; Nilsson, L. Rapid Determination of the Chloride Diffusivity in Concrete by Applying an Electric Field. *ACI Mater. J.* **1992**, *89*, 49–53. [CrossRef]
30. *NT BUILD 443*; Concrete, Hardened: Accelerated Chloride Penetration. Nordtest: Taastrup, Denmark, 1995.
31. *ASTM, C1556-11a*; Standard Test Method for Determining the Apparent Chloride Diffusion Coefficient of Cementitious Mixtures by Bulk Diffusion. ASTM International: West Conshohocken, PA, USA, 2016.
32. Nguyen, Q.D.; Castel, A.; Kim, T.; Khan, M.S.H. Performance of Fly Ash Concrete with Ferronickel Slag Fine Aggregate against Alkali-Silica Reaction and Chloride Diffusion. *Cem. Concr. Res.* **2021**, *139*, 106265. [CrossRef]
33. Shafikhani, M.; Chidiac, S.E. A Holistic Model for Cement Paste and Concrete Chloride Diffusion Coefficient. *Cem. Concr. Res.* **2020**, *133*, 106049. [CrossRef]
34. Pillai, R.G.; Gettu, R.; Santhanam, M.; Rengaraju, S.; Dhandapani, Y.; Rathnarajan, S.; Basavaraj, A.S. Service Life and Life Cycle Assessment of Reinforced Concrete Systems with Limestone Calcined Clay Cement (LC3). *Cem. Concr. Res.* **2019**, *118*, 111–119. [CrossRef]
35. Angst, U.M. Predicting the Time to Corrosion Initiation in Reinforced Concrete Structures Exposed to Chlorides. *Cem. Concr. Res.* **2019**, *115*, 559–567. [CrossRef]
36. Nanukuttan, S.V.; Basheer, L.; McCarter, W.J.; Robinson, D.J.; Basheer, P.A.M. Full-Scale Marine Exposure Tests on Treated and Untreated Concretes Initial 7-Year Results. *ACI Mater. J.* **2008**, *105*, 81. [CrossRef]
37. Khanzadeh Moradllo, M.; Shekarchi, M.; Hoseini, M. Time-Dependent Performance of Concrete Surface Coatings in Tidal Zone of Marine Environment. *Constr. Build. Mater.* **2012**, *30*, 198–205. [CrossRef]
38. Li, Y.; Xu, H.; He, Y.; Xu, Q.; Fan, X.; Liu, C. Study on the Correlation between Indoor Accelerated Corrosion Testing of Concrete Epoxy Polysiloxane Coating and Real-Sea Environmental Testing in the East China Sea. *Case Stud. Constr. Mater.* **2024**, *20*, e02905. [CrossRef]
39. Zhang, Z.; Yao, X.; Wang, H. Potential Application of Geopolymers as Protection Coatings for Marine Concrete III. *Field Experiment. Appl. Clay Sci.* **2012**, *67–68*, 57–60. [CrossRef]
40. Thompson, J.L.; Silsbee, M.R.; Gill, P.M.; Scheetz, B.E. Characterization of Silicate Sealers on Concrete. *Cem. Concr. Res.* **1997**, *27*, 1561–1567. [CrossRef]
41. Higgins, R.C. *Waterproofing for Concrete. Reference and Guide*; Sinak Corp.: San Diego, CA, USA, 1985.
42. Franzoni, E.; Pigino, B.; Pistolesi, C. Ethyl Silicate for Surface Protection of Concrete: Performance in Comparison with Other Inorganic Surface Treatments. *Cem. Concr. Compos.* **2013**, *44*, 69–76. [CrossRef]
43. Pigino, B.; Leemann, A.; Franzoni, E.; Lura, P. Ethyl Silicate for Surface Treatment of Concrete—Part II: Characteristics and Performance. *Cem. Concr. Compos.* **2012**, *34*, 313–321. [CrossRef]
44. Sandrolini, F.; Franzoni, E.; Pigino, B. Ethyl Silicate for Surface Treatment of Concrete—Part I: Pozzolanic Effect of Ethyl Silicate. *Cem. Concr. Compos.* **2012**, *34*, 306–312. [CrossRef]
45. Medeiros, M.; Helene, P. Efficacy of Surface Hydrophobic Agents in Reducing Water and Chloride Ion Penetration in Concrete. *Mater. Struct.* **2007**, *41*, 59–71. [CrossRef]
46. Sakr, M.R.; Bassuoni, M.T. Silane and Methyl-Methacrylate Based Nanocomposites as Coatings for Concrete Exposed to Salt Solutions and Cyclic Environments. *Cem. Concr. Compos.* **2021**, *115*, 103841. [CrossRef]
47. Ibrahim, A.M.; Bassuoni, M.T.; Carroll, J.; Ghazy, A. Performance of Concrete Superficially Treated with Nano-Modified Coatings under Sulfuric Acid Exposures. *J. Build. Eng.* **2024**, *86*, 108957. [CrossRef]
48. Gu, W.; Liu, R.; Zhang, Y.; Yu, X.; Feng, P.; Ran, Q.; Zhang, Y.; Zhang, Y. Robust Water-Borne Multi-Layered Superhydrophobic Coating on Concrete with Ultra-Low Permeability. *Constr. Build. Mater.* **2024**, *411*, 134573. [CrossRef]
49. Schueremans, L.; Van Gemert, D.; Giessler, S. Chloride Penetration in RC-Structures in Marine Environment-Long Term Assessment of a Preventive Hydrophobic Treatment. *Constr. Build. Mater.* **2007**, *21*, 1238–1249. [CrossRef]

50. Zhang, Z.; Yao, X.; Zhu, H. Potential Application of Geopolymers as Protection Coatings for Marine concrete I. Basic Properties. *Appl. Clay Sci.* **2010**, *49*, 7–12. [CrossRef]
51. Zhang, Z.; Yao, X.; Zhu, H. Potential Application of Geopolymers as Protection Coatings for Marine concrete II. Microstructure and Anticorrosion Mechanism. *Appl. Clay Sci.* **2010**, *49*, 7–12. [CrossRef]
52. Lv, S.; Ma, Y.; Qiu, C.; Sun, T.; Liu, J.; Zhou, Q. Effect of Graphene Oxide Nanosheets of Microstructure and Mechanical Properties of Cement Composites. *Constr. Build. Mater.* **2013**, *49*, 121–127. [CrossRef]
53. Pan, Z.; He, L.; Qiu, L.; Korayem, A.H.; Li, G.; Zhu, J.W.; Collins, F.; Li, D.; Duan, W.H.; Wang, M.C. Mechanical Properties and Microstructure of a Graphene Oxide–Cement Composite. *Cem. Concr. Compos.* **2015**, *58*, 140–147. [CrossRef]
54. Saafi, M.; Tang, L.; Fung, J.; Rahman, M.; Liggat, J. Enhanced Properties of Graphene/Fly Ash Geopolymeric Composite Cement. *Cem. Concr. Res.* **2015**, *67*, 292–299. [CrossRef]
55. Mohammed, A.; Sanjayan, J.G.; Duan, W.H.; Nazari, A. Incorporating Graphene Oxide in Cement Composites: A Study of Transport Properties. *Constr. Build. Mater.* **2015**, *84*, 341–347. [CrossRef]
56. Wang, Q.; Qi, G.; Wang, Y.; Zheng, H.; Shan, S.; Lu, C. Research Progress on the Effect of Graphene Oxide on the Properties of Cement-Based Composites. *New Carbon Mater.* **2021**, *36*, 729–750. [CrossRef]
57. Liu, C.; Huang, X.; Wu, Y.-Y.; Deng, X.; Zheng, Z.; Xu, Z.; Hui, D. Advance on the Dispersion Treatment of Graphene Oxide and the Graphene Oxide Modified Cement-Based Materials. *Nanotechnol. Rev.* **2021**, *10*, 34–49. [CrossRef]
58. Antolín-Rodríguez, A.; Merino-Maldonado, D.; Fernández-Raga, M.; González-Domínguez, J.M.; Morán-del Pozo, J.M.; Pozo, M.; García-González, J.; Juan-Valdés, A. Microstructural, Durability and Colorimetric Properties of Concrete Coated with a Controlled Application of Graphene Oxide. *J. Build. Eng.* **2024**, *86*, 108920. [CrossRef]
59. Antolín-Rodríguez, A.; Merino-Maldonado, D.; González-Domínguez, J.M.; Fernández-Raga, M.; Morán-del Pozo, J.M.; García-González, J.; Juan-Valdés, A. Performance of Graphene Oxide as a Water-Repellent Coating Nanomaterial to Extend the Service Life of Concrete Structures. *Heliyon* **2024**, *10*, e23969. [CrossRef]
60. Antolín-Rodríguez, A.; Merino-Maldonado, D.; Rodríguez-González, Á.; Fernández-Raga, M.; González-Domínguez, J.M.; Juan-Valdés, A.; García-González, J. Statistical Study of the Effectiveness of Surface Application of Graphene Oxide as a Coating for Concrete Protection. *Coatings* **2023**, *13*, 213. [CrossRef]
61. Gao, J.; Li, C.; Lv, Z.; Wang, R.; Wu, D.; Li, X. Correlation between the Surface Aging of Acrylic Polyurethane Coatings and Environmental Factors. *Prog. Org. Coat.* **2019**, *132*, 362–369. [CrossRef]
62. Zhang, J.-Z.; Buenfeld, N.R. Chloride Profiles in Surface-Treated Mortar Specimens. *Constr. Build. Mater.* **2000**, *14*, 359–364. [CrossRef]
63. Aguiar, J.B.; Camões, A.; Moreira, P.M. Coatings for Concrete Protection against Aggressive Environments. *J. Adv. Concr. Technol.* **2008**, *6*, 243–250. [CrossRef]
64. Almusallam, A.A.; Khan, F.M.; Dulaijan, S.U.; Al-Amoudi, O.S.B. Effectiveness of Surface Coatings in Improving Concrete Durability. *Cem. Concr. Compos.* **2003**, *25*, 473–481. [CrossRef]
65. Qu, H.; Feng, M.; Li, M.; Tian, D.; Zhang, Y.; Chen, X.; Li, G. Enhancing the Carbonation and Chloride Resistance of Concrete by Nano-Modified Eco-Friendly Water-Based Organic Coatings. *Mater. Today Commun.* **2023**, *37*, 107284. [CrossRef]
66. Basha, S.I.; Aziz, M.A.; Ahmad, S.; Al-Zahrani, M.M.; Shameem, M.; Maslehuddin, M. Improvement of Concrete Durability Using Nanocomposite Coating Prepared by Mixing Epoxy Coating with Submicron/Nano-Carbon Obtained from Heavy Fuel Oil Ash. *Constr. Build. Mater.* **2022**, *325*, 126812. [CrossRef]
67. Shi, L.; Liu, J.; Liu, J. Effect of Polymer Coating on the Properties of Surface Layer Concrete. *Procedia Eng.* **2012**, *27*, 291–300. [CrossRef]
68. Xiao, Z.; Liu, Y.; Wang, Y.; Shi, J. TA/Fe(III) Anti-Chloride Coating to Protect Concrete. *J. Clean. Prod.* **2020**, *259*, 120922. [CrossRef]
69. Han, S.; Han, T.H.; Kim, J.H. Surface Coating Method for Cement-Based Materials to Improve Chloride Ion Penetration Resistance Using Amine-Based CO_2 Solvent. *J. CO_2 Util.* **2024**, *83*, 102823. [CrossRef]
70. Kong, X.; Shen, Y.; Shi, J.; Zhang, N.; Kang, R.; Fu, Y. Superhydrophobic Concrete Coating with Excellent Mechanical Robustness and Anti-Corrosion Performances. *Colloids Surf. A Physicochem. Eng. Asp.* **2024**, *684*, 133157. [CrossRef]
71. Luo, X.; Yin, H.; Li, T.; Ran, J.; Fu, X.; Hu, C. Fabrication of Robust Superhydrophobic Myristic Acid/ZrO2 Coating with Improved Anti-Icing and Anti-Corrosion for Concrete via a Two-Step Method. *Colloids Surf. A Physicochem. Eng. Asp.* **2024**, *693*, 134040. [CrossRef]
72. Diamanti, M.V.; Brenna, A.; Bolzoni, F.; Berra, M.; Pastore, T.; Ormellese, M. Effect of Polymer Modified Cementitious Coatings on Water and Chloride Permeability in Concrete. *Constr. Build. Mater.* **2013**, *49*, 720–728. [CrossRef]
73. Liang, C.; Zhao, P.; Zou, H.; Song, Q.; Hou, P.; Huang, Y.; Wang, S.; Lu, L. Introducing Fiber to Enhance the Mechanical Properties and Durability of Polymer-Modified Cement-Based Coating. *Constr. Build. Mater.* **2023**, *372*, 130842. [CrossRef]
74. Zhang, M.; Xu, H.; Phalé Zeze, A.L.; Liu, X.; Tao, M. Coating Performance, Durability and Anti-Corrosion Mechanism of Organic Modified Geopolymer Composite for Marine Concrete Protection. *Cem. Concr. Compos.* **2022**, *129*, 104495. [CrossRef]
75. Wang, T.; Li, C.; Zheng, J.; Hackl, J.; Luan, Y.; Ishida, T.; Medepalli, S. Consideration of Coupling of Crack Development and Corrosion in Assessing the Reliability of Reinforced Concrete Beams Subjected to Bending. *Reliab. Eng. Syst. Saf.* **2023**, *233*, 109095. [CrossRef]

76. Zhang, J.-Z.; McLoughlin, I.M.; Buenfeld, N.R. Modelling of Chloride Diffusion into Surface-Treated Concrete. *Cem. Concr. Compos.* **1998**, *20*, 253–261. [CrossRef]
77. Wang, T.; Zheng, J.-J.; Dai, J.-G. Analysis of Time-Dependent Chloride Diffusion in Surface-Treated Concrete Based on a Rapid Numerical Approach. *Struct. Infrastruct. Eng.* **2023**, *19*, 332–344. [CrossRef]

materials

Article

Effects of Fineness and Morphology of Quartz in Siliceous Limestone on the Calcination Process and Quality of Cement Clinker

Donggen Nie [1], Wei Li [1,2,*], Lilan Xie [3], Min Deng [1,*], Hao Ding [4] and Kaiwei Liu [2]

1 College of Material Science and Engineering, Nanjing Tech University, Nanjing 211816, China
2 Anhui Province Key Laboratory of Advanced Building Materials, Anhui Jianzhu University, Hefei 230000, China
3 School of Materials and Architectural Engineering, Guizhou Normal University, Guiyang 550001, China
4 Hefei Cement Research and Design Institute Corporation Ltd., Hefei 230000, China
* Correspondence: stevenlee187@outlook.com (W.L.); dengmin@njtech.edu.cn (M.D.)

Abstract: With the increasing depletion of high-quality raw materials, siliceous limestone, sandstone and other hard-to-burn raw materials containing crystalline SiO_2 are gradually being used to produce clinker. This study investigates the influence of the quartz content and particle size in siliceous limestone on the calcination process and the resultant quality of cement clinker. Two different siliceous limestones were grinded to different fineness, and calcinated with some other materials. The content of the clinkers was analyzed with the XRD–Rietveld method and the microstructure of the clinkers was observed with laser scanning confocal microscopy (LSCM) and field emission scanning electron microscopy (FESEM). Three key outcomes of this study provide new insights on the use of siliceous limestone in cement production, namely that (i) reducing the fineness values of siliceous limestone from 15% to 0% of residue on a 0.08 mm sieve decreases the quantity of these larger quartz particles, resulting in an increase in C_3S content by up to 8% and an increase in 28d compressive strength by up to 4.4 Mpa, which is 62.30 Mpa; (ii) the morphology of quartz—either as chert nodules or single crystals—affects the microstructure of C_2S clusters in clinker, finding that chert nodules result in clusters with more intermediate phases, whereas large single crystals lead to denser clusters; (iii) the sufficient fineness values of siliceous limestone SL1 and SL2 are 5% and 7% of residue on a 0.08 mm sieve, respectively, which can produce a clinker with a 28d compressive strength greater than 60 Mpa, indicating that for different kinds of quartz in siliceous limestone, there is an optimum grinding solution that can achieve a balance between clinker quality and energy consumption without having to grind siliceous limestone to very fine grades.

Keywords: siliceous limestone; clinker; quartz; microstructure; fineness

Citation: Nie, D.; Li, W.; Xie, L.; Deng, M.; Ding, H.; Liu, K. Effects of Fineness and Morphology of Quartz in Siliceous Limestone on the Calcination Process and Quality of Cement Clinker. *Materials* **2024**, *17*, 3601. https://doi.org/10.3390/ma17143601

Academic Editor: Dimitrios Papoulis

Received: 17 June 2024
Revised: 18 July 2024
Accepted: 18 July 2024
Published: 21 July 2024

1. Introduction

Carbon emission reduction in the cement industry is a key and difficult issue in the implementation of the "double carbon" goal. SiO_2 is the main component of cement raw meal [1], and the crystalline SiO_2 in raw meal has a decisive influence on the energy consumption and clinker quality during the process of calcination. With the increasing depletion of high-quality raw meal, siliceous limestone, sandstone, shale and other hard-to-burn raw meal containing crystalline quartz are gradually being used to produce cement clinker. Studies have shown that the large-size quartz in the raw meal are weak to abrasion, and may influence the burnability of raw meal and the resultant quality of cement clinker [2,3]. The question of how to solve the problem of clinker performance degradation caused by large-size quartz is of great importance to improve clinker strength and reduce the energy consumption of clinker production.

SiO_2, as the main component of Portland cement raw meal, mainly exists in the form of quartz minerals and silicate minerals in the raw meal. Quartz is a stable tetrahedral Si-O structure, with a melting point higher than the clinker calcination temperature, making it difficult to depolymerize into more reactive $[SiO_4]^{4-}$ structure; hence, its ability to combine with CaO to form C_2S and C_3S is poor compared to silicate minerals, which is unfavorable for clinker calcination [4]. Quartz undergoes polymorphic transformations at different temperatures, but its tetrahedral Si-O structure does not change; thus, its impact on clinker calcination is still noticeable. However, research has found that the reactivity of different polymorphs of quartz crystals with CaO does show some differences, typically increasing in the following order: α-quartz < chalcedony < α-tridymite < α-cristobalite < amorphous SiO_2 [5,6]. Besides, the particle size and content of quartz are other key factors affecting the calcination of raw meal [7–11].

Currently, it is widely believed that the fineness of quartz particles in raw meal significantly affects the calcination of clinker [7,8,12–16]. Quartz particles mainly consist of quartz crystals of various shapes and sizes. Fundal and Christensen et al. [7,17,18] have proposed that quartz particle sizes exceeding 45 μm can adversely affect clinker calcination, whereas quartz particles smaller than 45 μm do not impact the reactivity of raw meal. Jiang et al. [19] investigated the distribution of quartz particles in different sandstone powders after grinding and their relationship with the content of f-CaO in the clinker. They found that the coarse quartz particle (125–160 μm) content has a correlation of about 95% with the content of f-CaO, while quartz particles smaller than 30 μm only have a correlation of about 50% with the content of f-CaO, indicating that the larger the quartz particles, the greater the impact on the content of f-CaO. Zhang [9] has used quartz-based sandstone of different levels of fineness to produce clinker, studying the effect of quartz particle fineness on the content of f-CaO in the clinker. It was pointed out that when the sieve residue of quartz particles exceeds 0.89% at 80 μm, the content of f-CaO will exceed 1.5%. It is generally believed that the impact of quartz crystals on calcination is that quartz crystals affect the formation of belite, thereby affecting calcination and leading to a decline in quality [15]. However, some studies have pointed out that coarse quartz particles react with CaO to form different belite clusters. The surface of these belite cluster is dense, which hinders the reaction between belite and CaO, thus affecting the formation of alite [20,21] and leading to an increase in the content of f-CaO [22–24].

The current research on the impact of quartz fineness on calcination is primarily focused on quartz particles, which are mixtures composed of quartz crystals of various shapes and sizes, and other minerals. This complex and diverse structure cannot accurately reflect the reaction processes of different quartz crystals and its influence on the calcination of clinker [25]. Therefore, further studies are needed, and so the purpose of this paper is to investigate the effect of the fineness and morphology of quartz in siliceous limestones on the calcination process and quality of the clinker. Two siliceous limestones with different morphologies of quartz from China were used. The internal standard methods of X-ray diffraction (XRD) and polarizing microscope analysis were chosen to analyze the distribution of quartz in raw meal. The XRD–Rietveld method [26] was chosen to determine the quantity of clinker minerals, and laser scanning confocal microscopy (LSCM) was used to analyze the microstructure of the clinker. Field emission scanning electron microscopy (FESEM) equipped with EDS was used to analyze the elemental distribution of clinker minerals. This investigation is anticipated to advance the knowledge on quartz's role in clinker calcination, and offer practical insights for improving the efficiency of clinker calcination using siliceous limestone.

2. Materials and Methods

2.1. Materials

Siliceous limestones SL1 and SL2, derived from Hongshi Cement Company located in Sanming, Fujian, China and Southwest Cement Company located in Guiyang, Guizhou, China, were used. Limestone L1, sandstone S1, aluminum corrective material FA, iron corrective material IS and coal derived from South Cement Company located in Hefei, Anhui, China were used. Table 1 shows the chemical compositions of the raw materials analyzed according to the standard GB/T 176-2017 [27], with Thermo ARL 9900 X-ray fluorescence (XRF) (Waltham, America). Siliceous limestones SL1 and SL2 contained 27.02% and 6.48% SiO_2, respectively. While SL1, SL2 and limestone L1 contained 37.00%, 50.91% and 53.84% CaO, respectively. Siliceous sandstone S1 contained 73.08% SiO_2. Aluminum corrective material FA contained 24.94% Al_2O_3, and iron corrective material IS contained 45.57% Fe_2O_3. Figure 1 shows the XRD patterns of three types of limestone determined according to JY/T 0587-2020. The ICSD card numbers indicating the types of calcite and quartz minerals in SL1, SL2, and L1 are 079674 and 083849, respectively [28]. The mineral content of three limestones were collected on a Rigaku Smartlab 3 kW diffractometer (Tokyo, Japan) using Cu Kα1 radiation (λ = 0.154 nm), operated in the reflection geometry ($\theta/2\theta$) at room temperature, and the X-ray tube was operated at 40 kV and 30 mA. For the sample powders, data were collected between 5° and 70°, with a scanning speed of 5°/min. Limestones SL1, SL2 and L1 were mainly composed of calcite. Siliceous limestones SL1 and SL2 contained some quartz, while limestone L1 contained a small amount of quartz. The quartz content in siliceous limestones SL1 and SL2 is 21.7 wt% and 5.4 wt%, obtained by the XRD internal standard method.

Table 1. Chemical composition of the raw materials.

Samples	Chemical Composition/wt.%									
	LOI	SiO_2	Fe_2O_3	Al_2O_3	CaO	MgO	K_2O	Na_2O	SO_3	Total
Siliceous limestone SL1	28.33	27.02	2.16	0.48	37.00	3.80	0.17	0.07	0.07	99.10
Siliceous limestone SL2	40.08	6.48	1.08	0.24	50.91	0.24	0.16	0.05	0.07	99.31
Limestone L1	42.83	0.45	0.33	0.12	53.84	0.26	0.03	0.03	0.05	97.94
Sandstone S1	7.25	73.08	3.84	7.21	5.12	0.49	0.71	0.06	0.09	97.85
Fly ash FA	6.29	57.42	5.73	24.94	1.83	1.45	0.67	0.51	0.07	98.91
Iron slag IS	19.34	6.48	45.57	2.24	15.24	3.99	0.51	0.55	0.11	94.03
Coal ash CA	0.45	50.39	5.48	36.30	4.22	0.24	-	-	0.08	97.16

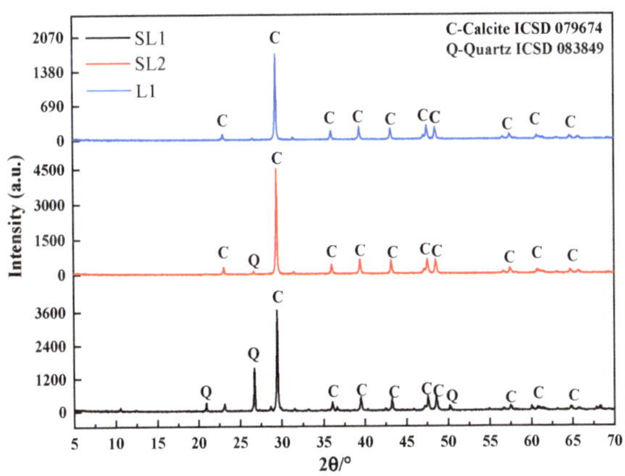

Figure 1. X-ray diffraction patterns of siliceous limestone SL1, SL2 and limestone L1.

Figure 2 demonstrates cross-polarized images of the siliceous limestones from thin sections. The morphology of quartz in SL1 and SL2 varies considerably. Siliceous limestone SL1 is mainly composed of chert nodules and calcite, and the chert nodules range from 300 to 2500 μm, composed of less than 110 μm quartz crystal. Siliceous limestone SL2 mainly consists of calcite and quartz crystals, and the crystal size of quartz in SL2 ranges from 50 to 5500 μm.

(**a**) SL1 (**b**) SL2

Figure 2. Petrographic microstructure of siliceous limestones (**a**) SL1 and (**b**) SL2.

2.2. Analytical Methods

2.2.1. Product Fineness

Cement test ball mill SM-500 is used to produce different fineness raw meal. Siliceous limestones SL1 and SL2 are ground to 20 ± 2%, 15 ± 2%, 10 ± 2%, 7% ± 2% and 0% of residue on an 80 μm sieve, while limestone L1 is ground to 10 ± 1% of residue on an 80 μm sieve. Aluminum corrective material and iron corrective material are ground to 20 ± 1% of residue on an 80 μm sieve. Pulverized coal is calcined to coal ash and then mixed at 1.5 wt.% with other raw meal.

Based on the three modulus values of the Southwest Cement Company, the lime saturation factor (KH), silica modulus (SM), and alumina modulus (IM) are set to the following values: KH: 0.90 ± 0.02, SM: 2.20 ± 0.02 and IM: 1.50 ± 0.02. As shown in Figure 3, siliceous limestones SL1 and SL2 of different fineness were mixed with other corrective materials and pressed to form a 60 mm × 60 mm × 10 mm rectangular specimen. The raw mix design is presented in Table 2. The specimens were then thermally treated in an electric furnace at a heating rate of 5 °C/min to 900 °C and then held for 30 min, and continued to be heated at a rate of 5 °C/min to 1450 °C and held for 30 min, and were then cooled rapidly in air.

Figure 3. A 60 mm × 60 mm × 10 mm rectangular specimen schematic diagram for raw meal.

Table 2. Raw mix design.

No.	Raw Materials/%						
	Siliceous Limestone SL1	Siliceous Limestone SL2	Limestone L1	Sandstone S1	Fly Ash FA	Iron Slag IS	Coal Ash CA
1	-	83.5	-	2.6	7.2	5.4	1.5
2	23.6	-	61.5	-	8.3	5.1	1.5

2.2.2. Measurement of Quartz Content in Siliceous Limestones

The internal standard method of X-ray diffraction was used for the quantitative measurement of quartz in siliceous limestones. $CaCO_3$ and SiO_2, with analytical purity, were used to prepare the mixtures used for the determination of the work curve. The batch of the mixture is listed in Table 3. A total of 50% TiO_2 with analytical purity was used as the internal standard substance, and mixed with the prepared mixtures. The X-ray powder diffraction data were collected by the Rigaku Smartlab 3kW diffractometer at 25 °C. X-ray scanning was carried out at a diffraction angle of 24° to 28° in steps of 1° per minute. The obtained data include the strongest peak of TiO_2 and the strongest peak of SiO_2, as shown by Figure 4. The established relationship between the content of quartz and the ratio of the peak intensity of quartz to TiO_2 in the mixtures of calcite, quartz, and TiO_2 is shown in Figure 5. Figure 5 presents the y-axis of the content of quartz in mixtures as a function of the (x-axis) ratio of the peak intensity of quartz and TiO_2. A best fit linear equation and the corresponding R^2 for the eight data points are shown in Figure 5.

Table 3. Composition of mixtures for determination of work curve.

Sample	1	2	3	4	5	6	7	8
$CaCO_3$	10%	20%	40%	50%	60%	70%	80%	90%
SiO_2	90%	80%	60%	50%	40%	30%	20%	10%
Total	100%	100%	100%	100%	100%	100%	100%	100%

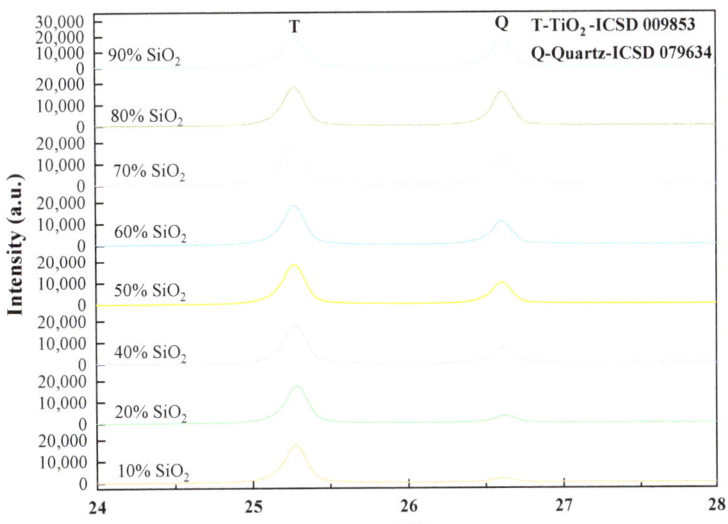

Figure 4. Peaks for quartz and TiO_2 in the mixtures of Calcite-Quartz-TiO_2.

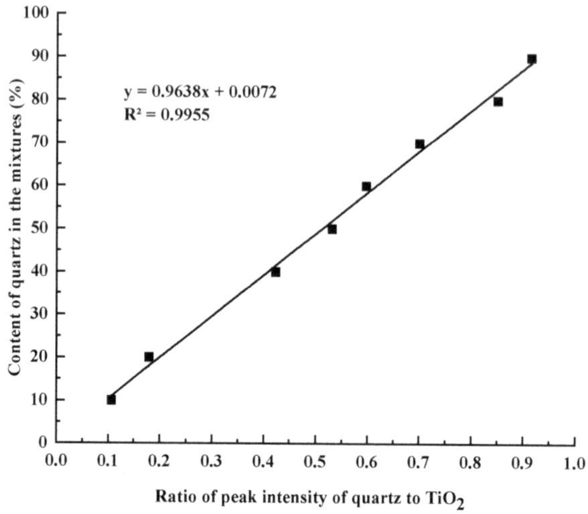

Figure 5. The relationship between the content of quartz and ratio of the peak intensity of quartz to TiO_2 in the mixtures of calcite, quartz and TiO_2.

Siliceous limestones with different fineness were dissolved by 1:1 dilute hydrochloric acid solution, filtered, and then dried in an oven at 105 °C for 12 h to obtain the acid insoluble content of different siliceous limestones. After that, the acid insoluble was mixed with 50% analytically pure TiO_2 powder, and was analyzed with Rigaku Smartlab 3 kW diffractometer in the 2θ range from 24° to 28° in steps of 1° per minute. The peak intensity of quartz and TiO_2 was calculated, and the content of quartz in siliceous limestones was determined according to Figure 5.

2.2.3. Quantification of Clinker Minerals

The calcined clinker was finely ground by a vibrating mill to pass through a 0.08 mm sieve. The free lime content in clinker was detected according to the glycol method in GB/T 176-2017. The XRD–Rietveld method was used to determine the content of C_3S, C_2S, C_3A and C_4AF, respectively. The data were collected by a Rigaku Smartlab 3 kW diffractometer from 5° to 65° (2θ) at a rate of 1°/min. The X-ray tube worked at 40 kV and 30 mA, and the quantitative analysis was performed with the Rietveld method using HighScore Plus 3.0e software.

2.2.4. Microstructure Analysis

The calcined clinker was crushed and sieved to a particle size range of 2.5–5.0 mm. The selected clinker particles were inlaid with epoxy resin and cured in an oven at 40 °C to prevent excessive epoxy from penetrating into the clinker pores. Then, the samples were finely polished with the Buehler AutoMet 250 Grinder Polisher System (Lake Country, America) to ensure that no scratches were visible on the polished surface. A 1 wt% NH_4Cl solution was selected to etch the polished surface to make each mineral show different colors in the optical microscope. The polished surface was immersed in 1 wt% NH_4Cl solution for 4–6 s and dried quickly with a hair dryer. An OLS4000 3D measurement laser confocal microscope (LSCM) (Tokyo, Japan) was used to observe the microstructure of the clinker phases and the distribution of the clinker minerals. Zeiss Ultra-55 field emission scanning electron microscopy (FESEM) (Jena, Germany), equipped with Oxford EDS in working condition with a 10–15 kV accelerating voltage and a vacuum environment of 1×10^{-5} bar, was used to analyze the elemental distribution in the clinker calcinated by siliceous limestones SL1 and SL2.

2.2.5. Mechanical Property

The flexural and compressive strength of the cement clinker were tested according to the standard GB/T 17671-2021 test method of cement mortar strength [29]. The different clinkers were mixed with 5 wt.% gypsum and ground in a ball mill to form a cement with a specific surface area of 345 m^2/kg. The mortar samples were molded to a size of 40 mm × 40 mm × 160 mm with a water to binder ratio of 0.5, and cured in a curing chamber at a relative humidity of more than 98% and a temperature of 20 ± 1 °C. The flexural and compressive strength of the specimen were measured at 28 days, and the instrument used was an ETM Series F electronic universal testing machine from Shenzhen WANCE Test Equipment Co., Ltd (Shenzhen, China).

3. Results

3.1. Quartz Distribution in Different Particle Size of Siliceous Limestones

3.1.1. Content of Quartz in Different Particle Size of Siliceous Limestones

Based on the internal standard method, the content of quartz in siliceous limestones SL1 and SL2 were 21.7% and 5.4%, respectively. After grinding, the fineness values of siliceous limestone SL1 were 19.41%, 14.28%, 9.32%, 5.10% and 0% of residue on an 80 μm sieve, respectively. While the fineness values of SL2 were 20.22%, 16.34%, 11.39%, 6.86% and 0% of residue on an 80 μm sieve, respectively. Figure 6 shows the content of quartz in different particle sizes of siliceous limestone SL1 and SL2. It can be seen from Figure 6a,b that for different fineness values of siliceous limestone, when the particle size is large than 0.045 mm, the content of quartz in different particle size series is larger than the average quartz content in the siliceous limestone, which means large grain size quartz crystals and chert are more difficult to grind than calcite. Additionally, for particle sizes greater than 200 μm, the content of quartz continues to decrease as the fineness of the siliceous limestone decreases from 20% to 7%, while for other particle size ranges, the trend is not consistent.

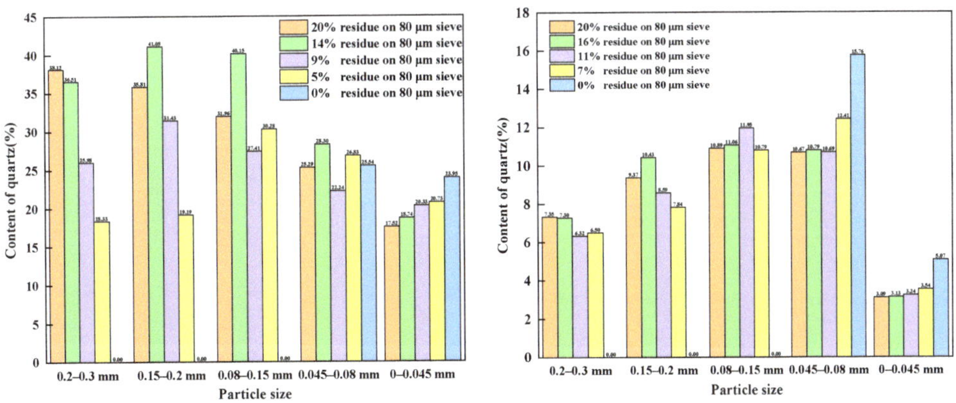

(**a**) SL1 (**b**) SL2

Figure 6. Content of quartz in different particle sizes of siliceous limestones: (**a**) SL1 and (**b**) SL2.

Figure 7 shows particle size distribution of different fineness values for siliceous limestones. It can be seen from Figure 7a,b that for siliceous limestone with different fineness values, when the particle size is greater than 0.08 mm, the powder content of each particle size gradually decreases. However, for particles with sizes between 0.045 mm and 0.08 mm, the powder content increases, indicating that with the decrease in fineness, the siliceous limestone powder is mainly concentrated below 45 μm.

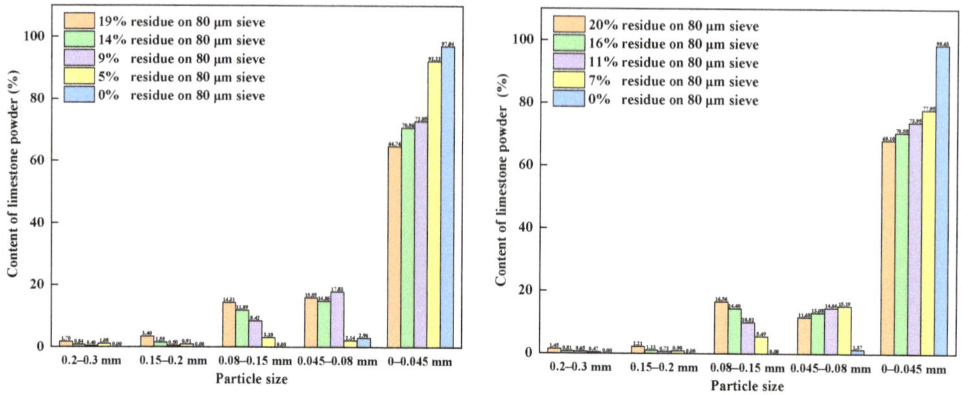

(**a**) SL1 (**b**) SL2

Figure 7. Particle size distribution of different fineness values of siliceous limestone: (**a**) SL1 and (**b**) SL2.

Figure 8 shows the distribution of quartz content by the particle size introduced by different fineness values of siliceous limestones in the raw meal. As shown in Figure 8, the content of quartz decreases with decreasing particle size for siliceous limestones larger than 0.08 mm, with the exception of SL1 in the 0.08–0.15 mm range, which has a fineness of 15% on an 80 μm sieve. It is also shown that even with a fineness of 10%, there are still some quartz crystals large than 0.08 mm. However, when continuing to grind the siliceous limestone to 0% of residue on a 0.08 mm sieve, the quartz crystals are mainly located in the particle size range of 0 to 0.045 mm. Siliceous limestones SL1 and SL2 account for 96.82% and 95.86% of the quartz content in the raw material, respectively.

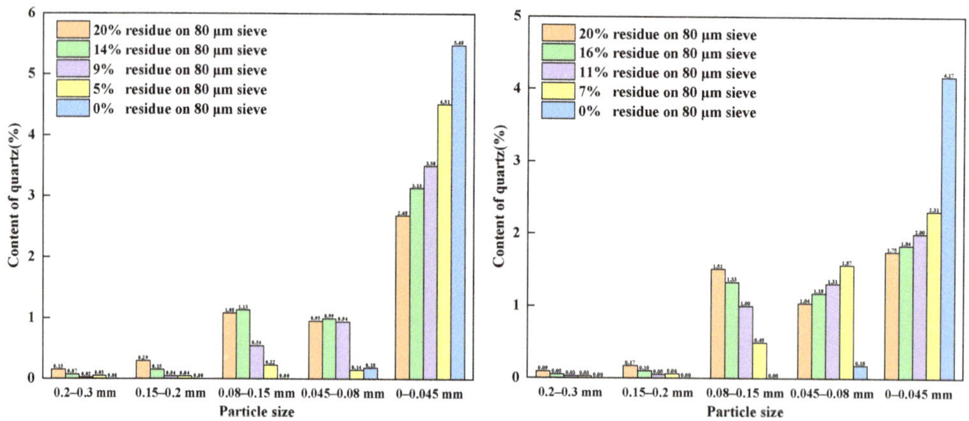

(**a**) SL1 (**b**) SL2

Figure 8. Distribution of quartz content by particle size introduced by different fineness values of siliceous limestones in the raw meal: (**a**) SL1 and (**b**) SL2.

3.1.2. Microstructure of Quartz with Different Particle Size

The petrographic method was used to analyze the distribution of quartz crystals in siliceous limestones SL1 and SL2. Figure 9 shows the distribution of quartz with different particle sizes in SL1. As can be seen from Figure 9, quartz crystals exist independently as chert nodules in powders with a particle size greater than 0.2 mm, within 0.15–0.2 mm and within 0.045–0.080 mm. When the particle size was between 0.080 and 0.15 mm, some quartz crystals were found to be distributed in the calcite matrix, with only a small amount of quartz distributed independently.

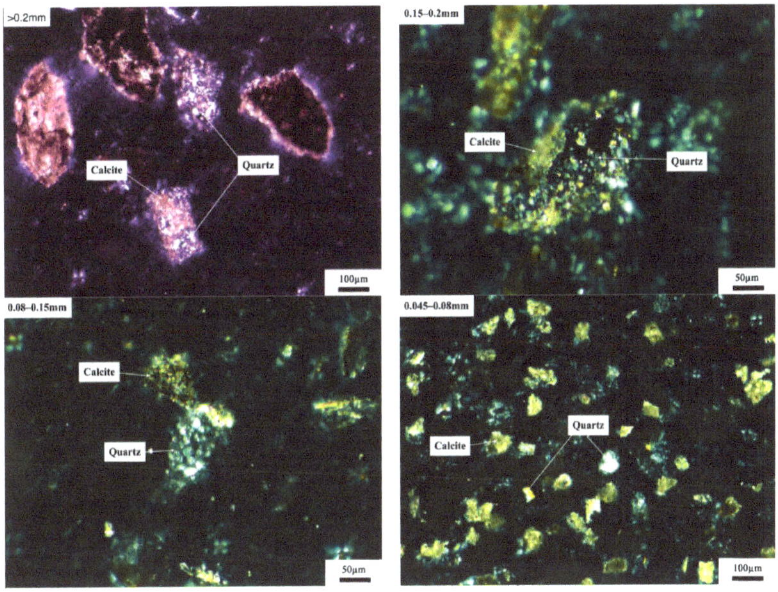

Figure 9. Distribution of quartz crystals with different particle sizes in siliceous limestone SL1.

Figure 10 shows the distribution of quartz with different particle sizes in SL2. As shown in Figure 10, for all particle sizes of siliceous limestone SL2, quartz crystals were found to be distributed independently in the powders, which is a little different from that of SL1. And this may due to the different morphology of quartz between SL1 and SL2.

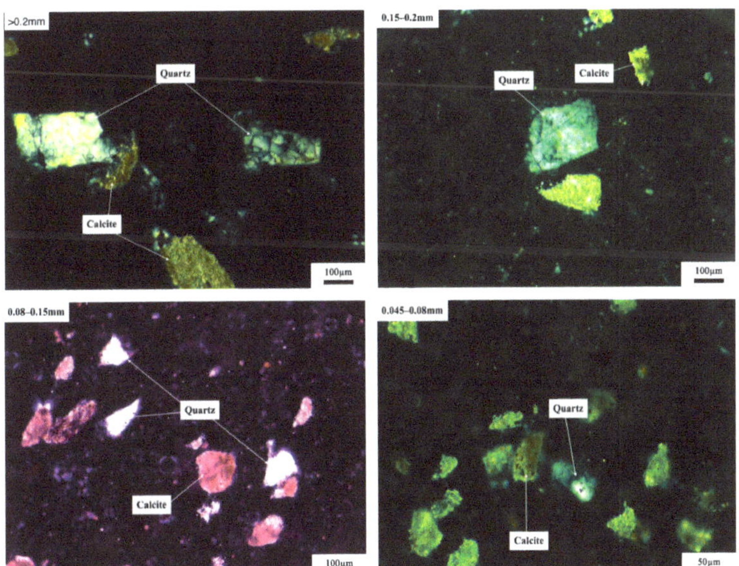

Figure 10. Distribution of quartz crystals with different particle sizes in siliceous limestone SL2.

3.2. *Effects of Siliceous Limestone Fineness on Clinker Minerals*

3.2.1. Burnability and Mineral Composition

Free lime content is one of the most important parameters for clinker quality. Figure 11 shows the free lime content of different clinkers calcined with siliceous limestone of different fineness. As shown in Figure 11, as the fineness of siliceous limestone in the raw meal decreases, the free lime content of clinker decreases, which indicates that the decrease in the fineness of siliceous limestone favors the burnability of clinker. With the same fineness of siliceous limestone, the free lime content of SL1 is lower than that of SL2, which indicates the burnability of SL1 is better than that of SL2.

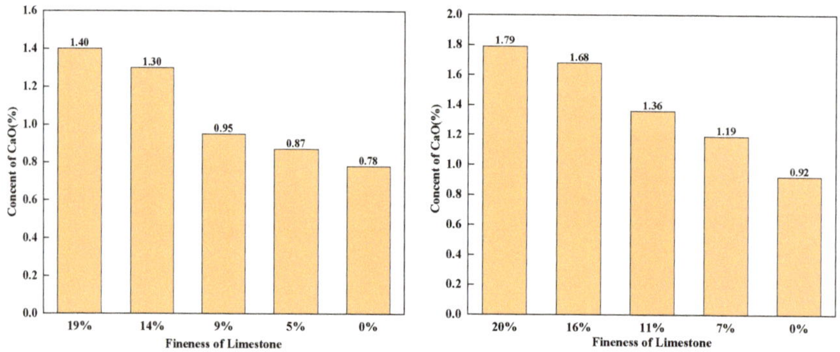

(**a**) SL1 (**b**) SL2

Figure 11. Effect of different fineness values of siliceous limestone on free lime content. (**a**) SL1 and (**b**) SL2.

Figure 12 shows the mineral content of calcined clinker for different fineness values of siliceous limestones SL1 and SL2. As shown in Figure 12a, as the fineness of the siliceous limestone decreases from 20% to 15%, the content of C_3S increases from 51.2% to 56.7%, while the content of C_2S decreases from 26.7% to 20.7%. However, as the fineness of SL1 decreases from 15% to 10%, the content of C_3S only increases by 1%. It can also be seen from Figure 12a that the content of C_3A is much lower than that of SL2. As shown in Figure 12b, with different fineness values of siliceous limestone SL2 in the raw meal, the content of C_3S and C_2S only vary a little. When the fineness of siliceous limestone decreases from 20% to 10%, the content of C_3S only increases from 60% to 63%, while the content of C_2S decreases from 18.0% to 16.7%.

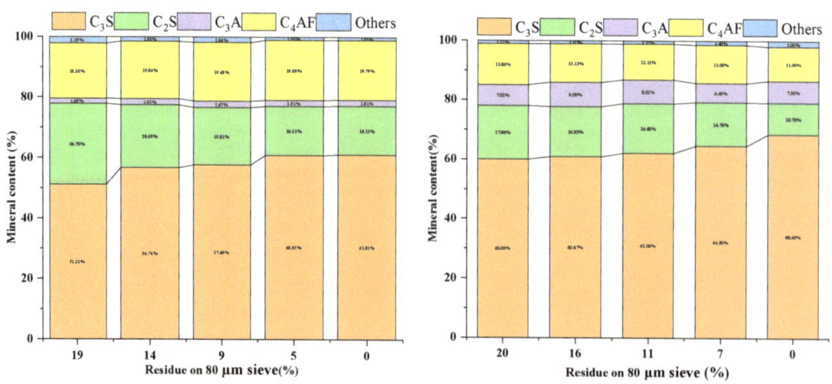

(**a**) SL1 (**b**) SL2

Figure 12. Mineral content of calcined clinker for different fineness values of siliceous limestone. (**a**) SL1 and (**b**) SL2.

3.2.2. Microstructure of Clinker Minerals

Figure 13 shows the microstructure of the clinker calcined with siliceous limestone SL1 with fineness values of 19%, 14%, 9% and 5% of residue on an 0.08 mm sieve. It can be seen from Figure 13 that there are obvious C_2S clusters in the clinker calcined with different fineness values of siliceous limestone, and most of the C_2S clusters are within the size range of 350–550 μm. In addition, the clinker morphology does not vary much when the fineness of the siliceous limestone SL1 decreases from 19% to 5% of residue on a 0.08 mm sieve. As shown in Figure 14, when the fineness of siliceous limestone SL1 continues to decrease to 0% of residue on a 0.08 mm sieve, the size of C_2S clusters obviously decrease, and most of the C_2S clusters are within the size range of 50–200 μm.

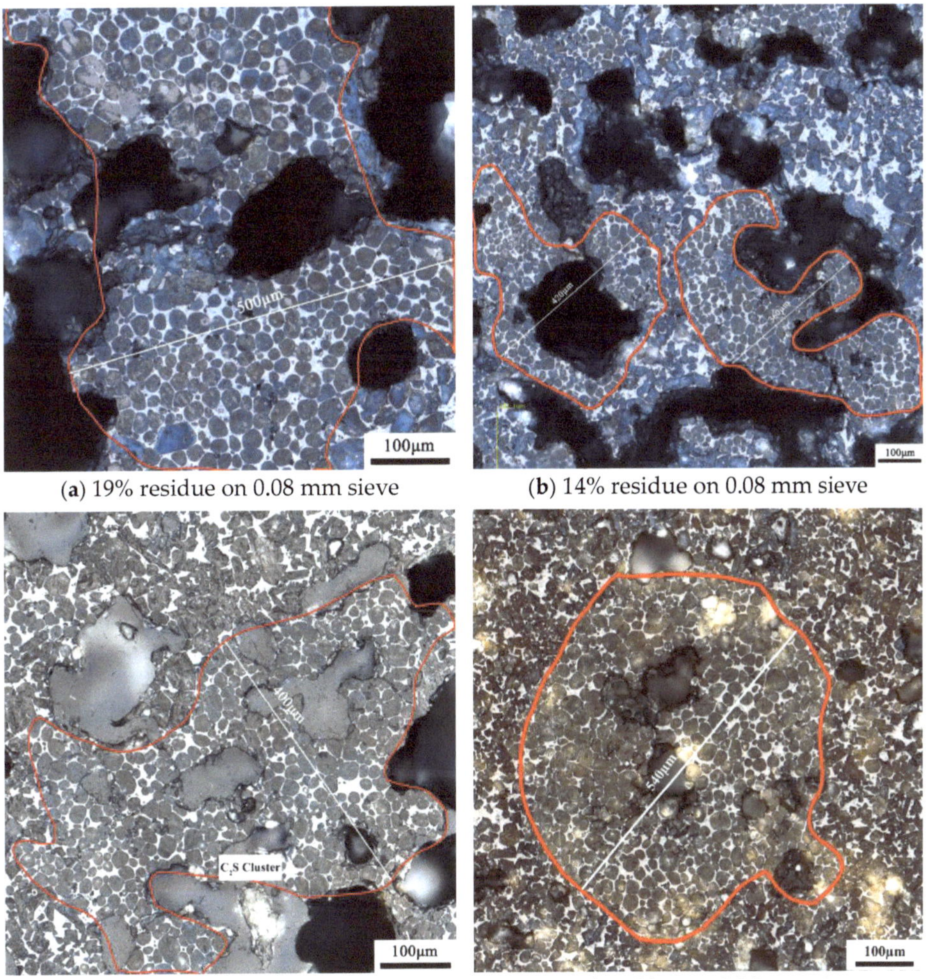

(**a**) 19% residue on 0.08 mm sieve (**b**) 14% residue on 0.08 mm sieve

(**c**) 9% residue on 0.08 mm sieve (**d**) 5% residue on 0.08 mm sieve

Figure 13. Petrographic images of clinker after calcination with different fineness values of siliceous limestone SL1. (**a**) 19% residue on 0.08 mm sieve, (**b**) 14% residue on 0.08 mm sieve, (**c**) 9% residue on 0.08 mm sieve and (**d**) 5% residue on 0.08 mm sieve.

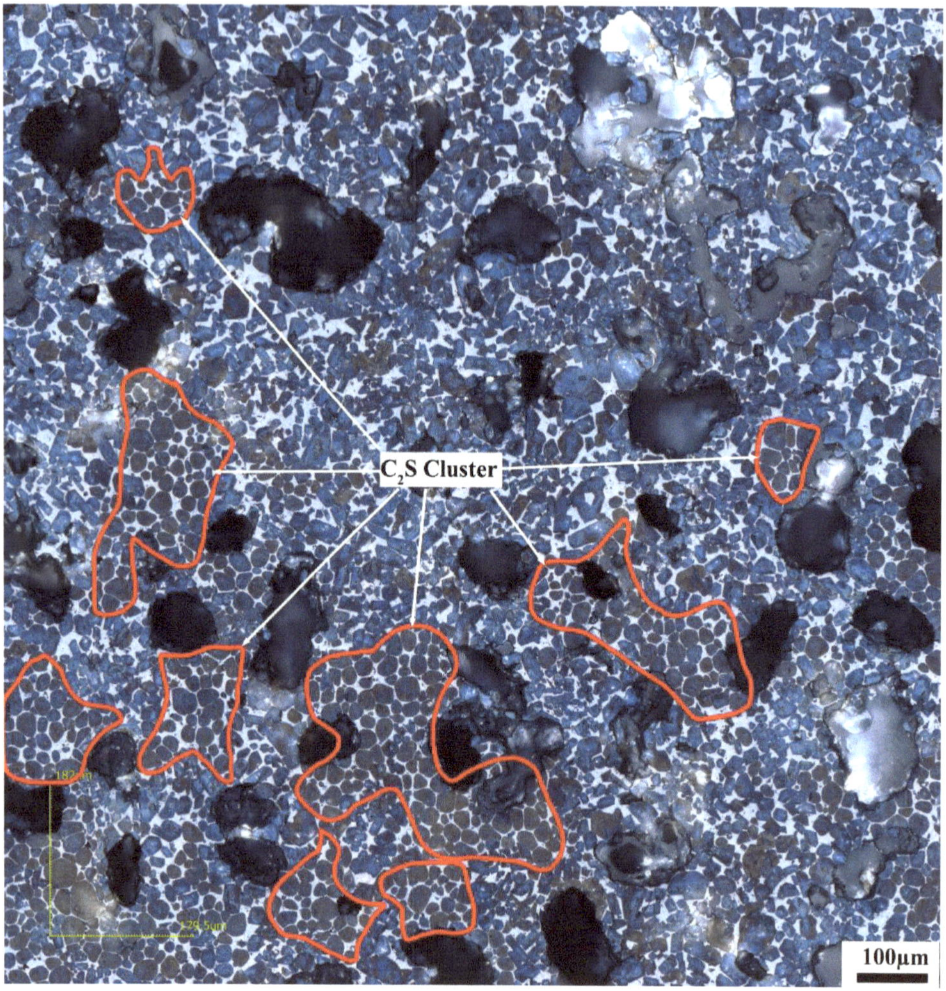

Figure 14. Petrographic images of clinker after calcination with siliceous limestone SL1 with a fineness of 0% of residue on 0.08 mm sieve.

Figure 15 shows the microstructure of the clinker calcined with siliceous limestone SL2 with different fineness values. Same as the clinker calcined with SL1, there are also significant C_2S clusters for different fineness values of siliceous limestone. Most of the C_2S clusters are within the size range of 200–600 μm, and the clinker morphology does not vary much when the fineness of the siliceous limestone SL2 decreases from 20% to 7% residue on a 0.08 mm sieve. As shown in Figure 16, when the fineness of siliceous limestone SL2 continued to decrease to 0% of residue on a 0.08 mm sieve, a little C_2S cluster was found in the matrix of C_3S, indicating that most of the C_2S crystals transformed to C_3S, with most of the quartz crystals grinded to less than 0.045 mm.

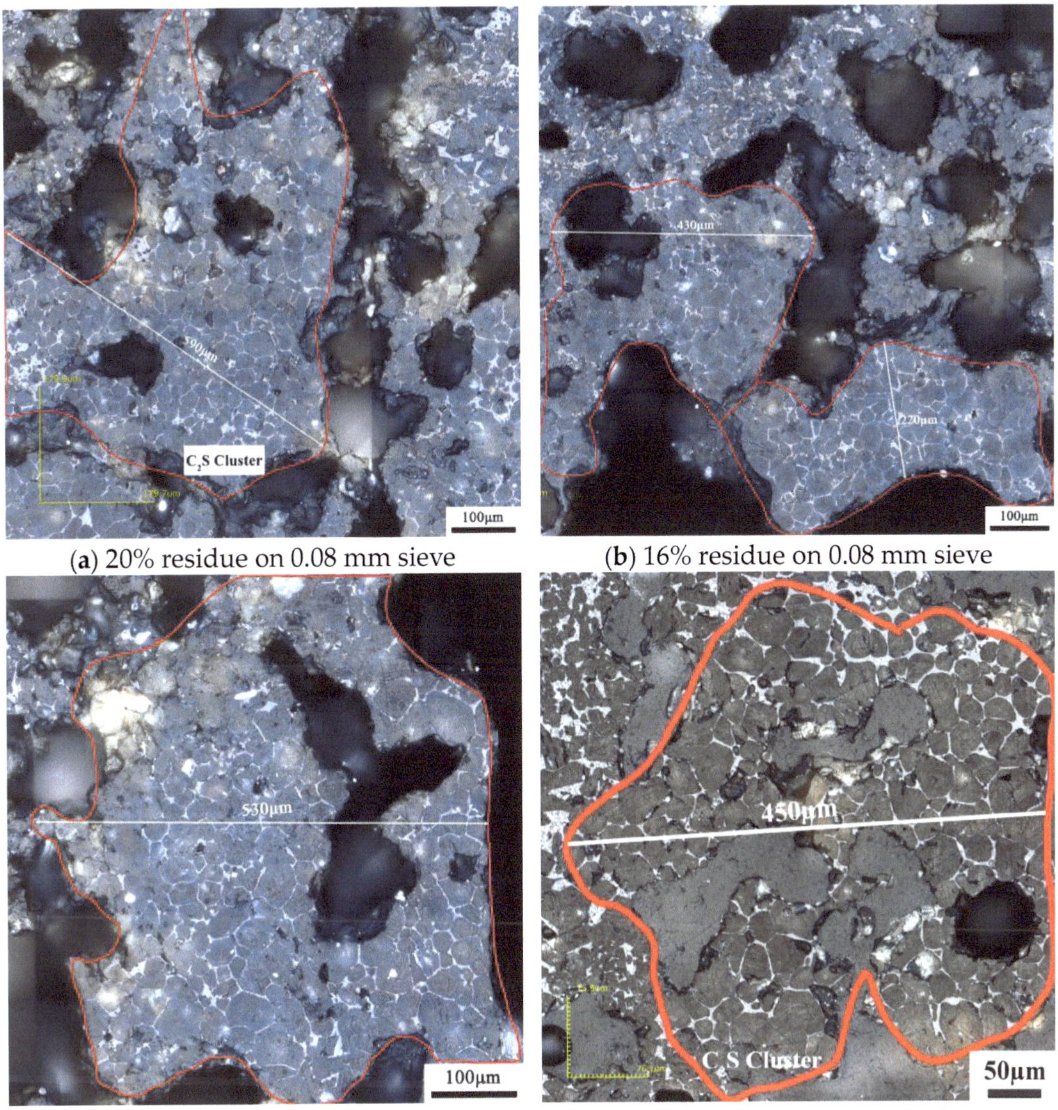

Figure 15. Microstructure of C_2S cluster of different clinkers after calcination with different fineness values of siliceous limestone SL2, (**a**) 20% residue on 0.08 mm sieve, (**b**) 16% residue on 0.08 mm sieve, (**c**) 11% residue on 0.08 mm sieve and (**d**) 7% residue on 0.08 mm sieve.

Figure 16. Petrographic images of clinker after calcination with siliceous limestone SL2 with a fineness of 0% of residue on 0.08 mm sieve.

3.3. *Effects of the Morphology of Quartz Crystal on the Microstructure of C_2S Clusters*

As shown in Figure 2, siliceous limestones SL1 and SL2 are composed of different morphologies of quartz crystal. Quartz in siliceous limestone SL1 is mainly in the form of chert nodules, while in SL2 it is mainly in the form of a single quartz crystal. Different forms of quartz in the raw meal will lead to different microstructure of C_2S.

Figure 17 shows the microstructure of a C_2S cluster formed by the calcination of the chert nodule-dominated siliceous limestone SL1. As can be seen in Figure 17, more intermediate phases are distributed in the C_2S cluster, except for the C_2S clustered in a limiting region. And as shown in Figure 17b, a small amount of C_3S can be found inside the C_2S cluster. Figure 18 shows the microstructure of a dense C_2S cluster formed by the calcination of large grain size of single quartz crystal-dominated siliceous limestone SL2. It can be seen from Figure 18 that there is very little of the intermediate phase inside the C_2S cluster regions, and the shape of C_2S is not as rounded as the traditional one. The clinker calcined by SL1 shows more intermediate phase in C_2S cluster than that of SL2, which leads to a better burnability, and this result is consistent with the content of free lime, as shown in Figure 11. However, even though the chert module-dominated siliceous limestone will lead to more of the intermediate phase inside the C_2S cluster, many C_2S clusters will still be present when some large-grained chert nodules are present in the raw meal, and it will still affect the quality of the clinker.

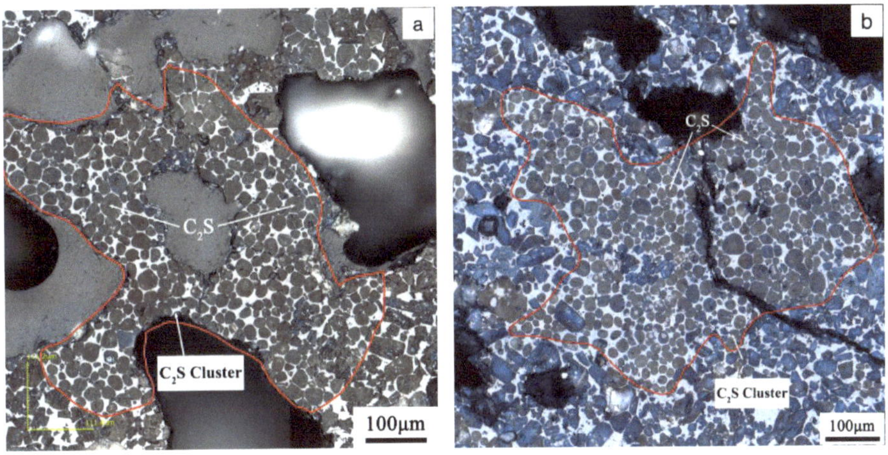

Figure 17. C_2S cluster with more intermediate phase formed by calcination of chert nodule-dominated siliceous limestone SL1, (**a**) 19% residue on 0.08 mm sieve and (**b**) 9% residue on 0.08 mm sieve.

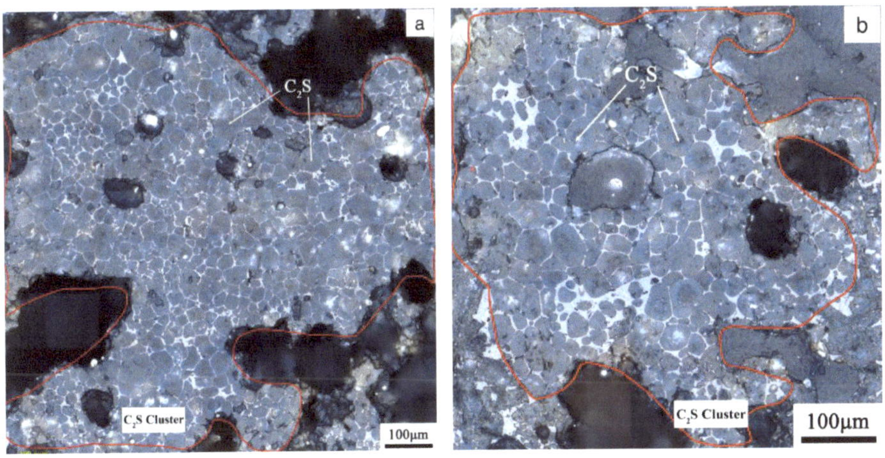

Figure 18. Dense C_2S cluster with little intermediate phase formed by calcination of large grain size of quartz crystal-dominated siliceous limestone SL2, (**a**) 20% residue on 0.08 mm sieve and (**b**) 11% residue on 0.08 mm sieve.

3.4. Effect of Fineness of Siliceous Limestones on the Mechanical Properties of Cement

Figure 19 shows the compressive and flexural strength of cement mortar samples molded with clinkers calcined by different fineness values of siliceous limestones SL1 and SL2 at 28 days. It can be seen from Figure 19 that with the decrease in the fineness of siliceous limestones SL1 and SL2, the comprehensive and flexural strength increase. When the residue on an 0.08 mm sieve is about 7%, the 28d comprehensive strength of SL1 and SL2 both exceed 60 MPa. When the residue on an 0.08 mm sieve continues to decrease to 0%, the 28d comprehensive strength of siliceous limestone SL1 and SL2 is 61.7 MPa and 62.3 MPa, respectively, which is 2.8 MPa and 4.4 MPa higher than that of clinkers calcined with 15% residue on an 0.08 mm sieve.

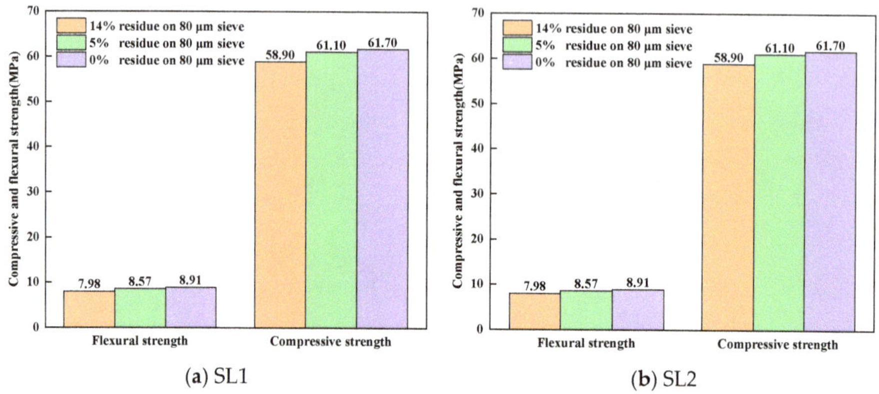

Figure 19. Compressive and flexural strength of the clinker calcined by different fineness values of siliceous limestone SL1 and SL2 at 28 days: (**a**) SL1 and (**b**) SL2.

4. Discussion

The purpose of this study is to find out how quartz in siliceous limestone affects the calcination of raw meal and the quality of cement clinker. Therefore, the present discussion focuses on the explanation and limitations of the results obtained in this study.

4.1. Effect of Particle Size of Quartz on the Formation of Clinker Minerals

It is generally believed that coarse particles will affect the burnability of clinker, especially for quartz crystals. In our experiment, siliceous limestones were crushed and ground to a fineness of 20%, 15%, 10%, 7% and 0% of residue on an 80 μm sieve, and the content of quartz in different particle size ranges were measured. As shown in Figures 6 and 8, the content of quartz in particles larger than 0.08 mm continues to decrease as the fineness of siliceous limestones decreases. However, even with a fineness of 10%, the content of quartz in particles larger than 0.08 mm is 0.60–1.08%. To produce 1 mol of C_2S, 1 mol of SiO_2 and 2 mol of CaO are required. The density of SiO_2 is 2.65 g/cm^3 and the density of C_2S is 3.28 g/cm^3. Thus, the volume of 1 mol of SiO_2 is 22.67 cm^3, while the volume of 1 mol of C_2S is 52.46 cm^3, which is 2.3 times than that of SiO_2. As a result, 1 mol of SiO_2 combined with CaO produces 1 mol of C_2S with a volume at least 2.3 times the original volume, which does not include the intermediate phases and holes present between the C_2S crystals. Thus, when the particle size of quartz larger than 80 μm, the size of the C_2S cluster is at least larger than 184 μm, which is consistent with the result that most of the C_2S clusters are within the size range of 200–600 μm. The presence of a C_2S cluster will break the equilibrium between C_2S and C_3S, and as a result the C_2S crystal in the C_2S cluster is hard to convert to C_3S, which leads to a reduction of C_3S in the clinker.

In order to minimize the drawbacks of quartz, siliceous limestones SL1 and SL2 were ground to a fineness of 0% of residue on an 80 μm sieve. The microstructure of the clinker calcinated with the ground siliceous limestones SL1 and SL2 is shown in Figures 14 and 16, where the size of the C_2S cluster decreases significantly, and even almost disappears in SL2. This means that if siliceous limestone is sufficiently fine, the adverse effects will be eliminated. However, it is unlikely that siliceous limestone can be ground that fine in plant production. For siliceous limestone SL1, when the fineness of it decreases to 5% of residue on an 80 μm sieve, the content of C_3S in the clinker is 60.85%, and the 28d compressive strength is 61.10 MPa, which is sufficient for cement production. While for siliceous limestone SL2, when the fineness of it decreases to 7% of residue on an 80 μm sieve, the content of C_3S in the clinker is 64.50%, and the 28d compressive strength is 60.30 MPa, which is also sufficient for cement production. This means, for different siliceous limestones, there must be an equilibrium where the amount of C_2S clusters does not significantly affect the quality of the clinker, and the siliceous limestone does not need to be that fine.

4.2. Effects of the Morphology of Quartz Crystal on the Microstructure of C_2S Clusters

As shown in Figures 17 and 18, different morphologies of quartz will lead to a different forms of C_2S clusters. A chert nodule in raw meal may lead to a C_2S cluster with more intermediate phase, while large-grained quartz may lead to a dense C_2S cluster. A chert nodule is composed of microcrystalline quartz that exhibits a mosaic texture. The gap between the microcrystalline quartz may be a path for the diffusion of f-CaO and intermediate phase, thus, there is much intermediate phase inside the C_2S cluster. As the size of the chert nodule composed of microcrystalline quartz is still too large, the C_2S produced in the chert nodule region is still hard to convert to C_3S in a limited calcination time.

Large-grained quartz is mainly in the form of a single quartz crystal in the raw meal, and there are few possible paths inside a single quartz crystal, through which f-CaO and intermediate phase could diffuse. This may explain the fact that little intermediate phase was found in the clinker calcinated by large-grained quartz-dominated siliceous limestone. Due to there being very little intermediate phase in the newly formed C_2S cluster, no C_3S was found inside the cluster, and the shape of C_2S is not as rounded as the traditional one.

Fundal [17] highlighted the finding that a 63 μm C_2S cluster was caused by a 44 μm quartz crystal, and then a series experiments concerning the effects of a quartz size larger than 44 μm on the burnability of raw meal were conducted. Based on Fundal's results, Christensen [7] deduced a relationship between free lime and quartz size, and pointed out that a quartz size larger than 44 μm will have a bad effect on the burnability of raw meal. Some other researchers have also made some progress on the effects of quartz size on the burnability of raw meal [8,9,12–16,20], but all the results were based on the free lime content, which could not accurately reflect the quality of the clinker. In the latest study by Joseé [25], the content of different phases in the clinker calcinated by different raw meal were determined by a high-temperature X-ray diffraction. In their study, raw meal with a relatively high amount of coarse quartz (>45 μm) shows a relatively high amount of free lime and a low amount of C_3S, which in their conclusion was due to the high content of coarse quartz in the raw meal. However, that conclusion was deduced from Fundal's research and no other evidence was presented. While in our research, more microstructural analysis was carried out to show how different morphologies of quartz and different sizes of quartz crystal affect the quality of clinker. Besides, the microstructure of the size of C_2S clusters and calculations of the volume of C_2S clusters caused by quartz show that the volume of the C_2S cluster is at least 2.3 times that of quartz. It is not appropriate to state that a quartz particle size threshold of less than 44 μm or some other size does not affect clinker quality. With the quantifications of different phases of clinker calcinated by different fineness values of raw meal and the study of the morphology of the clinker, the critical grain size of quartz varies for different types of quartz, which need further study.

4.3. The Reason for Low C_3A Content in Clinker Calcined from Siliceous Limestone SL1

In order to find out the reason for the low C_3A content in the clinker formed by calcination of siliceous limestone SL1, FESEM equipped with EDS was used to analyze the elemental distribution of C_2S and C_3S in the clinker formed by SL1 and SL2. Figures 20 and 21 show the FESEM images and EDS spectra of C_2S and C_3S formed by the calcination of siliceous limestones SL1 and SL2, respectively. As can be seen from Figure 20a,c, there exists 1.01 wt% Al in C_2S, and the chemical formula of C_2S based on EDS analysis is $CaO_{1.97}(SiO_2)(Al_2O_3)_{0.04}(Fe_2O_3)_{0.02}(MgO)_{0.04}(Na_2O)_{0.04}(K_2O)_{0.01}$. Figure 20b,d show the FESEM image and EDS pattern of C_3S; the content of Al is 1.18 wt% in C_3S, and the chemical formula of C_3S based on EDS analysis is $CaO_{2.89}(SiO_2)(Al_2O_3)_{0.06}(Fe_2O_3)_{0.03}(MgO)_{0.09}(Na_2O)_{0.01}$. There is a portion of Al_2O_3 solidly dissolved in C_2S and C_3S, thus leading to the lack of C_3A in the clinker calcinated by SL1.

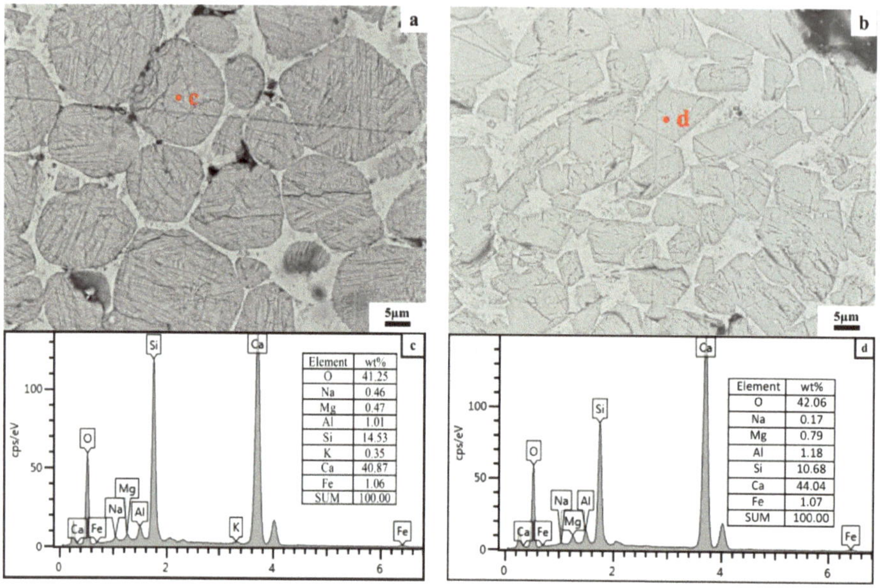

Figure 20. FESEM images and EDS patterns of C_2S and C_3S formed by calcination of siliceous limestone SL1. (**a**) C_2S, (**b**) C_3S, (**c**) EDS patterns for point c and (**d**) EDS patterns for point d.

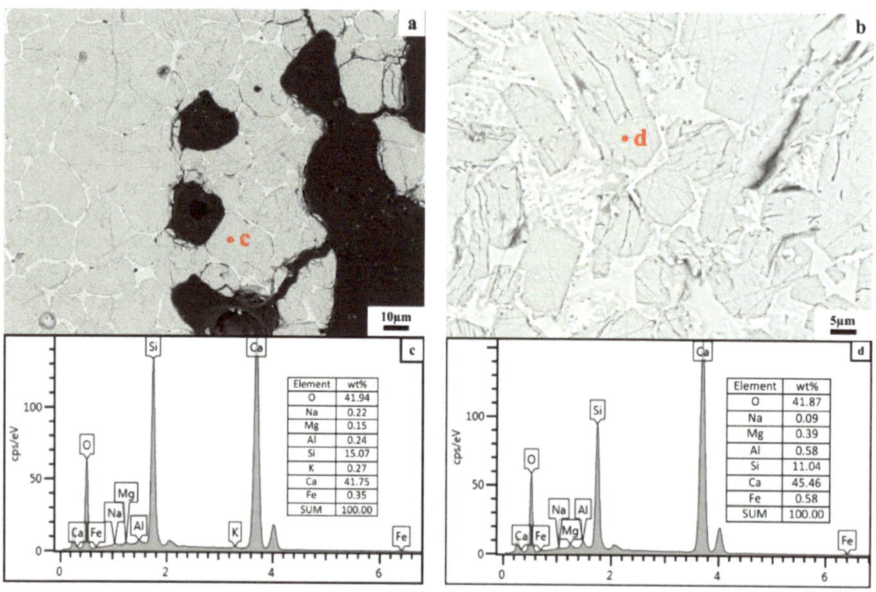

Figure 21. FESEM image and EDS pattern of C_2S and C_3S formed by calcination of siliceous limestone SL2. (**a**) C_2S, (**b**) C_3S, (**c**) EDS patterns for point c and (**d**) EDS patterns for point d.

Figure 21a,c show the microstructure and EDS pattern of C_2S formed by the calcination of SL2; the content of Al is 0.24 wt% in C_2S, and the chemical formula of C_2S based on EDS analysis is $CaO_{1.94}(SiO_2)(Al_2O_3)_{0.01}(Fe_2O_3)_{0.01}(MgO)_{0.01}(Na_2O)_{0.01}(K_2O)_{0.01}$. As shown in Figure 21 b,d, the content of Al in C_3S is 0.58%, and the chemical formula of C_2S based on EDS analysis is $CaO_{2.89}(SiO_2)(Al_2O_3)_{0.03}(Fe_2O_3)_{0.01}(MgO)_{0.04}$. Comparing the chemical formula of the clinker formed by the calcination of SL1 and SL2, the content of Al_2O_3 in C_2S and C_3S calcinated by SL1 is higher than that of SL2, which suggests that more Al_2O_3 is solidly dissolved in C_2S and C_3S calcinated by SL1 than SL2, leading to less Al_2O_3 to form C_3A, and consequently, a lower content of C_3A in the clinker.

5. Conclusions

This paper assesses the effects of siliceous limestone on clinker calcination. From our experiment work, the following major conclusions can be drawn:

1. The research reveals that larger quartz crystals and chert nodules in siliceous limestone are more difficult to grind than calcite. Decreasing the fineness of siliceous limestone reduces the quantity of large quartz particles, which is beneficial for the burnability of raw meal.
2. A direct correlation between the fineness of siliceous limestone and the quality of the clinker minerals produced is noted. The fineness affects the mineral content of the clinker, which in turn affects the mechanical properties. As the fineness of siliceous limestones SL1 and SL2 decreases from 15% to 0% of residue on a 0.08 mm sieve, the content of C_3S increases from 56.74% to 61.02% and 60.87% to 68.40%, respectively, while the content of C_2S decreases from 20.69% to 16.13% and 16.85% to 10.70%, respectively. The compressive strength increases from 58.90 MPa to 61.70 MPa and 57.90 MPa to 62.30 MPa, respectively.
3. The structure of quartz, whether as chert nodules or single crystals, can produce C_2S clusters during the calcination, affecting the microstructure and mineral content of the clinker. Chert nodules tend to produce C_2S clusters with more intermediate phases, whereas large-grained single quartz crystals lead to denser C_2S clusters with minimal intermediate phases.
4. It is important to optimize the fineness of siliceous limestone to mitigate the adverse effects of large quartz particles on clinker quality. The sufficient fineness values of siliceous limestones SL1 and SL2 are 5% and 7% of residue on a 0.08 mm sieve, respectively, which can produce a clinker with a 28d compressive strength greater than 60 MPa. Different grinding requirements are needed for the different morphologies of quartz in the raw meal, to strike a balance between clinker quality and energy consumption without having to grind siliceous limestone at very fine grades.

Author Contributions: D.N.: Methodology, Investigation, Writing—original draft. W.L.: Methodology, Investigation, Writing—review and editing. L.X.: Methodology, Investigation, Validation. M.D.: Supervision, Conceptualization, Resources, Writing—review and editing. H.D.: Resources, Methodology. K.L.: Writing—review and editing, Validation. All authors have read and agreed to the published version of the manuscript.

Funding: This research presented in this paper was supported by Opening Fund of Anhui Province Key Laboratory of Advanced Building Materials (Grant No. JZCL2407KF).

Institutional Review Board Statement: Not applicable.

Informed Consent Statement: Not applicable.

Data Availability Statement: Data is contained within the article.

Acknowledgments: Thanks to Jiangsu Funding Program for Excellent Postdoctoral Talent (2023ZB286) and Opening Fund of Anhui Province Key Laboratory of Advanced Building Materials (Grant No. JZCL2407KF).

Conflicts of Interest: Author Hao Ding was employed by the company Hefei Cement Research and Design Institute Corporation Ltd. The remaining authors declare that the research was conducted in the absence of any commercial or financial relationships that could be construed as a potential conflict of interest.

References

1. Ogli, J.Z.A. Chemical composition of Portland cement clinker. *Am. J. Pedagog. Educ. Res.* **2023**, *13*, 111–114.
2. Chatterjee, A.K. *Cement Production Technology: Principles and Practice*; CRC Press: Boca Raton, FL, USA, 2018.
3. Chatterjee, T. Burnability and clinkerization of cement raw mixes. In *Advances in Cement Technology*; Elsevier: Amsterdam, The Netherlands, 1983; pp. 69–113.
4. Zhang, W.; Liu, T.; Ren, X. Effects of siliceous raw materials on clinker formation. *China Build. Mater. Acad.* **2020**, *23*, 618–624.
5. Makashev, S. Effect of Raw Materials Physico-Chemical Properties on Reactivity of Raw Mix and Clinker Minerogenesis Processes. In Proceedings of the 6th ICCC, Moscow, Russia, 6–10 September 1976; pp. 1–17.
6. Mackenzie, K.J.D.; Hadipour, N. Formation kinetics of portland-cement clinker phases.3. beta-dicalcium silicate and tricalcium silicate. *Trans. J. Br. Ceram. Soc.* **1978**, *77*, 168–172.
7. Christensen, N.; Smidth, F. Burnability of cement raw mixes at 1400 °C II the effect of the fineness. *Cem. Concr. Res.* **1979**, *9*, 285–294. [CrossRef]
8. Dakang, Z.; Yuanbo, L. Influence and measures of large size SiO_2 in raw material on quality of clinker. *Cem. Eng.* **2016**, *5*, 27–29.
9. Jin, Z. *Research on the Impact of Minerals and Particle Size on the Content of Free Calcium Oxide and Petrographic Structure*; Guangxi University: Nanning, China, 2014.
10. Maki, I.; Ito, S.; Tanioka, T.; Ohno, Y.; Fukuda, K. Clinker grindability and textures of alite and belite. *Cem. Concr. Res.* **1993**, *23*, 1078–1084. [CrossRef]
11. Maki, I. Processing conditions of Portland cement clinker as viewed from the fine textures of the constituent minerals, Ceramic Transactions. *Am. Ceram. Soc.* **1994**, *40*, 3–17.
12. Tan, L.; Xuehong, R.; Wensheng, Z. Research Progress on the Influence of SiO_2 Crystalline Characteristics in Raw Materials on Clinker Sintering. *Bull. Chin. Ceram. Soc.* **2019**, *38*, 1777–1781.
13. Huagang, Z. Effect of the properties of siliceous material on the burnability of raw meal. *Cement* **1998**, *6*, 12–16.
14. He, S.; Ling, H.; Zhu, X. Influence of raw meal finess on burning of Portland cement clinker. *Cem. Eng.* **2012**, *5*, 23–26.
15. Ishak, K.; Hashim, S.F.S.; Azizli, K.A.M.; Palaniandy, S.; Hussin, H. Effect of Quartz in Clay on Grindability of Raw Mixes for Cement Production. *J. Phys. Sci.* **2022**, *33*, 45–59. [CrossRef]
16. Kakali, G.; Tsivilis, S. The effect of intergrinding and separate grinding of cement raw mix on the burning process. *Cem. Concr. Res.* **1993**, *23*, 651–662. [CrossRef]
17. Fundal, E. The burnability of cement raw mixes. *World Cem. Technol.* **1979**, *42*, 195–204.
18. Christensen, N. Burnability of cement raw mixes at 1400 °C I The effect of the chemical composition. *Cem. Concr. Res.* **1979**, *9*, 219–228. [CrossRef]
19. Jiang, Y.; Yuan, X.; Hu, J.; Xu, P.; Ni, X. The Influence of Grain Size and Crystal Structure of Quartz on Raw Meal Burnability. *Cem. Technol.* **2000**, *3*, 9–11.
20. Hou, G.; Shen, X.; Xu, Z. Composition design for high C_3S cement clinker and its mineral formation. *J. Wuhan Univ. Technol.-Mater. Sci. Ed.* **2007**, *22*, 56–60. [CrossRef]
21. Mathew, B.; Erikandath, S.; Lele, G.; Khadilkar, S. *Process Control & Clinker Quality Monitoring through Mineralogical and Micro-Structural Indices, Proceedings of the 14th NCB International Seminar on Cement and Building Materials, New Delhi, India, 1–4 December 2015*; National Council for Cement and Building Materials: New Delhi, India, 2016.
22. Maki, I.; Fukuda, K.; Imura, T.; Yoshida, H.; Ito, S. Formation of belite clusters from quartz grains in portland cement clinker. *Cem. Concr. Res.* **1995**, *25*, 835–840. [CrossRef]
23. Korkmaz, A.V. Evaluation of chemical, mineralogical and clinker burnability properties of mudstones as cement raw materials. *Case Stud. Constr. Mater.* **2019**, *11*, e00254. [CrossRef]
24. Castillo, J.A.; Broström, M.; Eriksson, M. Phase evolution and burnability of cement raw meal. *Adv. Cem. Res.* **2023**, *35*, 577–587. [CrossRef]
25. Shirahama, N.; Yamashita, M.; Tanaka, H. The influence of crystal properties of quartz in raw materials on the formation of clinker minerals. In Proceedings of the 14th International Congress on the Chemistry of Cement, Beijing, China, 13–16 October 2015; Volume 42.
26. De Matos, P.; Neto, J.A.; Sakata, R.; Kirchheim, A.; Rodríguez, E.; Campos, C. Strategies for XRD quantitative phase analysis of ordinary and blended Portland cements. *Cem. Concr. Compos.* **2022**, *131*, 104571. [CrossRef]
27. *GB/T 176-2017*; Methods for chemical analysis of cement, Standardization Administration of PRC. China National Standardization Administration: Beijing, China, 2017.
28. *JY/T 0587-2020*; General rules of polycrystalline X-ray diffraction method, Ministry of Education of PRC. Ministry of Education: Beijing, China, 2020.
29. *GB/T 17671-2021*; Test method of cement mortar strength, Standardization Administration of PRC. China National Standardization Administration: Beijing, China, 2021.

materials

MDPI

Article

Study on the Influence of Waste Rock Wool on the Properties of Cement Mortar under the Dual Fiber Effect of Polyvinyl Alcohol Fibers and Steel Fibers

Shijian Lu, Jiajia Cheng, Zhipeng Zhu, Luchao Yan, Yang Wang, Lingling Xu * and Min Deng *

College of Material Science and Engineering, Nanjing Tech University, Nanjing 211816, China
* Correspondence: xll@njtech.edu.cn (L.X.); dengmin@njtech.edu.cn (M.D.)

Abstract: In this paper, the effect of waste rock-wool dosage on the workability, mechanical strength, abrasion resistance, toughness and hydration products of PVA and steel fiber-reinforced mortars was investigated. The results showed that the fluidity of the mortar gradually decreased with the increase in the dosage of waste rock wool, with a maximum reduction of 10% at a dosage of 20%. The higher the dosage of waste rock wool, the greater the reduction in compressive strength. The effect of waste rock wool on strength reduction decreases with increasing age. When the dosage of waste rock wool was 10%, the 28 days of flexural and compressive strengths were reduced by 4.73% and 10.59%, respectively. As the dosage of waste rock wool increased, the flexural-to-compressive ratio increased, and at 20%, the maximum value of 28 days of flexural-to-compressive ratio was 0.210, which was increased by 28.05%. At a 5% dosage, the abraded volume was reduced from 500 mm^3 to 376 mm^3—a reduction of 24.8%. Waste rock wool only affects the hydration process and does not cause a change in the type of hydration products. It promotes the hydration of the cementitious material system at low dosages and exhibits an inhibitory effect at high dosages.

Keywords: waste rock wool; fiber-reinforced mortar; mechanical properties; hydration properties; microstructure

Citation: Lu, S.; Cheng, J.; Zhu, Z.; Yan, L.; Wang, Y.; Xu, L.; Deng, M. Study on the Influence of Waste Rock Wool on the Properties of Cement Mortar under the Dual Fiber Effect of Polyvinyl Alcohol Fibers and Steel Fibers. *Materials* **2024**, *17*, 3416. https://doi.org/10.3390/ma17143416

Academic Editor: Dolores Eliche Quesada

Received: 21 May 2024
Revised: 2 July 2024
Accepted: 8 July 2024
Published: 10 July 2024

1. Introduction

As a kind of building insulation material, rock wool produces a large number of waste rock wool in the demolition and construction process of buildings. Waste rock wool is expected to grow to 2.82 million tons by 2030 [1]. Taiwan produces over 100 thousand tons of waste rock wool per year [2]. The current treatment of waste rock wool is stacking and landfilling. A large amount of waste rock wool has produced a series of accumulations and environmental problems, so use of waste rock wool urgently needs to be resolved [3–5]. In order to reduce the amount of waste rock wool deposited in piles and landfills, Taiwan encourages industries to recycle and reuse waste rock wool with a price subsidy of TWD 1500 per ton [6].

Waste rock wool is an artificial inorganic fiber mainly produced by basalt under high-temperature melting, with high tensile strength, high modulus, chemical resistance and dimensional stability [7–9]. Waste rock wool consists of amorphous components such as silica, aluminum oxide and iron trioxide, which are potentially chemically active [10]. Unmilled waste rock wool is reported to have an average diameter of 6–8 microns and a length of 400 microns to 20 mm, and the length of the milled waste rock wool can be less than 34 microns [3,11–14]. Commonly used construction fibers are steel fibers and PVA fibers. Commonly used steel fibers have a diameter of 200 microns and a length of 13 mm, and commonly used PVA fibers have a diameter of 30 to 40 microns and a length of 6–12 mm. Both commonly used fibers are available in millimeter lengths [14,15]. Thus, waste rock wool, with a diameter of a few micrometers and a length of tens of micrometers, and its potential chemical activity were expected to play a hardening and enhanced role in cement-based materials at the micro-scale. Depending on the particle size and length of the waste rock wool, it can be used as a partial substitute for fine aggregates and as a supplementary cementitious material [16–20]. Many scholars [21–23] found that with the increase in mineral cotton content, the formation of large pores in mortar led to an increase in porosity and a general decrease in compressive strength. Kubiliute et al. [24], by collecting the dust from the melting process of a mineral wool raw material (mineral wool cupola dust) used as a cementitious material to replace cement, found that a dosage of washed dust below 20% was able to increase the compressive strength of specimens. Lin et al. [25] found that the incorporation of waste rock wool into cement matrix composites reduces its dry shrinkage and initial surface absorption and improves its compressive strength. These performance improvements were due to the filling effect of waste rock wool and the volcanic ash reaction.

In this paper, based on the optimum volumetric dosage of PVA fibers and steel fibers obtained by the research team, the effect of waste rock wool as a cementitious material and fibrous material on the workability, mechanical strength, toughness and hydration products of PVA fiber and steel fiber mortar was investigated by varying the dosage of waste rock wool. In this study, the volcanic ash properties of waste rock wool were analyzed by X-ray diffraction (XRD), heat of hydration, differential thermal analysis (DSC) and the thermogravimetric method (TG) to assess the effect of waste rock wool on the replacement of Portland cement from a microscopic point of view.

2. Experimental Program

2.1. Raw Materials

The cement used in the test was 52.5-grade ordinary silicate cement (Jiangnan Onoda Cement Co., Nanjing, China). Its 28 days of flexural strength and compressive strength were 8.15 MPa and 54.04 MPa, respectively. Fly ash and silica fume replaced 25% and 5% of cement, respectively. The waste rock wool came from an environmental company in Nanjing, and the particle size ranged from 45 μm to 80 μm. The physical properties of cement are shown in Table 1 [26]. The chemical composition of cement, fly ash and silica fume is shown in Table 2 [26]. Photographs of waste rock wool, fly ash and silica fume are shown in Figure 1. The X-ray diffraction patterns of cement, fly ash, silica fume and waste rock wool are shown in Figure 2. Figure 2b shows that the composition of fly ash was mainly mullite and quartz. Figure 2c shows that silica fume consists of amorphous SiO_2. Figure 2d shows that the waste rock wool consists of an amorphous substance and SiO_2. As shown in Figure 3, the waste rock-wool form was rod-shaped. Its diameter was below 20 μm. As shown in Table 3, most of the lengths of waste rock wool were less than 100 μm. The characteristic parameters of PVA fiber and steel fiber are shown in Table 4. The diameter and length of the PVA fiber were 50 μm and 6 mm, and the diameter and length of the steel fiber were 220 μm and 13 mm, respectively.

Table 1. Physico-mechanical properties of Portland cement.

Grade	Density (kg/m^3)	Specific Surface Area (m^2/kg)	Water Requirement of Normal Consistency (wt%)	Initial Setting Time (min)	Final Setting Time (min)	Flexural Strength (MPa)		Compressive Strength (MPa)	
						3 d	28 d	3 d	28 d
P·II 52.5	3060	372	28	135	261	5.10	8.15	30.75	54.04

Table 2. Chemical compositions of the three cementitious materials.

	SiO_2	CaO	Al_2O_3	Fe_2O_3	SO_3	MgO	K_2O	LOI
Fly ash (%)	50.87	3.06	32.42	7.56	0.37	0.78	1.84	1.84
Cement (%)	20.87	63.66	4.42	2.89	2.10	0.94	0.51	3.30
Silica fume (%)	92.88	0.06	0.62	0.28	-	0.02	0.03	3.85

Figure 1. Waste rock wool and supplementary cementitious materials: (**a**) Waste rock wool; (**b**) Fly ash; (**c**) Silica fume.

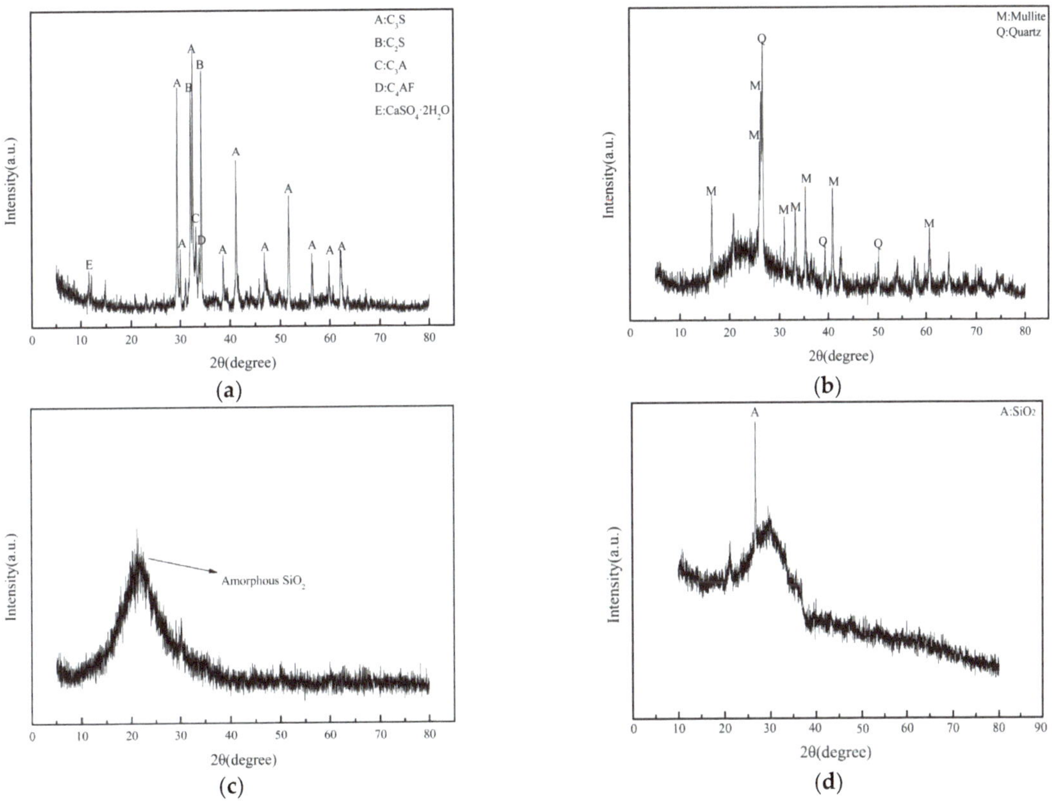

Figure 2. XRD pattern of raw materials: (**a**) Cement; (**b**) Fly ash; (**c**) Silica fume; (**d**) Waste rock wool.

Table 3. Composition of waste rock-wool particle size intervals.

Particle size range/μm	<3	3–16	16–32	32–45	45–80	80–100	>100
Waste rock wool	2.02	21.35	25.85	12.27	16.13	7.25	15.13

Table 4. Characteristic parameters of fibers.

Fiber	Diameter (μm)	Lengths (mm)	Tensile Strength (MPa)	Elongation at Break (%)	Modulus of Elasticity (GPa)	Densities (g/cm^3)
PVA	50	6	1600	7.0	35	1.27
Steel	220	13	≥2000	10	220	7.80

Figure 3. Waste rock-wool SEM image.

2.2. Mix Design

The sand-to-cement ratio for all mortars specimens was 3:1, with 1350 g of sand and 450 g of cementitious materials. As shown in Table 5, in 450 g of cementitious material, the dosages of fly ash and silica fume were fixed at 25% and 5%, respectively, and the remainder was the dosage of cement. When the dosage of waste rock wool increased by 5%, the dosage of cement was reduced by 5% accordingly. The water/cement ratio was 0.5 s The volumes of polyvinyl alcohol fiber and steel fiber were 0.09% and 0.13% of the mortar specimens volume, respectively. The preparation process of fiber-reinforced mortar is shown in Figure 4. After mixing cement, fly ash, silica fume, water and waste rock wool for 30 s, sand was added with low-speed mixing for 30 s and then high-speed mixing for 60 s, after which steel fibers and PVA fibers were added with low-speed mixing for 240 s to obtain the fiber-reinforced mortar. The mortar was packed into 40 mm × 40 mm × 160 mm molds and cured at 20 ± 2 °C and 95% relative humidity for 24 h, before being cured for 7, 28 and 90 days.

Table 5. Mortar ratios and fluidity.

Mix No	W/C	Cement (%)	Waste Rock Wool (%)	Fly Ash (%)	Silica Fume (%)	PVA Fiber (%)	Steel Fiber (%)	Fluidity (mm)
R0	0.5	70	0	25	5	0.09	0.13	210
R5	0.5	65	5	25	5			205
R10	0.5	60	10	25	5			199
R15	0.5	55	15	25	5			194
R20	0.5	50	20	25	5			189

Note: Rx denotes the replacement of cement by rock-wool with a mass fraction of x on the basis of the base proportion of cementitious materials; Fiber dosage is (mortar specimen) volume dosage.

Figure 4. Preparation process of fiber-reinforced mortar.

2.3. *Test Methods*

2.3.1. Slurry Fluidity

The fluidity test of mortar refers to GB/T 2419-2005 [27]. The fluidity test of mortar is shown in Figure 5. After 25 jumps on the table, we measured the diameter of the mortar in two directions perpendicular to each other and took the average value as the test result. The entire flow test process, from the mixing of the mortar with water to the end of the diameter measurement, should be controlled within 6 min.

Figure 5. The fluidity test of mortar.

2.3.2. Mechanical Strength

The test of mechanical strength in this experiment was in accordance with GB/T 17671-1999 [28]. The compressive strength and flexural strength of specimens at 7, 28 and 90 days were measured by a Universal Testing Machine (Wance, Shenzhen, China) at a loading speed of 2.4 kN/s and 0.05 kN/s, respectively. Specimen sizes for tests of flexural strength and compressive strength were 40 mm × 40 mm × 160 mm and 40 mm × 40 mm × 40 mm, respectively. There were three flexural strength test specimens in each group, and the average of the three specimen test results was calculated, which was the result of the flexural strength test. If the test result was more than 10% of the average value, the value was discarded and the average value of the remaining values was taken as the result. The flexural strength test was conducted using the fracture of the six test pieces as compressive strength test specimens, and we calculated the average value of the test results of the six test pieces, and the average value was the compressive strength test results. If there was more than 10% of the average value in the test results, the value was discarded, and the average of the remaining values was taken as the result.

2.3.3. Impact Resistance

The test method of impact resistance referred to GB/T 38494-2020 [29]. The specimen for the impact resistance test was a plate of 300 mm × 300 mm × 30 mm. When the specimen was cured for 28 days, the impact test was carried out on the specimen using the falling ball method. The schematic diagram of the impact test is shown in Figure 6. The height of the falling ball was fixed to fix the impact energy, and the impact strength of the specimen was used as an index for evaluating the impact resistance of the specimen. The first visible cracks that appeared in the impact process of the specimen were taken as the number of initial impacts N_0, and the number of impacts when the specimen was damaged or a crack width greater than 3 mm was taken as the number of final impacts N_X. Impact energy dissipation and impact strength were calculated according to Equations (1)–(3):

$$W_0 = N_0 mgh \quad (1)$$

$$W_x = N_x mgh \quad (2)$$

$$S = \frac{W_x}{t^2} \quad (3)$$

where W_0 and W_x refer to the impact energy consumption of initial and final crack(J), respectively; N_0 and N_x refer to the impact number of initial and final crack, respectively. m was the mass of the steel ball (g). In this test, m was 1040 g; h refers to the drop height (m). In this test, h was 0.35 m; g refers to the acceleration of gravity (m/s^2). In this test, g was 9.81 m/s^2; S refers to the impact strength (J/mm^2); and t refers to thickness of the specimen (mm).

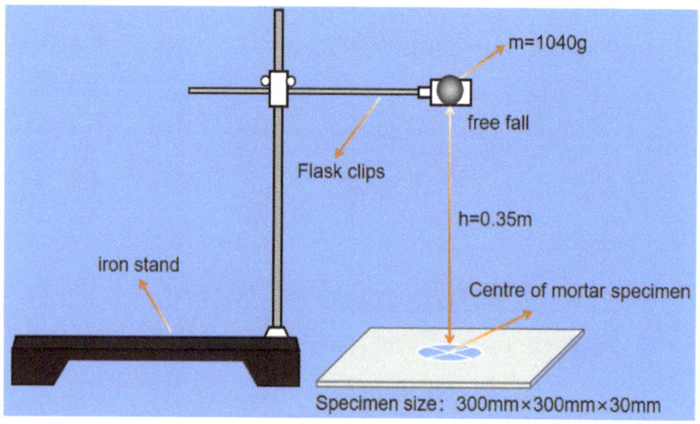

Figure 6. Schematic diagram of the impact test.

2.3.4. Abrasion Resistance

The abrasion resistance test of the hardened mortar after 28 days was carried out with reference to GB/T 3810.6-2016 [30]. An abrasion resistance tester and test samples are shown in Figure 7, and the feeding rate of the abrasive was 100 ± 10 g/100 r. Vernier calipers were used to measure the chord length of the abrasion pit after 150 r of the friction steel wheel. The mass of the test block was measured before and after the abrasion test, and the mass loss of the test block before and after the experiment was used to express the abrasion mass. The abrasion resistance of the test block was characterized by the volume of the abrasion pit, which was calculated according to Equations (4) and (5):

$$V = \left(\frac{\pi \cdot \alpha}{180} - \sin\alpha\right) \times \left(\frac{h \times d^2}{8}\right) \quad (4)$$

$$\sin\frac{\alpha}{2} = \frac{L}{d} \quad (5)$$

where V refers to abrasion volume (mm^3); L refers to the chord length (mm); α refers to the angle formed by the chord length and the center of the rotating steel wheel (°); h refers to the thickness (mm), which in this test h was 10 mm; and d refers to the diameter of the rotating steel wheel (mm), which in this test h was 200 mm.

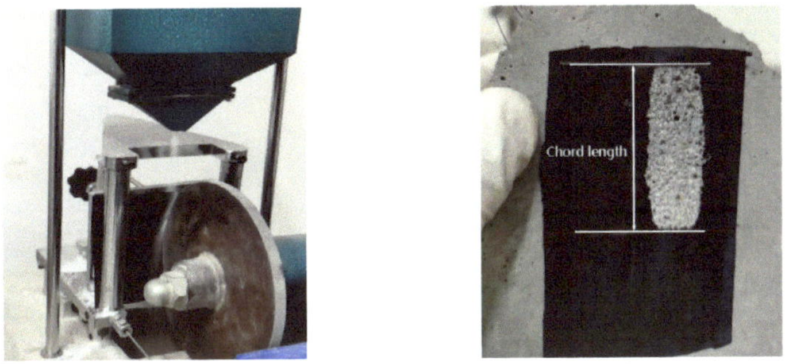

(**a**) (**b**)

Figure 7. Abrasion resistance tester and tested sample: (**a**) Abrasion resistance tester; (**b**) tested sample.

2.3.5. Microstructure Testing Methods

The microscopic morphology of mortar was obtained by scanning electron microscopy (SEM, JSM-5900, Japan Electronics Corporation, Osaka, Japan). The effect of waste rock wool on the hydration of cement was investigated by thermogravimetry-differential scanning calorimetry (TG-DSC, STA 449 F1, Netzsch, Selb, Germany). The effect of waste rock wool on cement hydration products was studied using an X-ray diffractometer (XRD, 0031D max/RB, Rigaku Corporation, Tokyo, Japan). The influence of waste rock wool on the total heat release and heat release rate of cement was studied using a cement isothermal calorimeter (Calmetrix, Boston, MA, USA).

3. Results and Discussion

3.1. Slurry Fluidity

As shown in Table 5, waste rock wool reduced the fluidity of the mortar, which gradually decreased with the increase in the dosage. When the dosage was 20%, the fluidity decreased from 210 mm to 189 mm, a decrease of 10.00%, but the fluidity of the mortar still met the requirements of the actual construction. Waste rock wool leads to the reduction of mortar fluidity mainly due to two factors: on the one hand, the waste rock wool has a high water-absorption rate, and with the increase in the dosage, its water-absorption ability was more and more obvious [31]. On the other hand, the fiber morphology of the waste rock wool makes it interlaced, forming a three-dimensional network system, which increases the resistance to the flow of the slurry.

3.2. Mechanical Strength

Figure 8 shows the compressive strength and flexural strength of mortar with waste rock wool at different ages. Figure 8a reflects the decreasing compressive strength of mortar at 7, 28 and 90 days with the increase in waste rock-wool dosage. The compressive strength of the mortar decreased the most when the dosage of waste rock wool was 20%, and the decrease in compressive strength decreased with the extension of the age period. The 7 days of compressive strength of the mortar decreased from 37.17 MPa to 20.6 MPa (a decrease of 44.58%), the 28 days of compressive strength decreased from 55.32 MPa to

35.85 MPa (a decrease of 35.20%), and the 90 days of compressive strength decreased from 61.67 MPa to 42.15 MPa (a decrease of 31.66%).

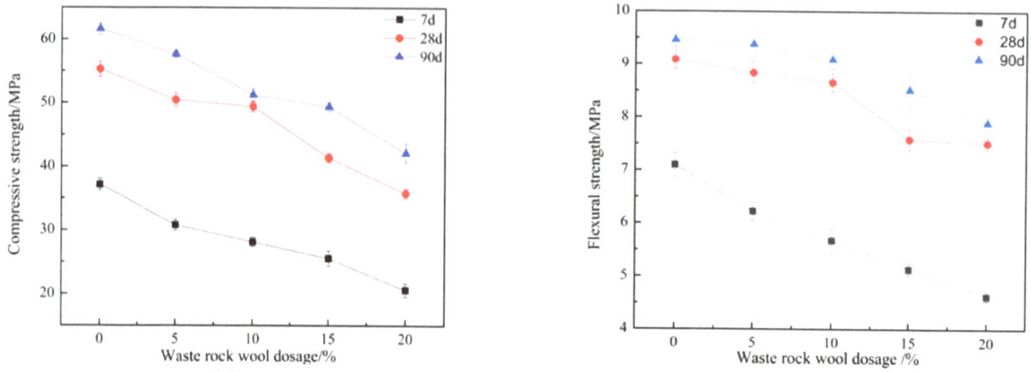

Figure 8. Effect of waste rock-wool dosage on mechanical strength: (**a**) Compressive strength; (**b**) Flexural strength.

In general, the increase in the packing density of cementitious materials in mortar has a positive effect on its mechanical strength. The compressive strength showed a decreasing trend, and as the dosage of waste rock wool increased, the proportion of cement decreased, the dosage of cement hydration products decreased, and the internal structure of the mortar was loosened. The negative effect on the compressive strength was more significant—the combined performance of the compressive strength of the mortar was reduced. However, the increase in compressive strength of the mortar increases with age, which indicates that the detrimental effect of spent rock wool on the later strength of the mortar decreases. This was due to the high water absorption of waste rock wool, which reduces the early hydration rate, and in the later stages of the hydration reaction, waste rock wool can release the absorbed water, which has the effect of internal curing. Klyuev [18] reported similar research results.

Figure 8b demonstrates similar results to the compressive strength, where the flexural strength of the mortar shows a decreasing trend with the increase in the dosage of waste rock wool. The greatest decrease in flexural strength was observed when the waste rock-wool dosage was 20%, and the 7, 28 and 90 days of flexural strength were reduced by 34.93%, 17.27% and 17.19%, respectively. At the age of 28 days, the reduction of the flexural strength of the mortar decreased, the reduction of the flexural strength of the mortar was smaller when the waste rock-wool dosage was less than 10%, and the reduction of the flexural strength continued to increase when the waste rock-wool dosage was more than 10%. When the dosage of waste rock wool was 10%, the flexural strength decreased from 9.09 MPa to 8.66 MPa, which was only 4.73% lower. In order to achieve the requirement of 28 days of flexural strength not less than 8 MPa, the dosage of waste rock wool should be controlled within 10%. At the age of 90 days, the change rule of mortar flexural strength with the dosage of waste rock wool was consistent with the change rule of 28 days of flexural strength with the dosage of waste rock wool. At the age of 90 days, the change rule of mortar flexural strength with the dosage of waste rock wool was consistent with the change rule of 28 days of flexural strength with the dosage of waste rock wool. When the dosage of waste rock wool was 10%, the flexural strength of mortar decreased from 9.54 MPa to 9.09 MPa, which was a decrease of 4.75%.

3.3. Toughness

3.3.1. Flexural to Compressive Ratio

In this paper, the flexural and compressive ratio was used to characterize the toughness of cementitious materials, and the flexural and compressive ratio data of 7-, 28- and 90-day ages of waste rock-wool mortar are shown in Table 6. Waste rock wool improves the 7, 28 and 90 days of flexural compression ratios of mortar, which indicates that the waste rock wool has a certain toughening effect. From the data in Table 6, it can be seen that the flexural and compressive ratio of the mortar increases with the dosage of waste rock wool, with an optimal dosage of 20%. At the age of 7 days, the flexural and compressive ratio of mortar increased from 0.191 to 0.224—an increase of 17.28%. At the age of 28 days, the flexural and compressive ratio of mortar increased from 0.164 to 0.210—an increase of 28.05%. At the age of 90 days, the flexural and compressive ratio of mortar increased from 0.155 to 0.187—an increase of 21.13%.

Table 6. Flexural and compressive ratio of mortars.

Mix No.	7 d	28 d	90 d
R0	0.191	0.164	0.155
R5	0.202	0.173	0.163
R10	0.201	0.175	0.177
R15	0.201	0.183	0.173
R20	0.224	0.210	0.187

3.3.2. Impact Resistance

The increase in the flexural and compressive ratio of the mortar was due to the fact that the reduction in the compressive strength of the mortar was greater than the reduction in the flexural strength. It was not reasonable to express the toughening effect of waste rock wool by the flexural and compressive ratio alone. Therefore, this research further explored the effect of waste rock wool on the toughness of mortar using the falling ball method.

As can be seen from the data in Table 7, waste rock wool reduced the number of initial and final cracking impacts on the mortar specimens. Increasing the dosage of waste rock wool reduced the impact of initial cracking effects from 16 to 2 (a reduction of 87.50%) and the number of final cracking effects from 60 to 25 (a reduction of 58.33%). The impact strength of the specimen with the increase in waste rock-wool dosage also has a similar rule of change: in 20% of the dosage, the impact strength decreased from 53.56 J/cm^2 to 22.32 J/cm^2 (a decrease of 58.33%), which indicates that the waste rock wool weakened the impact resistance of the mortar. Thus, the larger the dosage, the more significant the weakening effect. Waste rock wool plays a role in bridging cracks and preventing microcracks from sprouting and expanding inside the matrix, but its adverse effect on the internal structure of the material was more significant and therefore shows a negative effect on the impact resistance.

Table 7. Effect of waste rock wool on the impact resistance of mortar.

Mix No.	Number of First Crack Impacts N_0	Number of Final Crack Impacts N_1	Impact Strength S/(J/cm^2)
R0	16	60	53.56
R5	14	48	42.85
R10	10	38	33.92
R15	4	32	28.57
R20	2	25	22.32

Figure 9 shows the effect of waste rock wool on the total impact energy consumption of mortar. With the increase in waste rock-wool dosage, the total impact energy consumption of mortar at the first and final cracking decreases at the same time. The differences between the two also decreased; the differences were 157.12 J, 121.41 J, 99.98 J, 99.99 J, and 82.13 J. The total energy difference between the initial and final cracking impacts was reduced to the lowest value at a 20% waste rock-wool dosage, indicating that the addition of waste rock wool reduced the toughness of the mortar.

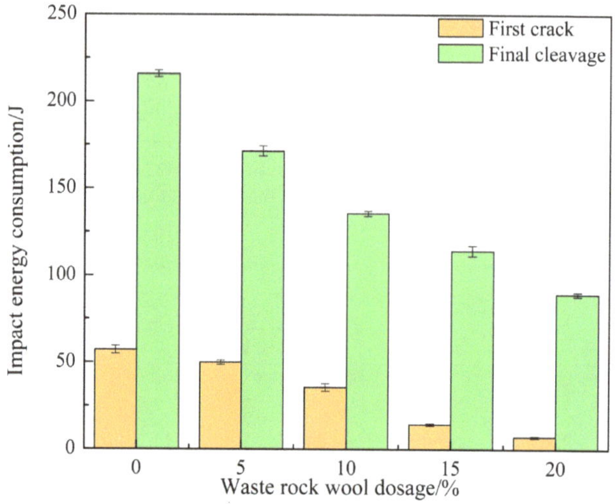

Figure 9. Effect of waste rock wool on total mortar impact energy consumption.

3.4. Abrasion Resistance

The incorporation of waste rock wool improved the abrasion resistance of the mortar to a certain extent, and the specific test data are shown in Table 8. The optimum dosage of waste rock wool was 5%, at which time the abrasion volume and abrasive mass of the mortar were 376 mm^3 and 0.85 g. This was reduced by 24.80% and 29.75%, respectively, compared with R0. The abrasion resistance of the specimens decreased continuously after the dosage of waste rock wool was higher than 5% (except for R15). Because R15 has a uniform distribution of fibers in the abrasion region without agglomeration, its abrasion volume and abrasion mass were instead lower compared to R10. In contrast, the phenomenon that the mortar's chord length of abrasion volume and abrasion volume were lower than that of the control group while the abrasion mass was higher than that of the control group at 10% dosing was attributed to the increase in abrasion mass due to the shedding of steel fiber in the blended fibers during the abrasion test.

Table 8. Effect of waste rock wool on abrasion resistance of mortar.

Mix No.	Chord Length of Abrasion Volume (mm)	Abrasion Volume (mm^3)	Mass Loss (g)
R0	39.0	500	1.21
R5	35.5	376	0.85
R10	37.5	444	1.29
R15	35.0	361	0.89
R20	40.5	561	1.34

3.5. Heat of Hydration

The hydration process of cement not only involved the generation of hydration products but was also accompanied by the release of a large dosage of heat. The dosage of heat released and the hydration rate of cement hydration were closely related to its hydration degree, so the hydration rate and the heat of hydration can be used to characterize the hydration degree of cement. As can be seen in Figure 10a, the waste rock wool delayed the emergence of the hydration-induced period of the composite slurry. Compared with R0, the dosage of 5%, 10%, 15% and 20% of waste rock wool delayed the appearance of the hydration exothermic peak by 0.55 h, 0.83 h, 1.04 h and 1.59 h, respectively. In the second hydration peak, the shoulder associated with the aluminate hydration reaction and the eventual formation of calculite was also observed [32]. This indicates that the role of waste rock wool in the system was dominated by physical filling, and the increase in the dosage of waste rock wool leads to a decrease in the percentage of cement in the cementitious material, and therefore an increase in the effective water/cement ratio in the system, leading to a decrease in the concentration of calcium ions in the pore solution and therefore an increase in the time required to reach the supersaturated state [33]. The peak of the hydration rate decreases continuously with the increase in waste rock-wool dosage, which was attributed to the dilution effect of waste rock wool that reduces the dosage of C_3A and C_3S in the cementitious system, thus slowing down the hydration process [34].

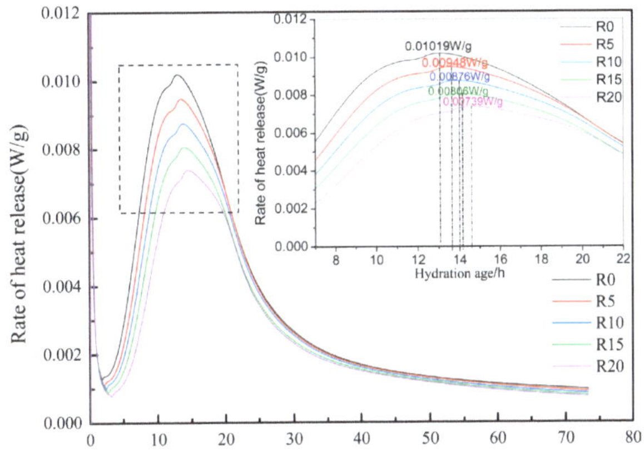

(**a**)

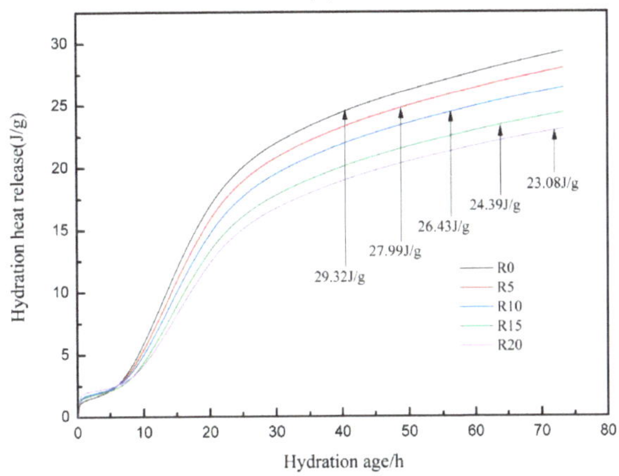

(**b**)

Figure 10. Effect of waste rock-wool dosing on heat of hydration: (**a**) Exothermic rate of hydration; (**b**) Total exothermic hydration.

The hydration exotherm of the cementitious material system with different dosages of waste rock wool is shown in Figure 10b, and the dosage of waste rock wool reduces the hydration exotherm of the system. Compared to R0, the exothermic heat of hydration was reduced by 4.54%, 9.86%, 16.81% and 21.28% at dosages of 5%, 10%, 15% and 20%, respectively. At 5% and 10% dosages, the decrease in exothermic capacity is lower than the dosage of waste rock wool, while the two dosages of 15% and 20% lead to a greater decrease in exothermic capacity than the dosage of waste rock wool, which indicates that waste rock wool has a promoting effect on hydration when the mass fraction is not greater than 10%. This was because the waste rock wool has very low activity and hardly reacts, which can increase the effective water/cement ratio of the system when it replaces cement in mortar with a low dosage, thus promoting the early hydration of cement. When the dosage is greater than 10%, waste rock wool will inhibit the hydration process of the cementitious system, which can be attributed to the high water absorption of waste rock wool. When the dosage is larger, it absorbs more water, resulting in a reduction in the effective water/cementitious ratio of the system, which results in a reduction in the degree of the early hydration of cement.

3.6. X-ray Diffraction

Figure 11 shows the X-ray diffraction patterns of the net cement paste at the age of 3 days and 28 days with different waste rock-wool dosages. From the figure, it can be seen that the main crystalline phases in the composite system were AFt, C_2S, C_3S, $CaCO_3$, $Ca(OH)_2$ and quartz, of which C_2S and C_3S were the main clinker phases of the cement, AFt and $Ca(OH)_2$ were the hydration products, and $CaCO_3$ was the result of the carbonation of $Ca(OH)_2$. Comparing the control and test groups in Figure 11a,b, it was found that there was no change in the position of the diffraction peaks of the hydration products, which indicated that no new hydration products were generated by the waste rock-wool admixture, and it only affected the relative intensity of the diffraction peaks. From Figure 11a, it can be seen that a 5% to 10% waste rock-wool dosage increased the relative intensity of the diffraction peaks of $Ca(OH)_2$ corresponding to $2\theta = 34.0°$, whereas the relative intensity of the $Ca(OH)_2$ diffraction peak gradually decreased at dosages of 15% to 20%. The results are consistent with those of the early hydration analysis of cement.

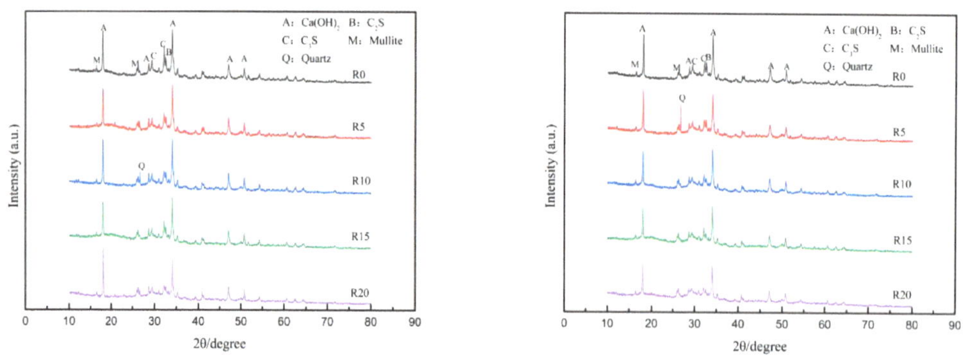

(a) (b)

Figure 11. XRD pattern of cement net paste mixed with waste rock wool: (**a**) 3 d; (**b**) 28 d.

Comparing Figure 11a,b, it can be found that the relative diffraction peaks of the clinker phases C_3S and C_2S decreased significantly with the increase in the age of curing, which indicates that the degree of cement hydration further increased. In contrast, the relative diffraction peak intensity of $Ca(OH)_2$ decreased, which indicates that the aluminous and siliceous components in fly ash and silica fume had a secondary hydration reaction with $Ca(OH)_2$ and consumed part of $Ca(OH)_2$.

3.7. TG-DSC Analysis

It was demonstrated that the temperature ranges of 60~150 °C, 390~460 °C and 540~800 °C corresponded to the volatilization of free water and the dehydration of C-H-S gels, the hydrolysis of calcium hydroxide and the decomposition of calcium carbonate, respectively [35–37].

As can be seen from Figure 12a, the mass loss of calcium hydroxide was 2.16%, 2.42%, 2.18%, 1.96% and 1.82% for R0, R5, R10, R15 and R20 at the age of 3 days, respectively. The content of calcium hydroxide showed an increase and then decrease with the increase in waste rock-wool dosage, which indicated that the waste rock wool promoted the hydration of cement at a low dosage, while the high dosage reduced the degree of cement hydration. This conclusion is consistent with the results of the analysis of the heat of hydration. From Figure 12b, it can be seen that the mass loss of calcium hydroxide was 2.44%, 2.44%, 1.93%, 2.04%, and 1.84% for R0, R5, R10, R15, and R20 at the age of 28 days, respectively. Comparing the mass loss of calcium hydroxide at the age of 3 days and 28 days, it can be seen that the content of calcium hydroxide increases less or even decreases with the increase in age, which indicates that in the hydration of the cementitious material system, fly ash, silica fume and calcium hydroxide undergo a secondary hydration reaction, which further improves the degree of hydration of the cementitious material system, in agreement with the analytical results of the XRD diffractograms.

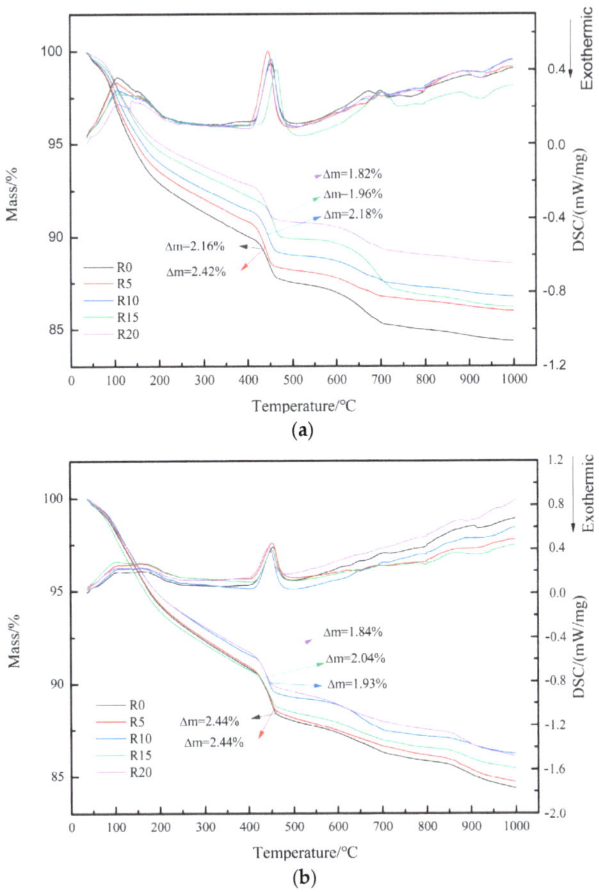

Figure 12. TG-DSC of cement paste mixed with waste rock wool: (**a**) 3 d; (**b**) 28 d.

3.8. SEM Analysis

Figure 13a shows the micro-morphology of hydration products of R0 (without waste rock wool) mortar, and Figure 13b shows the micro-morphology of hydration products of R5 (contains 5% waste rock wool) mortar. As can be seen in the figure, the hydration product type of the mortar remains the same, regardless of the presence or absence of waste rock wool, and C-S-H gel, layered $Ca(OH)_2$, ettringite and spherical fly ash particles could be clearly observed in the SEM image of both groups of mortars.

(**a**)

(**b**)

Figure 13. Microscopic morphology of waste rock-wool mortar: (**a**) R0; (**b**) R5.

The presence of defects such as microcracks inside the cement mortar as well as the self-contraction of the mortar can lead to the generation of microcracks, which are susceptible to stress concentration under loading, leading to the emergence of more and larger cracks [38–40]. In cementitious materials, fiber consumes energy through fiber bridging, fiber pullout, fiber breakage and crack deflection to improve the toughness of cement mortar. Different fibers have different effects on the material improvement, and a size of less than 100 μm waste rock wool fills the internal structure and inhibits the initiation and extension of the micro-cracking role. As shown in Figure 14a–c, the waste rock wool with a size less than 100 μm can be used as a filler to fill the internal pores of the mortar. It can also reduce the stress concentration at the crack tip through bridging actions, crack deformations and other action mechanisms. Energy is absorbed to inhibit microcrack expansion and improve the toughness of the mortar through the two destructive forms of tension and pullout (Figure 14a,b). However, when the dosage was too much, the phenomenon of fiber agglomeration occurs, as shown in Figure 14c, and the fiber agglomerates form defects, resulting in the weakening of the toughening effect. Figure 14d shows the effect of PVA fibers and steel fibers. The PVA fibers transmit stress and absorb energy in the form of twisting and deformation or even fracture, while the steel fibers consume energy by overcoming friction with the substrate and pulling it out of the substrate. In addition, in Figure 14b,c, the waste rock wool is randomly distributed around the PVA fibers, while in Figure 14d, it is randomly distributed around steel fibers. This indicates that the waste rock wool, PVA fibers and steel fibers play the roles of crack-blocking and toughening at different structural levels to achieve the layered reinforcement effect.

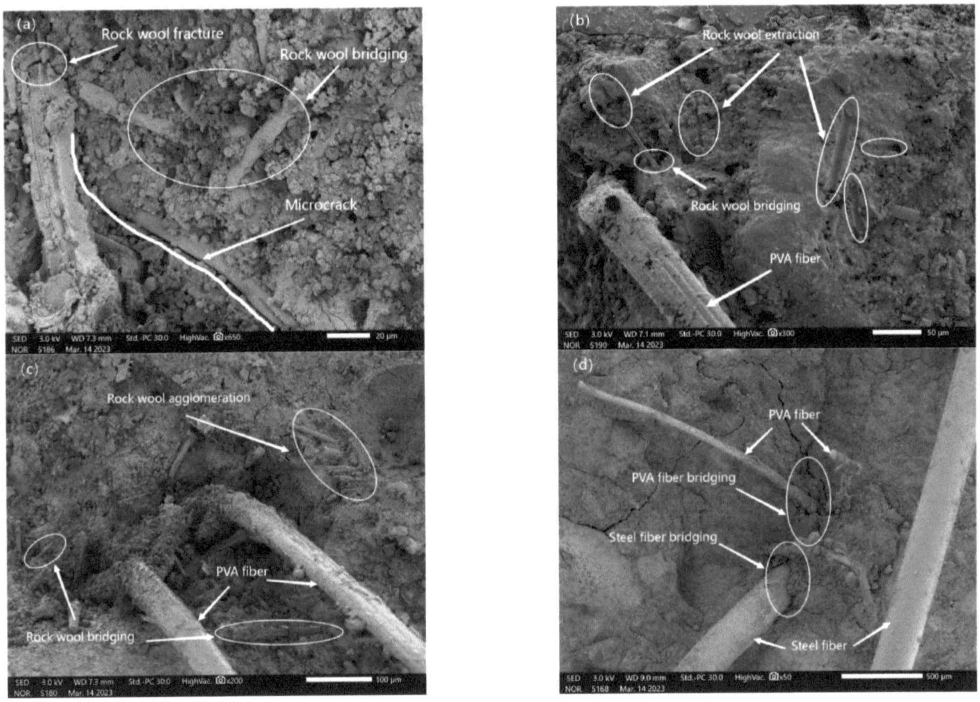

Figure 14. SEM image of multi-scale fiber-reinforced mortar. (**a**) Fibre fracture; (**b**) Fibre extraction; (**c**) Fibre agglomeration; (**d**) Fibre bridging.

4. Conclusions

In this paper, the effect of waste rock-wool dosage on the workability, strength, toughness, abrasion resistance, and hydration properties of PVA- and steel-fiber hybrid fiber mortar was investigated, and the mechanism of the effect of waste rock-wool dosage on the hybrid fiber mortar was summarized from a microscopic point of view. The research results were as follows:

1. Waste rock wool reduces the flexural strength and compressive strength of the mortar, and in the range of the selected dosage, the larger the dosage of waste rock wool, the greater the reduction in strength. However, with the increase in age, the effect of the reduction of strength was gradually reduced. When the dosage of waste rock wool was 10%, the 28 days of flexural and compressive strengths were reduced by 4.73% and 10.59%.
2. The effect of waste rock wool on the flexural-to-compressive ratio and impact toughness varied. With the increase in waste rock-wool dosage, the flexural-to-compressive ratio of the mortar at all ages increased, and the maximum value of the flexural-to-compressive ratio of the mortar at 28 days under 20% dosage was 0.210, which increased by 28.05%. The impact toughness decreased with the increase in the dosage, and the impact strength of the mortar at 20% doping was 22.32 J/cm^2, which decreased by 58.33%. Multi-scale hybrid combinations of waste rock wool, polyvinyl

alcohol fibers and steel fibers were able to achieve toughening effects at different structural levels.

3. When waste rock wool was used as cementitious material, it consumed waste rock wool and reduced the dosage of cement and auxiliary cementitious materials. This is in line with the concept of green development. Waste rock wool has low activity or even no activity and can only play the role of physical filler. Therefore, in order to maintain high mechanical strength, the amount of waste rock wool should not be higher than 10%.

Author Contributions: Conceptualization, L.X. and M.D.; data curation, J.C. and S.L.; validation, S.L., Z.Z., L.Y. and Y.W.; writing—original draft preparation, S.L.; writing—review and editing, L.X., S.L., Z.Z., L.Y. and Y.W. All authors have read and agreed to the published version of the manuscript.

Funding: This work was supported by the Priority Academic Program Development of Jiangsu Higher Education Institutions (PAPD).

Institutional Review Board Statement: Not applicable.

Informed Consent Statement: Not applicable.

Data Availability Statement: The original contributions presented in the study are included in the article, further inquiries can be directed to the corresponding authors.

Acknowledgments: The authors gratefully acknowledge the assistance from Lingling Xu, Min Deng and Jiajia Cheng.

Conflicts of Interest: The authors declare no conflicts of interest.

References

1. Yap, Z.S.; Khalid, N.H.A.; Haron, Z.; Mohamed, A.; Tahir, M.M.; Hasyim, S.; Saggaff, A. Waste Mineral Wool and Its Opportunities—A Review. *Materials* **2021**, *14*, 5777. [CrossRef] [PubMed]
2. Chen, C.; Huang, R.; Wu, J.; Yang, C. Waste E-glass particles used in cementitious mixtures. *Cem. Concr. Res.* **2006**, *36*, 449–456. [CrossRef]
3. Yliniemi, J.; Kinnunen, P.; Karinkanta, P.; Illikainen, M. Utilization of Mineral Wools as Alkali-Activated Material Precursor. *Materials* **2016**, *9*, 312. [CrossRef]
4. Ntimugura, F.; Vinai, R.; Harper, A.; Walker, P. Mechanical, thermal, hygroscopic and acoustic properties of bio-aggregates—Lime and alkali—Activated insulating composite materials: A review of current status and prospects for miscanthus as an innovative resource in the South West of England. *Sustain. Mater. Technol.* **2020**, *26*, e00211. [CrossRef]
5. Egnot, N.S.; Benson, S.M.; Vater, M.F.; Hazan, R.; Patel, O.; Marsh, G.M. Systematic review and meta-analysis of epidemiological literature evaluating the association between exposure to man-made vitreous fibers and respiratory tract cancers. *Regul. Toxicol. Pharmacol.* **2020**, *112*, 104585. [CrossRef] [PubMed]
6. Wei, M.-S.; Huang, K.-H. Recycling and reuse of industrial wastes in Taiwan. *Waste Manag.* **2001**, *21*, 93–97. [CrossRef] [PubMed]
7. Yliniemi, J.; Laitinen, O.; Kinnunen, P.; Illikainen, M. Pulverization of fibrous mineral wool waste. *J. Mater. Cycles Waste Manag.* **2018**, *20*, 1248–1256. [CrossRef]
8. Klima, K.; Luo, Y.; Brouwers, H.; Yu, Q. Effects of mineral wool waste in alkali activated-artificial aggregates for high-temperature applications. *Constr. Build. Mater.* **2023**, *401*, 132937. [CrossRef]
9. Lin, W.-T.; Cheng, A.; Huang, R.; Zou, S.-Y. Improved microstructure of cement-based composites through the addition of rock wool particles. *Mater. Charact.* **2013**, *84*, 1–9. [CrossRef]
10. Ramírez, C.P.; del Río Merino, M.; Arrebola, C.V.; Barriguete, A.V.; Kosior-Kazberuk, M. Analysis of the mechanical behaviour of the cement mortars with additives of mineral wool fibres from recycling of CDW. *Constr. Build. Mater.* **2019**, *210*, 56–62. [CrossRef]
11. Medeiros, M.G.; Nadaleti, W.C.; Rocha, J.C.; Cheriaf, M.; Gleise, P.J.P.; de Castilhos, A.B.D.C., Jr. A cleaner material production by the incorporation of the rockwool waste into portland cement matrices. *J. Clean. Prod.* **2021**, *293*, 126059. [CrossRef]
12. Yliniemi, J.; Walkley, B.; Provis, J.; Kinnunen, P.; Illikainen, M. Nanostructural evolution of alkali-activated mineral wools. *Cem. Concr. Compos.* **2020**, *106*, 103472. [CrossRef]
13. da Rocha, E.B.D.; de Sousa, A.M.F.; Furtado, C.R.G. Properties Investigation of novel nitrile rubber composites with rockwool fibers. *Polym. Test.* **2020**, *82*, 106291. [CrossRef]
14. Cheng, A.; Lin, W.-T.; Huang, R. Application of rock wool waste in cement-based composites. *Mater. Des.* **2011**, *32*, 636–642. [CrossRef]
15. Yoo, D.-Y.; Kim, J.; Zi, G.; Yoon, Y.-S. Effect of shrinkage-reducing admixture on biaxial flexural behavior of ultra-high-performance fiber-reinforced concrete. *Constr. Build. Mater.* **2015**, *89*, 67–75. [CrossRef]

16. Cheng, L.; Chen, S.; Chen, F.; Wang, C.; Chen, Q. Research Progress and Performance Evaluation of Polyvinyl Alcohol Fiber Engineered Cementitious Composites. *Sustainability* **2023**, *15*, 10991. [CrossRef]
17. Abed, Z.M.; Jaber, A.A.; Ghaeb, H.O.; Hasan, A.M. Investigating the Various Sustainable Variables of Composite Ferrocement Mortar Using Residual Rock Wool Fibers. *J. Mater. Civ. Eng.* **2020**, *32*, 04020381. [CrossRef]
18. Väntsi, O.; Kärki, T. Mineral wool waste in Europe: A review of mineral wool waste quantity, quality, and current recycling methods. *J. Mater. Cycles Waste Manag.* **2014**, *16*, 62–72. [CrossRef]
19. Klyuev, S.; Fediuk, R.; Ageeva, M.; Fomina, E.; Klyuev, A.; Shorstova, E.; Sabitov, L.; Radaykin, O.; Anciferov, S.; Kikalishvili, D.; et al. Technogenic Fiber Wastes for Optimizing Concrete. *Materials* **2022**, *15*, 5058. [CrossRef]
20. Gebremariam, A.T.; Vahidi, A.; Di Maio, F.; Moreno-Juez, J.; Vegas-Ramiro, I.; Łagosz, A.; Mróz, R.; Rem, P. Comprehensive study on the most sustainable concrete design made of recycled concrete, glass and mineral wool from C&D wastes. *Constr. Build. Mater.* **2021**, *273*, 121697. [CrossRef]
21. Luo, Y.; Yu, Q. Valorization of mineral wool waste in Class F fly ash geopolymer: Geopolymerization, macro properties, and high temperature behavior. *Cem. Concr. Compos.* **2024**, *145*, 105318. [CrossRef]
22. Awoyera, P.O.; Akinrinade, A.D.; de Sousa Galdino, A.G.; Althoey, F.; Kirgiz, M.S.; Tayeh, B.A. Thermal insulation and mechanical characteristics of cement mortar reinforced with mineral wool and rice straw fibers. *J. Build. Eng.* **2022**, *53*, 04568. [CrossRef]
23. Ramírez, C.P.; Sanchez, E.A.; del Río Merino, M.; Arrebola, C.V.; Barriguete, A.V. Feasibility of the use of mineral wool fibres recovered from CDW for the reinforcement of conglomerates by study of their porosity. *Constr. Build. Mater.* **2018**, *191*, 460–468. [CrossRef]
24. Kubiliute, R.; Kaminskas, R.; Kazlauskaite, A. Mineral wool production waste as an additive for Portland cement. *Cem. Concr. Compos.* **2018**, *88*, 130–138. [CrossRef]
25. Lin, W.-T.; Cheng, A.; Huang, R.; Wu, Y.-C.; Han, T.-Y. Rock wool wastes as a supplementary cementitious material replacement in cement-based composites. *Comput. Concr.* **2013**, *11*, 93–104. [CrossRef]
26. Cheng, J.; Shi, X.; Xu, L.; Zhang, P.; Zhu, Z.; Lu, S.; Yan, L. Investigation of the effects of styrene acrylate emulsion and vinyl acetate ethylene copolymer emulsion on the performance and microstructure of mortar. *J. Build. Eng.* **2023**, *75*, 106965. [CrossRef]
27. *GB/T 2419-2005*; Test Method for Fluidity of Cement Mortar. China National Standards: Beijing, China, 2005.
28. *GB/T 17671-1999*; Method of Testing Cements Determination of Strength. China National Standards: Beijing, China, 1999.
29. *GB/T 38494-2020*; Test Method for Impact Resistance of Ceramic Wares. China National Standards: Beijing, China, 2020.
30. *GB/T 3810.6-2016*; Test Methods of Ceramic Tiles-Part 6: Determiantion of Resistance to Deep Abrasion for Unglazed Tiles. China National Standards: Beijing, China, 2016.
31. Istenič, D.; Prosenc, F.; Zupanc, N.; Turel, M.; Holobar, A.; Milačič, R.; Marković, S.; Mihelič, R. Composting of recovered rock wool from hydroponics for the production of soil amendment. *Environ. Sci. Pollut. Res.* **2024**, *31*, 29280–29293. [CrossRef] [PubMed]
32. Hesse, C.; Goetz-Neunhoeffer, F.; Neubauer, J. A new approach in quantitative in-situ XRD of cement pastes: Correlation of heat flow curves with early hydration reactions. *Cem. Concr. Res.* **2011**, *41*, 123–128. [CrossRef]
33. Laibao, L.; Yunsheng, Z.; Wenhua, Z.; Zhiyong, L.; Lihua, Z. Investigating the influence of basalt as mineral admixture on hydration and microstructure formation mechanism of cement. *Constr. Build. Mater.* **2013**, *48*, 434–440. [CrossRef]
34. Ji, G.; Peng, X.; Wang, S.; Hu, C.; Ran, P.; Sun, K.; Zeng, L. Influence of magnesium slag as a mineral admixture on the performance of concrete. *Constr. Build. Mater.* **2021**, *295*, 123619. [CrossRef]
35. Chintalapudi, K.; Pannem, R.M.R. Enhanced microstructural characteristics of binary and ternary blended cements reinforced with graphene oxide. *Fuller. Nanotub. Carbon Nanostructures* **2022**, *30*, 987–1001. [CrossRef]
36. Tan, H.; Nie, K.; He, X.; Deng, X.; Zhang, X.; Su, Y.; Yang, J. Compressive strength and hydration of high-volume wet-grinded coal fly ash cementitious materials. *Constr. Build. Mater.* **2019**, *206*, 248–260. [CrossRef]
37. Yang, K.; Tang, Z.; Cheng, Z.; Zhao, H.; Feng, R.; Long, G. Mechanical properties of ultra-high strength cement-based materials (UHSC) incorporating metal powders and steel fibers. *Constr. Build. Mater.* **2022**, *318*, 125926. [CrossRef]
38. Cai, R.; Qi, H.; Mao, J.; Lv, J.; Jin, D. Improved Crack Resistance and Pore Structure of Cement-Based Materials by Adding EVA Powder. *J. Mater. Civ. Eng.* **2022**, *34*, 0004143. [CrossRef]
39. Wang, Y.; Ge, Y.; Wang, X.; Chen, X.; Li, Q. The effect of powder activated carbon on mechanical properties and pore structures of cement-based mortars. *Constr. Build. Mater.* **2022**, *316*, 125798. [CrossRef]
40. Zhang, Q.; Gong, S.; Zhao, Y.; Wu, Y.; Zhou, J. Mechanical properties of multi-scale fiber compound reinforced cement-based materials. *J. Civ. Environ. Eng.* **2021**, *43*, 123–129. [CrossRef]

materials

Article

Effects of Curing Temperature on Expansion of Concrete Due to ASR

Yongfu Yang [1,2], Min Deng [1,3,*], Liwu Mo [1] and Wei Li [1,*]

1 College of Materials Science and Engineering, Nanjing Tech University, Nanjing 210009, China; andymoliwu@njtech.edu.cn (L.M.)
2 School of Materials Science and Engineering, Luoyang Institute of Science and Technology, Luoyang 471023, China
3 State Key Laboratory of Materials-Oriented Chemical Engineering, Nanjing 210009, China
* Correspondence: dengmin@njtech.edu.cn (M.D.); well_lee@njtech.edu.cn (W.L.)

Abstract: In the laboratory study of alkali–silica reaction (ASR), models attempt to predict the service life of concrete due to ASR by correlating the performance of concrete at high and low temperatures. However, the consequences of elevating temperature are not so encouraging. In this paper, the influence of temperature on the expansion of 2-graded concrete and 3-graded concrete caused by ASR was investigated by curing the concrete under different temperatures ranging from 40 °C to 80 °C. Increased temperature resulted in rapid expansion at the early stages, but the expansion rate of concrete prisms cured at the higher temperatures (70 °C and 80 °C) was slowed down at the later stages, and concrete prisms cured at 50 °C or 60 °C showed the highest expansions during the experimental period. The chemical analysis results of the pore solution expressed from the concrete show that the ASR expansion is significantly influenced by the $[OH^-]$: the decrease in $[OH^-]$ leads to the retardation of the ASR expansion. The decrease in $[OH^-]$ is attributed to the consumption of OH^- ions for the alkali–silica reaction and the decrease in activity of NaOH(aq) influenced by the temperature. For large cross-section specimens, the OH^- within the concrete for alkali–silica reactions cannot be effectively compensated by the external alkali solution. In the accelerated test to evaluate ASR for large cross-section specimens, a curing temperature of less than 60 °C is suggested. This study provides critical insights into the temperature dependency of ASR expansion of concrete, offering a curing temperature range for developing predictive models of ASR expansion under varied environmental conditions.

Keywords: curing temperature; ASR expansion; pore solution; concentration of OH^- ions; activity of $NaOH_{(aq)}$

check for updates

Citation: Yang, Y.; Deng, M.; Mo, L.; Li, W. Effects of Curing Temperature on Expansion of Concrete Due to ASR. *Materials* **2024**, *17*, 3140. https://doi.org/10.3390/ma17133140

Academic Editor: Sukhoon Pyo

Received: 3 May 2024
Revised: 25 June 2024
Accepted: 25 June 2024
Published: 27 June 2024

1. Introduction

An alkali–silica reaction (ASR) is a chemical reaction between the reactive silica component contained in aggregate and hydroxyl ions (OH^-) from the pore solution of concrete [1]. This reaction produces alkali–silicate gel (ASR gel), which can absorb water and cause deleterious expansion, even cracking, in the concrete. Structures can be severely damaged by ASR expansion and are irreparable. It is, therefore, very important to assess the risk of ASR in concrete and to take appropriate measures in advance to prevent ASR expansion. The use of non-reactive aggregates is the most reliable measure to prevent concrete deterioration due to ASR. However, this option is often not practical due to the limited availability of non-reactive aggregates in many locations. In China, alkali-reactive aggregates are widely distributed [2], and their use may be an unavoidable choice, e.g., the dam concretes for the Jinping Hydroelectric Project located in Sichuan Province have no choice but to use alkali-reactive sandstone as a coarse aggregate.

Although ASR is adequately understood, it is still a challenge to assess the risk of deterioration in concrete containing alkali-reactive aggregates. In the laboratory study of ASR, in order to correlate the performance of concrete at high and low temperatures, experimental studies on concrete specimens are typically performed under accelerated conditions by increasing the curing temperature, and some ASR models are proposed to predict the future performance and progression of ASR in structures in service. Although many studies have investigated the influence of temperature on the expansion rate and the final expansion [3–5], the relationship between the final expansion and the temperature dependence remains unclear. Tang et al. [6] and Poole et al. [7] presented models based on an Arrhenius equation to extrapolate from laboratory experiments to the field. They found that the measured rate constants of ASR at different temperatures fit the Arrhenius equation well, and the reaction rate at normal temperature could be estimated by extrapolating the regression line through the results at higher temperatures. According to Larive's work [8], temperature does not affect the final expansion, but it does affect the expansion kinetics [9], and the temperature dependence of the kinetics of ASR expansion follows Arrhenius' law. The expansion curve at a certain reference temperature has been required to estimate some parameters, and models have been developed by Ulm et al. [10], Farage et al. [11], and Comi et al. [12]. However, Diamond et al. [13] reported that the final expansion was greater at 20 °C than at 40 °C. Kawabata et al. [14] proposed the alkali-wrapped concrete prism test (AW-CPT) to evaluate the effect of temperature on ASR expansion, avoiding the effect of alkali leaching and moisture loss, and the results showed that the final expansion was greater at 40 °C than at 60 °C. Chatterji et al. [15] studied the alkali–silica reaction of two types of sand, and the results showed that for any sand and salt concentration, the expansion increased with the decreasing temperature. Fournier et al. [3] discussed the results of an interlaboratory study on the accelerated concrete prism test (i.e., 60 °C, RH > 95%) for ASR. With increasing temperature, the main expansion phase of the test prisms at 60 °C is often complete after 3 months of testing. Lindgård et al. [16] discussed how exposure conditions influence prism expansion: for prisms exposed to 60 °C, the rate and amount of alkali leaching is the main controlling factor for the prism expansion. Ideker et al. [17] reported that expansions due to alkali–silica reaction (ASR) in the accelerated concrete prism test (ACPT-60 °C) showed a significant reduction at 13 weeks compared to 52 weeks testing in the standard concrete prism test (CPT-38 °C).

Temperature is a critical factor in the alkali–silica reaction, while the expansion pressure exerted by the ASR gel inside the reactive aggregates is a key factor in the expansion process of concrete. Some ASR expansion models describing the expansion behavior associated with the chemical reaction have been developed by Dunant et al. [18], Putatatsananon et al. [19], Multon et al. [20], Takahashi et al. [21], and Miura et al. [22]. However, the present evaluation methods are mainly based on the results of mortar bars or concrete specimens (75 mm × 75 mm × 250 mm, CPT prism), in which the aggregate size is smaller than that of practical concrete. The crack pattern, such as onion skin or sharp cracks [23] in the aggregate, is strongly dependent on the reactive rock type and the random distribution of the expansion sites in the aggregate. The size of the aggregate can strongly affect the alkali transfer pathways and the ASR gel trapped within the aggregate [24]. The results obtained from the mortar bar or CPT prism are not well correlated with the ASR expansion behavior of actual concrete.

The development of accurate and reliable performance tests for the durability of concrete is still a challenge. As stated by Thomas et al. [25], the appropriate benchmark of a laboratory performance test is against real concrete structures or, as a surrogate, against large concrete blocks exposed to natural weathering conditions. However, such field experience is extremely time-consuming. Therefore, an accelerated performance test to evaluate the ASR expansion of concrete is very necessary. According to Hobbs [26] and Glasser [27], the key parameters influencing the rate and extent of alkali–silica reactions are temperature, alkali content, and humidity. In this paper, the 3-graded concrete prism with 300 mm × 300 mm × 500 mm and the 2-graded concrete prism with 150 mm × 150 mm × 550 mm were cast using the practical aggregate, Jingping sandstone. Compared to the CPT prism, the cross-section of samples was enlarged, even for the

3-graded concrete cured in an alkali solution, to avoid alkali leaching. The influence of curing temperature on the ASR expansion of large concrete blocks was studied. It tends to evaluate the effect of temperature on ASR expansion and the temperature dependence of ASR expansion.

2. Materials and Methods

2.1. Materials

The coarse aggregate selected for this study was sandstone used in a dam located in Liangshan, Sichuan Province, China. The aggregate shows a mosaic structure consisting of quartz, mica, feldspar, calcite, and chlorite, according to the petrographic analysis. A total of 10% of microcrystalline quartz was considered to be the reactive phase, which is aggregated in the rock. The expansions according to the accelerated mortar bar method (ASTM C1260 [28]) in 14 days and the concrete prism test method (ASTM C1293 [29]) in 1 year are 0.19% and 0.04%, respectively. The coarse aggregate is potentially alkali reactive. The fine aggregate used in the concrete was non-reactive river sand. Portland cement was selected on the basis of its alkali content. Cement LA is a low-alkali cement (0.56% Na_2O_{eq} by mass) and the chemical composition of the cement is given in Table 1. Polycarboxylate superplasticizer (SP, 1.49% Na_2O_{eq} by mass) was added to the concrete to ensure good workability. Chemical-grade sodium hydroxide NaOH was used to adjust the alkali content of the concrete.

Table 1. Chemical composition of the cement.

Sample	Chemical Composition/wt.%									
	SiO_2	Al_2O_3	Fe_2O_3	CaO	MgO	SO_3	K_2O	Na_2O	L.O.I	Total
Cement LA	19.56	4.28	4.17	63.61	1.27	2.05	0.53	0.21	3.19	98.87

2.2. Sample Preparation and Curing Conditions

The 3-graded concrete and the 2-graded concrete prisms were cast, respectively. The mixes of the laboratory concrete were based on those of the field concretes, but some parameters were adjusted: the alkali content in the concrete was increased to 5.25 kg/m^3 by adding NaOH to the mixing water, and fly ash was replaced by cement. The water–cement ratio of concrete is 0.5. For the 3-graded concrete, the sand ratio is 39%, and the mass ratio of 5–20 mm aggregate/20–40 mm aggregate/40–80 mm aggregate is 3:3:4. For the 2-graded concrete and the reference concrete, the sand ratio is 30%, and the 5–20 mm aggregate/20–40 mm aggregate is 5:5 by mass. For the reference concrete, the alkali content was 1.60 kg/m^3, and alkali was supplied only by the cement and superplasticizer. The mixing proportions of the concretes are given in Table 2. Polycarboxylate superplasticizer was added to the fresh concretes to ensure they had the same slump.

Table 2. Mix proportions of the concretes.

Concrete Type	Na_2O_e /(kg/m^3)	Mix Proportions/(kg/m^3)						
		Cement	Water	Fine Aggregate	Coarse Aggregate			SP
					5–20 mm	20–40 mm	40–80 mm	
3-graded	5.25	204	102	605	468	468	624	1.8
2-graded	5.25	280	140	614	715	715	-	1.8
Reference	1.60	280	140	614	715	715	-	1.1

The 3-graded concrete prisms with a size of 300 mm × 300 mm × 500 mm, the 2-graded concrete prisms with a size of 150 mm × 150 mm × 550 mm, and the reference concrete prisms with a size of 150 mm × 150 mm × 550 mm were cast. A strain gauge was embedded in the concrete (Figure 1) to measure ASR expansion. In addition, 12 prisms of the 3-graded concrete with dimensions of 300 mm × 300 mm × 500 mm and 18 cubes of the 2-graded concrete with dimensions of 150 mm × 150 mm × 150 mm were cast for pore solution extraction at a specific period. Samples were cured for 5 days at 22 ± 1 °C and RH ≥ 95% and then stored in the curing boxes. Prisms of the 3-graded concrete were immersed in 0.7 mol/L NaOH solution. Prisms or cubes of the 2-graded concrete were stored above water, i.e., cured in the moisture. All concrete prisms were stored vertically in the curing boxes, and the curing temperature was set at 40, 50, 60, 70, and 80 °C, respectively.

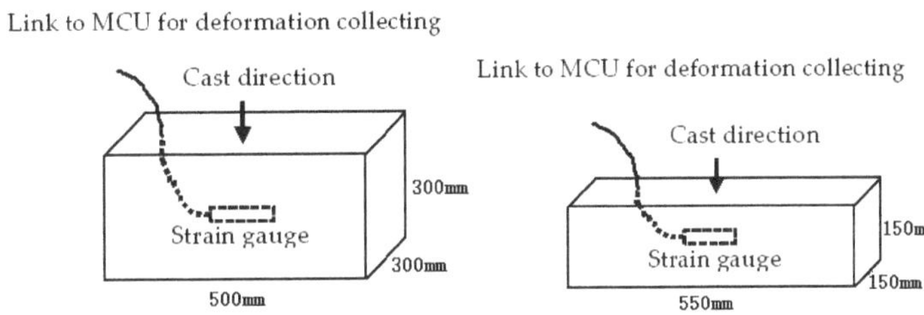

(**a**) prism with 300 mm × 300 mm × 500 mm (**b**) prism with 150 mm × 150 mm × 550 mm

Figure 1. Schematic of stain gauge embedded in sample.

2.3. Expansion Measurement

Strain gauges embedded in the concrete were used to measure the deformation due to ASR expansion. The sensors were linked to the microcontroller unit (MCU) system, and deformation data was collected at 12 h intervals.

2.4. Internal Moisture Content and Open Porosity of Concrete

The evaporable water content was determined to represent the internal humidity of the concrete, and the open porosity of the concrete was calculated. Concrete cubes with sizes of 100 mm × 100 mm × 100 mm were cast and cured in moisture at 40 °C, 60 °C, and 80 °C, respectively. After a long-term (1 year) saturation by the moisture in the curing boxes, the mass of the specimens was measured, then they were finally dried at 105 °C and weighed. The internal moisture content (C_m) and the concrete open porosity (p) are calculated using the following equations:

$$C_m = \frac{m_0 - m_i}{m_i} \times 100\% \tag{1}$$

$$\mathrm{p} = \frac{V_{water}}{V_{concrete}} = \frac{(m_0 - m_i)\cdot\rho_{concrete}}{m_0\cdot\rho_{water}} \times 100\% \tag{2}$$

where V_{water} is the pore volume in concrete, $V_{concrete}$ is the volume of concrete, m_0 is the mass of the moisture-saturated specimen, m_i is the mass of the dry specimen, $\rho_{concrete}$ is the density of concrete, and ρ_{water} is the density of water.

2.5. Ion Concentration of Pore Solution

The concretes for pore solution were cured at 40, 60, and 80 °C, respectively. The 3-graded concrete prisms were immersed in 0.7 mol/L NaOH solution, while the 2-graded concrete cubes were cured in moisture. At certain ages, the power supply to the curing chambers was deactivated, allowing the containers to cool to ambient temperature. Cores with the diameter of 160 mm and the length of 200 mm (Φ160 mm × 200 mm) were taken out from the 3-graded concrete and divided into two parts: the inner and the outer with the size of Φ160 mm × 100 mm. Samples for pore solution were crushed and then subjected to high-pressure expression (1200 MPa), following the method outlined by Barneyback [30]. Following expression, approximately 200 μL of pore solution (with an accuracy of 5 μL) was transferred to a 50 mL volumetric flask and diluted with deionized water to reach volume. The concentration of Na and K were determined using a flame photometer. The concentration of OH^- ions was assessed through direct titration with 0.01 mol/L HCl (with an accuracy of 0.0001 mol/L) using the phenolphthalein endpoint. Chemical analysis was carried out at 22 ± 2 °C. In the experimental process, some measures were taken to keep the pore solution from carbonating, such as fresh deionized water was always used.

3. Results and Discussion

3.1. Deformation of the Reference Concrete

The deformation of the concretes includes the expansion due to ASR and the shrinkage due to the cement hydration. Figure 2 presents the deformation at different curing temperatures obtained on the reference concrete. It is evident that the concrete experiences shrinkage and this shrinkage becomes more pronounced with higher temperatures. The rapid hydration of cement at elevated temperatures leads to increased shrinkage, particularly during the early age. The reference concrete with low alkali content (1.60 kg/m^3 Na_2O_{eq}) is regarded as non-reactive concrete, though they were cast by reactive aggregate.

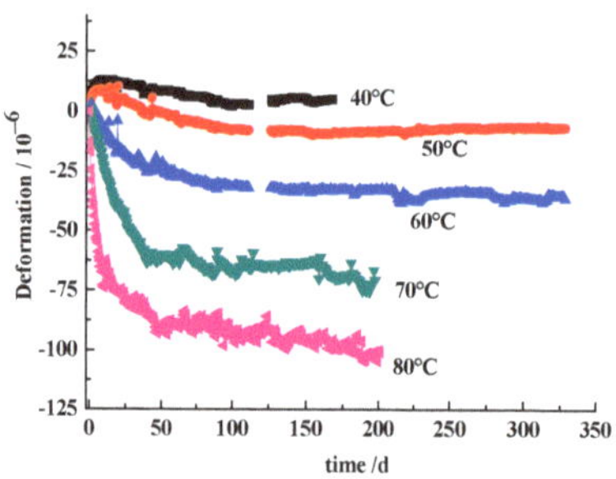

Figure 2. Deformation of the reference concrete stored over water.

3.2. ASR Expansion

Figures 3 and 4 show the deformation at different curing temperatures obtained on the 3-graded concrete and the 2-graded concrete, respectively. The induction period of the expansion depends on the curing temperature. At a low temperature (40 °C), the curves show an S-shape, indicating a very slow expansion rate at first, then a rapid increase, and finally, a gradual slowing down. As the temperature increased, the specimens expanded rapidly after a short latent period, followed by a low rate of expansion. For the 3-graded concrete prisms immersed in alkali solution, the expansions in 1 year under five curing temperatures of 40, 50, 60, 70, and 80 °C are 0.050%, 0.084%, 0.099%, 0.057%, and 0.046%, respectively. The prism cured at 60 °C produced the highest expansion during the experimental period. For the 2-graded concrete prism cured in moisture at 70 °C, the curve indicates a very rapid ASR expansion rate in the first 25 days, then a gradual slowing down in the period of 25–150 days. After 150 days, the expansion remains almost unchanged, and the expansion up to a final asymptote is 0.027%. For the 2-graded concrete prism cured at 60 °C, it expands rapidly in the first 40 days, then a gradual slowing down in the period of 40–250 days. The specimen remains stable after 250 days, with an expansion of 0.035% at 420 days. For the prism cured at 50 °C, the expansion rate remains relatively fast before 100 days, then slows down. The expansion at 420 days is 0.061%. For the prism cured at 40 °C, after a latent period of about 25 days, the specimen begins to expand with the slowest expansion rate. However, in the later age, its expansion exceeds that of specimens cured at 60 °C and 70 °C. At 420 days, the expansion is 0.040%. The prism cured at 50 °C in moisture shows the highest expansion. The expansion of the 2-graded concrete prism cured in moisture is always lower than that of the 3-graded concrete prisms immersed in alkali solution at the same curing temperature. Whether the concrete is cured in moisture or alkaline solution, within the curing temperature range of 40–80 °C, the higher the curing temperature, the faster the initial ASR expansion of concrete prisms. Increasing the curing temperature significantly accelerates the ASR expansion process of concrete. However, increasing the curing temperature will also shorten the duration of the rapid expansion stage of concrete, and the earlier the ASR expansion of concrete tends to stabilize. The prisms cured at lower temperatures have a slower initial expansion rate, but they maintain stable growth in the later ages. The expansion rate of concrete prisms cured at the higher temperatures (70 °C and 80 °C) was slowed down at the later ages, and concrete prisms cured at 50 °C or 60 °C showed the highest expansions during the experimental period. ASR expansion of prism cured at the higher temperature is not necessarily larger.

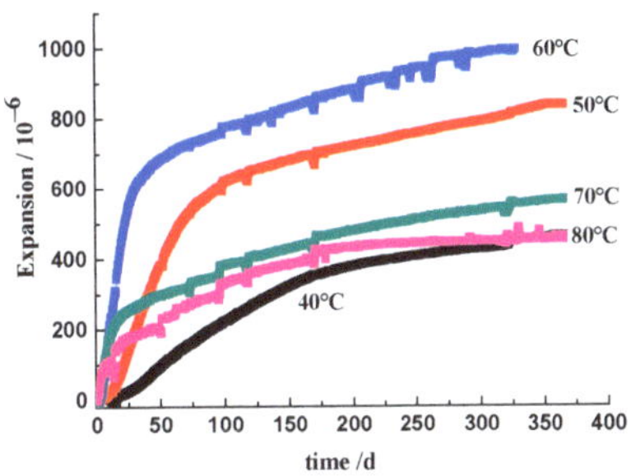

Figure 3. Deformation of the 3-graded concrete immersed in 0.7 mol/L NaOH solution.

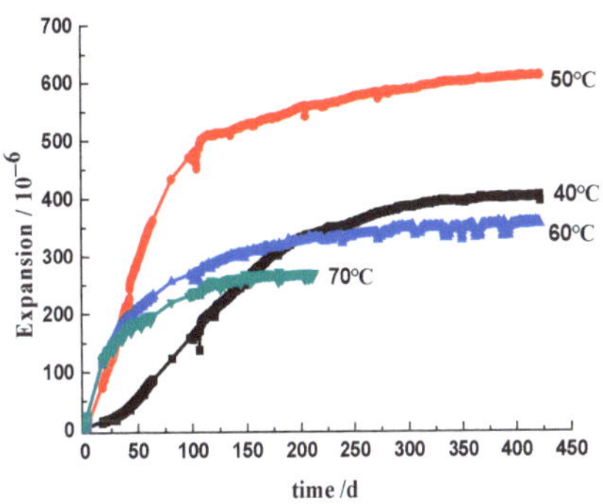

Figure 4. Deformation of the 2-graded concrete stored over water.

3.3. Internal Moisture Content and Porosity

The internal moisture content and the open porosity of concrete samples cured at 40 °C, 60 °C, and 80 °C were tested, and the results are shown in Table 3. The internal moisture content within concrete increases with increasing curing temperature, indicating that concrete samples can maintain a high RH level even at an elevated temperature to ensure that ASR expansion proceeds [16]. The open porosity of the concrete also increased with increasing temperature.

Table 3. Internal moisture content and porosity of the 2-graded concrete cured in moisture.

Samples Cured at	Internal Moisture Content/%	Porosity/%
40 °C	6.41	7.81
60 °C	6.47	9.30
80 °C	6.55	9.72

3.4. Alkali Content in Pore Solution

Table 4 shows the ion concentrations of the pore solution in the 2-graded concrete cured in moisture. Table 5 shows the ion concentrations of the pore solution in the 3-graded concrete cured in 0.7 mol/L NaOH solution. For each storage temperature, the [Na^+], [K^+], and [OH^-] of the pore solution in concretes, whether cured in moisture or in alkali solution, tend to decrease with the development of ASR for up to 1 year, except at 40 °C in 7 days, where an increase in the [Na^+] and [K^+] is observed. Moreover, the ion concentrations in the pore solution decreased more rapidly as the curing temperature increased. Compared to the concrete cured in moisture, the ion concentrations of concrete cured in alkali solution decreased more slowly. For concrete cured in moisture at 60 °C and 80 °C for 1 year, the [OH^-] of the pore solution is too low to be detected. For concrete cured in alkali solution, the [K^+] in the pore solution of the inner concrete at any curing temperature is always higher than that of the outer concrete. The [OH^-] in the pore solution of the inner concrete cured at 40 °C, 60 °C, and 80 °C for 28 days and at 40 °C for 60 days is higher than that of the outer concrete, but it inverses for concrete cured for 60 days at 60 °C, 80 °C, and for 1 year at each curing temperature. Similar results to the [Na^+] are also observed in Table 5.

Table 4. Ion concentration of the pore solution in concrete cured in moisture.

Curing Time/Days		0	7	14	28	180	365
[K^+]/(mmol/L)	40 °C	212.3	231.6	193.0	155.7	130.7	39.8
	60 °C		141.5	111.2	102.5	70.8	26.5
	80 °C		128.9	106.2	56.6	25.0	23.6
[Na^+]/(mmol/L)	40 °C	523.5	616.0	557.4	451.8	372.6	262.2
	60 °C		443.7	353.4	300.4	179.3	201.7
	80 °C		403.4	419.5	236.6	94.9	143.4
([K^+] + [Na^+])/(mmol/L)	40 °C	735.8	847.6	750.4	607.5	503.3	302.0
	60 °C		585.2	464.6	402.9	250.1	228.2
	80 °C		532.3	525.7	293.2	119.9	167.0
[OH^-]/(mmol/L)	40 °C	515.8	496.2	441.8	394.9	346.8	220.3
	60 °C		375.3	336.0	233.6	161.4	-
	80 °C		302.3	293.6	112.9	78.2	-

0 d: the sample for pore solution cured at 22 ± 2 °C, RH 100% for 1 day. -: the OH^- concentration is too low to be detected.

Table 5. Ion concentration of the pore solution in concrete cured in 0.7 mol/L NaOH solution.

Curing Time/Days		Ion Concentration of Pore Solution/(mmol/L)								
		40 °C			60 °C			80 °C		
		[K^+]	[Na^+]	[OH^-]	[K^+]	[Na^+]	[OH^-]	[K^+]	[Na^+]	[OH^-]
0		331.5	786.7	658.4	331.5	786.7	658.4	331.5	786.7	658.4
28	Outer	212.3	676.6	484.0	252.5	558.2	379.4	185.1	455.1	285.4
	Inner	297.3	748.6	503.4	261.0	564.7	453.8	245.4	530.6	301.2
60	Outer	225.6	595.0	378.1	177.9	505.8	287.8	109.2	451.8	165.9
	Inner	253.2	608.1	409.5	160.7	427.3	268.2	148.6	441.0	153.3
365	Outer	88.9	630.4	348.9	56.3	412.8	213.9	26.5	363.0	60.5
	Inner	100.3	546.8	322.7	70.2	402.7	179.4	45.0	388.9	29.3

0 d: the sample for pore solution cured at 22 ± 2 °C, 100% RH for 1 day.

3.5. Discussion

The results described in this paper show that the expansion of concrete either cured in moisture or cured in alkali solution and the [OH^-] in concrete pore solution are greatly influenced by temperature. In general, increasing the curing temperature from 40 °C to 80 °C can accelerate the rate of ASR expansion at early ages. However, in this experiment, the higher temperature of 80 °C did not result in higher final expansion. The results are different from those of Liu [31] and Ke [32]. In their experiments, mortar bars were cured in 1 mol/L NaOH solution, and the results showed that the ASR expansion increased with increasing temperature, i.e., the mortar bar cured at 80 °C showed the highest expansion in the experimental periods.

3.5.1. Contradiction to the Alkali Leaching

As mentioned in the Introduction, the reduction in the final expansion of concrete stored over water at elevated temperatures has been attributed to the alkali leaching. Many researchers explained that alkali leaching would reduce the alkalinity of the pore solution in the concrete, thus significantly reducing the final expansion [16,20,33]. In this experiment, alkali leaching is inevitable in the 2-graded concrete stored over water, though some measures were taken to avoid or reduce the leaching of alkali from the concrete, such as increasing the cross-section of the concrete prism, prolonging the pre-storage period at room temperature. It cannot be denied that alkali leaching made an important contribution to the reduction in ASR expansion of concrete cured at an elevated temperature in moisture, even when immersed in water. However, for the 3-graded concrete immersed in 0.7 mol/L NaOH solution, the prisms were surrounded by the alkali ions, and alkali leaching from the concrete to the simulated solution was negligible. In addition, the concentration of NaOH solution decreased to 0.6 mol/L after 1 year at 80 °C, indicating that the alkali had penetrated into the concrete to compensate for alkali consumption. At this storage condition, which is similar to that of Ke and Liu's experiments, the prism cured at 60 °C showed the highest expansion, unlike Ke and Liu's results: the higher the curing temperature, the higher the expansion obtained. This phenomenon, which occurred in the 3-graded concrete, cannot be explained by alkali leaching.

3.5.2. The [OH^-] Influenced by Temperature

The OH^- ions in the pore solution were produced by the hydration of the cement and the dissolution of NaOH in the mixing water. Lothenbach et al. [34] investigated the change in concentration of OH^- ions in Portland cement paste cured at temperatures ranging from 5 °C to 50 °C. Their results showed that the [OH^-] increased with increasing temperature and the evolution of hydration. Durand et al. [35] showed a similar result that the [OH^-] of the pore solution in Portland cement paste cured at 38 °C was maintained at a high level at later ages. It can be concluded that the [OH^-] will not vary dramatically in concrete if no OH^- ions are consumed. In our experiments, comparing Figure 4 and Table 4, the 2-graded concrete prisms cured in moisture expanded rapidly at the early ages with the decline in [OH^-] in concrete pore solution, followed by a decrease in the rate until a long-term asymptote was reached, accompanying the [OH^-] decreased to a very low level. Comparing Figure 3 and Table 5, for the 3-graded concrete, prisms expanded continuously after the rapid expansion at the early ages. Simultaneously, the [OH^-] decreased due to the consumption of OH^- ions in concrete, though the concrete prisms were immersed in alkali solution. For concrete prism cured at 80 °C for 1 year, the concentration of OH^- ions in the inner concrete decreased to a very low level of 0.029 mol/L, far from the threshold [OH^-] of 0.25 mol/L suggested by Diamond [36]. In our laboratory, the dissolution kinetics of reactive silica in Jingping sandstone exposed to NaOH solution was studied, and the experimental results indicate that ASR behaves as a first-order reaction, accompanied by the exponential decrease in the [OH^-] with time [37]. The alkali content and the concentration of OH^- in the concrete pore solution play a major role in the development of ASR, especially the role of [OH^-] in the evolution of ASR [9]. Therefore, the change in alkali content and OH^- concentration in the pore solution can well interpret the rate and extent of ASR during laboratory performance testing. The alkali and OH^- ions permeate into sandstone aggregates and react with the microcrystal silica phase, producing expansive ASR gels. At the early ages, the alkali–silica reaction proceeds rapidly for the high [OH^-], and the reaction rate increases with the increasing temperature. Accordingly, the concrete prisms expand rapidly. As the alkali–silica reaction proceeds, OH^- ions are continuously consumed, resulting in a decrease in $[OH^-]$. When the $[OH^-]$ decreases to the threshold for initiating ASR expansion, the reaction becomes very slow, and no apparent expansion is observed. For the mortar bar immersed in the alkali solution, ASR expansion of the specimen with small scale section occurred in an alkali-supply environment, but for the prism in 300 mm × 300 mm × 500 mm cured in 0.7 mol/L NaOH solution, the alkali ions in pore solution cannot be compensated effectively from the outside alkali solution.

3.5.3. The Effective $[OH^-]$ Influenced by Temperature

The effective $[OH^-]$ in the pore solution of concrete is also influenced by the curing temperature. NaOH in concrete pore solution will dissociate into Na^+ and OH^-, as $NaOH_{(aq)}$ is a strong electrolyte. The degree of ionization depends on the concentration and temperature of the solution. According to the results of Pabalan et al. [38], activity coefficients of $NaOH_{(aq)}$ at 1 bar or saturation pressure are shown in Figure 5. When the concentration of sodium hydroxide is greater than 0.25 mol/L, the activity coefficient increased slightly with the temperature increase from 0 °C to 50 °C, then it declined gradually. When the concentration is less than 0.25 mol/L, the activity coefficient decreases with the increasing temperature. This means that the amount of effective OH^- ions in the pore solution of concrete cured at a lower temperature (40 °C) is greater than that of concrete cured at a higher temperature (80 °C) if the concrete has the same content of NaOH.

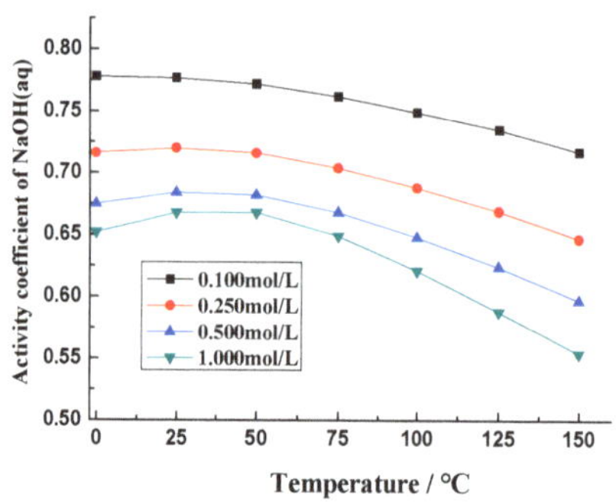

Figure 5. Activity coefficients of $NaOH_{(aq)}$ [38].

The effective OH^- ions play a major role in the ASR expansion of concrete. When the specimen with the 2-graded aggregate cured at 70 °C in moisture for 220 days (the expansion had reached a long-term asymptote) was transferred to the conditions of 40 °C, it re-expanded rapidly and continuously (Figure 6). As mentioned in Table 4, moisture content in concrete decreased with the decrease in temperature, so the re-expansion of the concrete prism is not attributed to the wet swelling of concrete. Notably, samples cured at a high temperature show a high porosity, which means that the re-expansion needs more ASR gel to fill the confined space compared to the prism cured at 40 °C all the time [39]. With the temperature decreased, the amount of effective OH^- ions increased, and ASR proceeded again, leading to the re-expansion of the concrete prism.

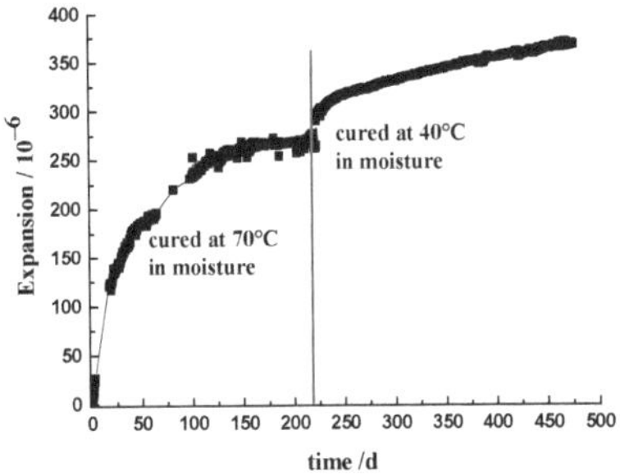

Figure 6. Re-expansion of the prism when transferred from 70 °C to 40 °C of curing condition.

3.5.4. Temperature Dependency of ASR Expansion

The ASR expansion of concrete is influenced by the effective alkaline ion concentration of the concrete pore solution. At the initial curing stage of concrete, the concentration of alkali ions in the pore solution is high, and the alkali–silica reaction rate is rapid, leading to more ASR gel generation and prisms expanding rapidly at various temperatures. As the curing age prolongs, the concentrations of K^+, Na^+, and OH^- ions continue to decrease, and the rate of alkali–silicate reaction slows down; as a result, the expansion of the prism cured at various temperatures also slows down.

The influence of elevating temperature on ASR expansion may manifest in the following ways: (1) When the alkali content in the concrete pore solution is adequate, elevating temperature will accelerate the process of alkali–silica reaction, and ASR gels are mostly confined in a limited space within the aggregate, which induces the rapid ASR expansion, i.e., in the early ages, the higher curing temperature, the greater ASR expansion; (2) As the temperature increases, the $[OH^-]$ in the concrete pore solution decreases faster to the threshold for initiating ASR expansion, which slows down the rate of ASR expansion; (3) The activity coefficients of OH^- ion may decrease with the increasing temperature, which means that the amount of effective OH^- ions in pore solution of concrete cured at a higher temperature is less than a lower temperature if the concrete has the same content of NaOH, and this may decrease ASR expansion; (4) Elevating temperature boosts the exudation of ASR gels from the sandstone into the pore solution, which reduces ASR expansion [5,37]. Thus, ASR expansion is comprehensively affected by the curing temperature.

4. Conclusions

This study investigated the influence of curing temperature on the ASR expansion of concrete containing reactive sandstone as the coarse aggregate. Two sets of tests were carried out, namely, (1) 3-graded concrete with 300 mm × 300 mm × 500 mm cured in alkali solution and (2) 2-graded concrete with 150 mm × 150 mm × 550 mm cured in moisture. The following conclusions are obtained:

- ASR expansion of concrete shows a strong temperature dependency. Increasing the curing temperature accelerates the expansion rate of ASR in the early ages, which exhibits a positive temperature dependency, while increasing the temperature also accelerates the $[OH^-]$ reduction, decreases the effective $[OH^-]$, and boosts the exudation of ASR gels from aggregates, which shows a reduced, negative temperature dependency.
- The ASR expansion is significantly influenced by the $[OH^-]$ in the pore solution of concrete. Increasing the curing temperature may accelerate the decrease in $[OH^-]$ and shorten the duration of the rapid expansion stage of concrete. Specimens cured at 50 °C or 60 °C showed the highest expansion during the experimental period. The ASR expansion of specimens cured at higher temperatures was not necessarily greater.
- For the 300 mm × 300 mm × 500 mm specimens, the OH^- within concrete for ASR cannot be compensated effectively from the outside alkali solution.

- The temperature dependency of ASR expansion in the large cross-section concrete prisms should be considered when correlating concrete performance at high and low temperatures. In the accelerated test to evaluate ASR for large cross-section specimens, a curing temperature of less than 60 °C is suggested.

Author Contributions: Conceptualization, Y.Y. and M.D.; Methodology, Y.Y.; Software, Y.Y.; Validation, Y.Y., M.D. and L.M.; Formal analysis, Y.Y.; Investigation, Y.Y.; Resources, Y.Y. and W.L.; Data curation, Y.Y.; Writing—original draft, Y.Y.; Writing—review and editing, Y.Y., W.L. and L.M.; Visualization, Y.Y.; Supervision, M.D.; Project administration, M.D.; Funding acquisition, M.D. All authors have read and agreed to the published version of the manuscript.

Funding: This research was financially supported by the National Key Research and Development Plan of China (2017YFB0309903-01).

Institutional Review Board Statement: Not applicable.

Informed Consent Statement: Not applicable.

Data Availability Statement: Data are contained within the article.

Conflicts of Interest: The authors declare no conflicts of interest.

References

1. Wang, H.; Gillott, J. Mechanism of alkali-silica reaction and the significance of calcium hydroxide. *Cem. Concr. Res.* **1991**, *21*, 647–654. [CrossRef]
2. Deng, M.; Lan, X.; Xu, Z.; Tang, M. Petrographic characteristic and distribution of reactive aggregates in China. In Proceedings of the 12th International Conference on Alkali-Aggregate Reaction in Concrete, Beijing, China, 15–19 October 2004.
3. Fournier, B.; Chevrier, R.; Grosbois, M.; Lisella, R.; Folliard, K.; Ideker, J.; Shehata, M.; Thomas, M.; Baxter, S. The accelerated concrete prism test (60 °C): Variability of the test method and proposed expansion limits. In Proceedings of the 12th International Conference on Alkali-Aggregate Reaction in Concrete, Beijing, China, 15–19 October 2004.
4. Lindgård, J.; Nixon, P.; Borchers, I.; Schouenborg, B.; Wigum, B.J.; Haugen, M.; Åkesson, U. The EU "PARTNER" Project—European standard tests to prevent alkali reactions in aggregate: Final results and recommendations. *Cem. Concr. Res.* **2010**, *40*, 611–635. [CrossRef]
5. Kawabata, Y.; Dunant, C.; Yamada, K.; Scrivener, K. Impact of temperature on expansive behavior of concrete with a highly reactive andesite due to the alkali–silica reaction. *Cem. Concr. Res.* **2019**, *125*, 105888. [CrossRef]
6. Tang, M.; Han, S. Kinetics of alkali-silica reaction. In Proceedings of the 6th International Congress on Alkalis in Concrete, Copenhagen, Denmark, 22–25 June 1983.
7. Poole, T.; Wong, G. Service life prediction, ASR deterioration. In Proceedings of the Materials Engineering Conference, Infrastructure: New Materials and Methods of Repair, San Diego, CA, USA, 13–16 November 1994.
8. Larive, C. *Apports Combinés de l'expérimentation et de la Modélisation à la Compréhension de l'alcali-réaction et de ses Effets Mécaniques*; Ecole Nationale des Ponts et Chaussées: Paris, France, 1997.
9. Kim, T.; Olek, J.; Jeong, H. Alkali-silica reaction: Kinetics of chemistry of pore solution and calcium hydroxide content in cementitious system. *Cem. Concr. Res.* **2015**, *71*, 36–45. [CrossRef]
10. Ulm, F.; Coussy, O.; Li, K.; Larive, C. Thermo-chemo-mechanics of ASR expansion in concrete structures. *J. Eng. Mech.* **2000**, *126*, 233–242. [CrossRef]
11. Farage, M.; Alves, J.; Fairbairn, E. Macroscopic model of concrete subjected to alkali-aggregate Reaction. *Cem. Concr. Res.* **2004**, *34*, 495–505. [CrossRef]
12. Comi, C.; Fedele, R.; Perego, U. A chemo-thermo-damage model for the analysis of concrete dams affected by alkali-silica reaction. *Mech. Mater.* **2009**, *41*, 210–230. [CrossRef]
13. Diamond, S.; Barneyback, R.; Struble, L. Physics and chemistry of alkali-silica reactions. In Proceedings of the 5th International Conference on Alkali-Aggregate Reaction in Concrete, Cape Town, South Africa, 30 March–3 April 1981.
14. Kawabata, Y.; Yamada, K.; Sagawa, Y.; Ogawa, S. Alkali-Wrapped Concrete Prism Test (AW-CPT)—New testing protocol toward a performance test against alkali-silica reaction. *J. Adv. Concr. Technol.* **2018**, *16*, 441–460. [CrossRef]
15. Chatterji, S.; Christensen, P. Studies of alkali-silica reaction. Part 7. Modelling of expansion. *Cem. Concr. Res.* **1990**, *20*, 285–290. [CrossRef]
16. Lindgård, J.; Thomas, M.; Sellevold, E.; Pedersen, B.; Andiç-Çakır, Ö.; Justnes, H.; Rønning, T.F. Alkali-silica reaction (ASR)—Performance testing: Influence of specimen pre-treatment, exposure conditions and prism size on alkali leaching and prism expansion. *Cem. Concr. Res.* **2013**, *53*, 68–90. [CrossRef]
17. Ideker, J.; East, B.; Folliard, K.; Thomas, M.D.; Fournier, B. The current state of the accelerated concrete prism test. *Cem. Concr. Res.* **2010**, *40*, 550–555. [CrossRef]

18. Dunant, C.; Scrivener, K. Micro-mechanical modelling of alkali-silica-reaction-induced degradation using the AMIE framework. *Cem. Concr. Res.* **2010**, *40*, 517–525. [CrossRef]
19. Putatatsananon, W.; Saouma, V. Chemo-mechanical micromodel for alkali-silica reaction. *ACI Mater. J.* **2013**, *110*, 67–77.
20. Multon, S.; Sellier, A. Multi-scale analysis of alkali–silica reaction (ASR): Impact of alkali leaching on scale effects affecting expansion tests. *Cem. Concr. Res.* **2016**, *81*, 122–123. [CrossRef]
21. Takahashi, Y.; Ogawa, S.; Tanaka, Y.; Maekawa, K. Scale-dependent ASR expansion of concrete and its prediction coupled with silica gel generation and migration. *J. Adv. Concr. Technol.* **2016**, *14*, 444–463. [CrossRef]
22. Miura, T.; Multon, S.; Kawabata, Y. Influence of the distribution of expansive sites in aggregates on the microscopic damage due to alkali-silica reaction (ASR)—Insights into the mechanical origin of expansion. *Cem. Concr. Res.* **2021**, *142*, 1–18. [CrossRef]
23. Sanchez, L.; Fournier, B.; Jolin, M.; Duchesne, J. Reliable quantification of AAR damage through assessment of the Damage Rating Index (DRI). *Cem. Concr. Res.* **2015**, *67*, 74–92. [CrossRef]
24. Dunant, C.; Scrivener, K. Effects of aggregate size on alkali-silica reaction induced expansion. *Cem. Concr. Res.* **2012**, *42*, 745–751. [CrossRef]
25. Thomas, M.; Fournier, B.; Folliard, K.; Ideker, J.; Shehata, M. Tests methods for evaluating preventive measures for controlling expansion due to alkali-silica reaction in concrete. *Cem. Concr. Res.* **2006**, *36*, 1842–1856. [CrossRef]
26. Hobbs, D. *Alkali-Silica Reaction in Concrete*; Thomas Telford Ltd.: London, UK, 1988.
27. Glasser, F. Chemistry of the alkali-aggregate reaction. In *The Alkali Silica Reaction in Concrete*; Swamy, R., Ed.; Van Nostrand Reinhold: New York, NY, USA, 1992.
28. *ASTM C1260*; Standard Test Method for Potential Alkali Reactivity of Aggregates (Mortar-Bar Method). ASTM International: West Conshohocken, PA, USA, 1994.
29. *ASTM C1293*; Standard Test Method for Determination of Length Change of Concrete Due to AlkaliSilica Reaction. ASTM International: West Conshohocken, PA, USA, 2009.
30. Barneyback, R.; Diamond, S. Expression and analysis of pore fluids from hardened cement paste and mortars. *Cem. Concr. Res.* **1981**, *11*, 279–285. [CrossRef]
31. Liu, C. Experimental Investigation on Suppression and Expansion Prediction of Alkali-Silica Reaction in Concrete. Master's Thesis, China Institute of Water Resources and Hydropower Research, Beijing, China, 2006. (In Chinese).
32. Ke, J. Kinetics of Expansion due to Alkali-Aggregate Reaction and Expansion Prediction. Master's Thesis, Nanjing Tech University, Nanjing, China, 2008. (In Chinese).
33. Rogers, C.; Hooton, R. Reduction in mortar and concrete expansion with reactive aggregates due to alkali leaching. *Cem. Concr. Aggreg.* **1991**, *13*, 42–49. [CrossRef]
34. Lothenbach, B.; Winnefeld, F.; Alder, C.; Wieland, E.; Lunk, P. Effect of temperature on the pore solution microstructure and hydration products of Portland cement pastes. *Cem. Concr. Res.* **2007**, *37*, 483–491. [CrossRef]
35. Durand, B.; Bérand, J.; Roux, R.; Soles, J. Alkali-silica reaction: The relation between pore solution characteristics and expansion test results. *Cem. Concr. Res.* **1990**, *20*, 419–428. [CrossRef]
36. Diamond, S. Alkali reactions in concrete—Pore solution effects. In Proceedings of the Sixth International Conference, Copenhagen, Denmark, 22–25 June 1983.
37. Yang, Y.; Deng, M.; Mo, L.; Li, W. Kinetics of alkali-silica reaction: Application to the sandstone. *Materials* **2024**, *17*, 2956. [CrossRef]
38. Pabalan, R.; Pitzer, K. Thermodynamics of NaOH(aq) in hydrothermal solutions. *Geochim. Cosmochim. Ac.* **1987**, *51*, 829–837. [CrossRef]
39. Lindgård, J.; Sellevold, E.; Thomas, M.; Pedersen, B.; Justnes, H.; Rønning, T.F. Alkali–silica reaction (ASR)-performance testing: Influence of specimen pre-treatment, exposure conditions and prism size on concrete porosity, moisture state and transport properties. *Cem. Concr. Res.* **2013**, *53*, 145–167. [CrossRef]

materials

MDPI

Article

Kinetics of Alkali–Silica Reaction: Application to Sandstone

Yongfu Yang [1,2], Min Deng [1,3,*], Liwu Mo [1] and Wei Li [1,*]

[1] College of Materials Science and Engineering, Nanjing Tech University, Nanjing 210009, China; hn_yyf@163.com (Y.Y.)
[2] School of Materials Science and Engineering, Luoyang Institute of Science and Technology, Luoyang 471023, China
[3] State Key Laboratory of Materials-Oriented Chemical Engineering, Nanjing 210009, China
* Correspondence: dengmin@njtech.edu.cn (M.D.); stevenlee187@outlook.com (W.L.)

Abstract: Despite extensive research, the relationship between the progression of the alkali–silica reaction (ASR) and the expansion of concrete due to ASR, particularly for the heterogeneous aggregate with slow reactivity, is not thoroughly understood. In this paper, the dissolution kinetics of reactive silica present in sandstone when exposed to NaOH solutions, alongside the expansion characteristics of rock prisms under ASR conditions, were studied. The experimental results indicate that ASR behaves as a first-order reaction, accompanied by an exponential decrease in the concentration of OH^- over time, and the dissolution rate of silica is predominantly governed by diffusion dynamics. Notably, increasing the temperature accelerates ASR, which augments the expansive pressure in a confined and limited space, leading to more significant aggregate expansion. Conversely, higher temperatures also result in a diminished retention of ASR gels within the aggregate, leading to the mitigation of ASR expansion. Our findings underscore that larger aggregates retain a greater quantity of gels, resulting in more pronounced expansion. To establish an ASR prediction model based on the relationship of the ASR expansion of concrete to high and low temperatures, the parameters such as the range of curing temperatures and the grading size of aggregates should be carefully considered for the experiments.

Keywords: alkali–silica reaction; kinetics; sandstone; rock prism; mechanism of ASR expansion

Citation: Yang, Y.; Deng, M.; Mo, L.; Li, W. Kinetics of Alkali–Silica Reaction: Application to Sandstone. *Materials* **2024**, *17*, 2956. https://doi.org/10.3390/ma17122956

Academic Editor: Weiting Xu

Received: 18 April 2024
Revised: 28 May 2024
Accepted: 11 June 2024
Published: 17 June 2024

1. Introduction

The alkali–silica reaction (ASR) is a complex chemical reaction taking place between the reactive silica in the aggregate and hydroxyl ions (OH^-) from the pore solution in concrete [1,2]. ASR gel may adsorb water and cause the expansion and cracking of concrete. In general, two steps are involved in the ASR process: (1) the chemical reaction to form ASR gel, and (2) the expansion of gel due to the uptake of water [3]. Correspondingly, the ASR expansive behavior of concrete depends on: (1) chemical factors influencing the chemical reaction rate and degree, such as temperature [4,5], pore solution chemistry [6,7], and the reactivity of silica; (2) the response of physical factors to expansive pressure, such as the confined conditions of ASR gel [8–11], the distribution of reactive silica in the aggregate [12], and the rheological behavior of the gel [13]. Therefore, it is necessary and important to comprehensively understand the influence factors and expansion mechanism of ASR for establishing the ASR prediction model.

1.1. Chemistry of Alkali–Silica Reaction

The kinetics is equally important because the rate at which silica dissolves in the pore solution can largely control the overall rate of ASR. The explanation of the chemical process was given by Glasser et al. [14], and in more detail, by Bulteel et al. [15]. The topochemical rupture of the siloxane bonds possibly acts as follows:

$$\equiv \text{Si-O-Si} \equiv + 2OH^- \Leftrightarrow 2(\equiv \text{Si-O}^-) + H_2O \quad (1)$$

But it may occur in two steps [15]:

(1) The formation of Q3 sites due to siloxane bonds first breaking up by hydroxide ion attack, as follows:

$$2SiO_2 + OH^- \rightarrow SiO_{5/2}^- + SiO_{5/2}H \quad (2)$$

(2) The dissolution of silica due to continued hydroxide ion attack on the Q3 sites to form silica ions, as follows:

$$SiO_{5/2}^- + OH^- + \frac{1}{2}HO_2 \rightarrow H_2SiO_4^{2-} \quad (3)$$

As the reactive silica structures (≡Si-OH, ≡Si-O-Si≡) gradually break down, they also attract the soluble alkali hydroxides (sodium hydroxide, NaOH, and/or potassium hydroxide, KOH), then form the hydrated alkali–silicate gel within the aggregate. Dron et al. [16] argued that ASR involves mainly ionic reactions in the pore solution. Different chemical equilibriums are possible, depending on the pH of the pore solution and the ionic species in the solution. Balachandran et al. [17] proposed a three-stage behavioral model to explain the dynamic relationship between ASR progression and pore solution composition. A high calcium ASR product (C–S–H- type gel) or a low calcium ASR product (ASR-type gel) was dictated by their accessibility to the pore solution.

The dissolution rate of silica is influenced by various factors, such as the pH, temperature, surface area of the reactive silica, the reactivity of silica, and the concentrations of K^+, Na^+, and Ca^{2+} present in concrete [18,19]. Dron et al. [20] studied the kinetics of the dissolution of opal in NaOH solution according to the conditions described in ASTM C289 [21]. Bulteel [15] and Verstraete et al. [22] quantified the reaction degree of flint dissolved in 0.79 mol/L KOH solution at different temperatures. Maraghechi [23] studied the dissolution rate of soda–lime–silica glass, and the results showed that it is highly related to both pH and temperature. Up to pH = 14 and at 60 °C, the dissolution rate of glass was found to be a function of $[OH^-]^{0.2}$. Kim et al. [19] investigated the chemical reaction sequence and kinetics of ASR by exposing α-cristobalite to 0.7 mol/L KOH or NaOH with the presence of solid calcium hydroxide ($Ca(OH)_2$). They subsequently presented the results on the kinetics of the chemistry of the pore solution in a cementitious system containing the highly reactive aggregate [24].

1.2. *Effect of Aggregate Properties*

Whether or not concrete structures suffer ASR damage during service is fundamentally determined by the content of reactive silica in the aggregates and their properties. Aggregates are commonly classified as: non-reactive, moderately reactive, highly reactive, or very highly reactive, according to the results of standardized tests (e.g., the accelerated mortar bar test (AMBT, ASTM C1260 [25]) or the concrete prism test (CPT, ASTM C1293 [26])). Amorphous SiO_2 (e.g., opal, natural, or synthetic glass) is known to be the most reactive, followed by metastable crystals (e.g., cristobalite and tridymite), microcrystalline silica, and other crystalline forms containing many lattice defects, residual strains, or internal microcracks. If the aggregate particle is composed entirely of reactive silica, such as opal or glasses, which have uniform composition throughout their volume, the aggregate surface may react quickly with the alkaline pore solution to form ASR gel at the surface. Onion skin cracks can be observed when alkali ions diffuse uniformly into the highly reactive aggregate [27,28]. Ichikawa et al. [27] proposed a reaction rim model to describe the ASR evolution. However, if the reactive silica is absent at the surface and is mostly contained as fine reactive particles distributed within a non-reactive matrix of the aggregate, such as many sandstones, greywackes and siliceous limestones, it takes time for the alkaline pore solution to penetrate into and contact with the reactive silica. As a result, the reaction proceeds more slowly, and the gel forms within the aggregate, leading to sharp cracks throughout the aggregate. As a type of slow-reacting aggregate, the mineralogy and fabric of rocks are responsible for different manifestations of the alkali–silica reaction [29,30]. Based on the results, Dunant et al. proposed a gel pocket model [31]. In the gel pocket model, expansive pressure forms at the random location of a reactive site (gel pocket), i.e., the origin of expansion is randomly distributed inside the aggregate [28,32].

The porosity degree of aggregate particles also plays a vital role in the rate of ASR expansion. According to the study by Ben et al. [30], it was reported that highly porous aggregates suffered more ASR expansion as compared to the aggregates with lower porosity. It was found that the alkali ions are more easily diffused in the porous aggregate particles than in the less porous aggregates, which then initiates aggregate dissolution leading to ASR.

The kinetics of the highly reactive aggregates, such as opal, glass, flint, and α-cristobalite, has been studied by many researchers, while for the slow-reacting aggregates, the reactive silica randomly distributed within the non-reactive matrix of the aggregates, few studies have focused on the kinetics of silica dissolution. On the other hand, for the same aggregate, the expansion of ASR gel at different temperatures and aggregate sizes varies. The cause of the expansion induced by ASR gel remains controversial. In this paper, the kinetics of ASR in sandstone was investigated, and the corresponding induced expansion was discussed.

2. Materials and Methods

2.1. *Materials*

Sandstone, a clastic sedimentary rock, was used in this study, which also adopted the dam concrete for the Jinping Hydroelectric Project located at Sichuan Province, China, as the coarse aggregate. The sandstone is mainly composed of 72.0 wt.% SiO_2, 11.2 wt.% Al_2O_3, 4.5 wt.% Fe_2O_3, 1.4 wt.% CaO, 3.6 wt.% MgO, 2.0 wt.% K_2O, and 1.0 wt.% Na_2O; and the ignition loss is 4.1 wt.%. The rock exhibits a mosaic structure consisting of quartz, mica, feldspar, calcite, and chlorite, according to the petrographic analysis using a LEICA DM750P Polarization Microscope (Leica Microsystems, Tokyo, Japan) (Figure 1a). Microcrystalline quartz is considered to be the reactive phase, which distributes as an aggregate in the rock (Figure 1b). The amount of microcrystalline quartz in the rock is approximately 10 wt.%. A continuous foliation can be distinguished, and the direction of the foliation planes is shown by a black arrow in Figure 1c. Minerals in the rock were also confirmed by X-ray diffraction (XRD, Bruker D8 Focus, Billerica, MA, USA, Cu, 40 kV, 40 mA, 5°/min,), and the results are shown in Figure 2. The expansion of the mortar was 0.19% at 14 days according to the accelerated mortar bar method (ASTM C1260 [25]), and the expansion of the concrete prisms was 0.04% at 1 year according to ASTM C1293 [26]. This indicates that sandstone is potentially alkali-reactive, and that it is a slow-reacting aggregate. Chemical grade NaOH (NaOH $\geq$ 95%) was used to prepare alkali solutions with deionized water.

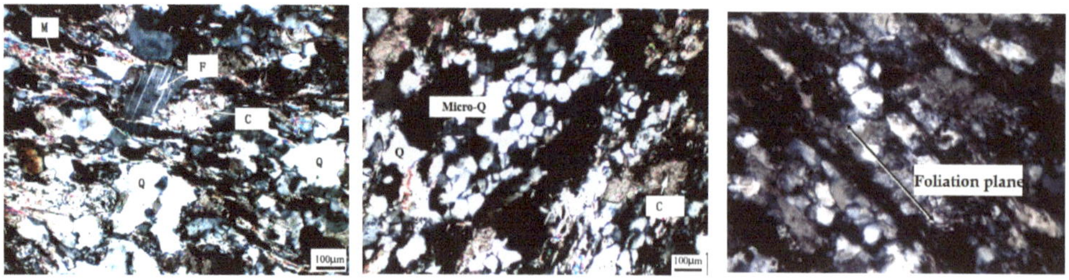

(a) (b) (c)

Figure 1. Petrographic analysis of sandstone with a crossed polarizer. (**a**) Clastic texture of sandstone; (**b**) microcrystalline quartz aggregated in sandstone; (**c**) direction of foliation planes. Q—quartz; C—calcite; F—feldspar; M—mica.

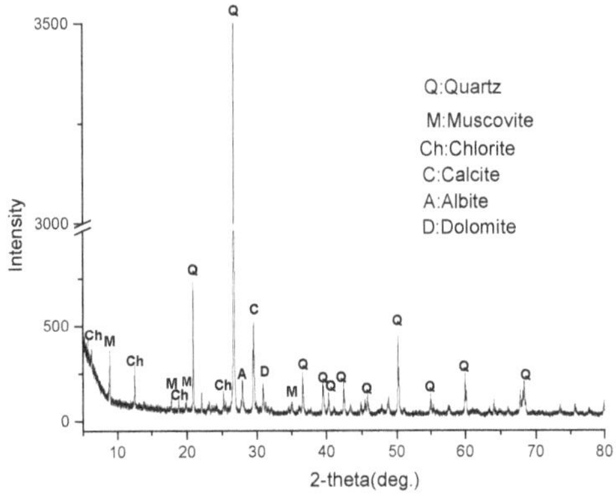

Figure 2. XRD pattern of the sandstone.

2.2. Experimental Methods

2.2.1. Kinetics of Silica Dissolution

The extent of the reaction between the NaOH solution and sandstone was determined using ASTM C289, while some conditions were adjusted: (1) The crushed sandstone particles with the size fraction of 0.315–0.630 mm instead of that of 0.15–0.3 mm, described in ASTM C289, were used in the experiment. (2) The reaction conditions, such as different temperatures (40 °C, 60 °C, and 80 °C) and the different concentration of NaOH solution (1.0 mol/L, 0.5 mol/L, 0.3 mol/L and 0.2 mol/L), were used in the experiment. (3) The duration of the reaction was prolonged to 90 days.

Two batches of test specimens were prepared for the experiment. One is used for investigating the influence of temperatures designated at 40 °C, 60 °C, and 80 °C, respectively, and the concentration of NaOH solution is 1.0 mol/L. Another batch of specimens is used for investigating the influences of NaOH concentration, for which the concentrations are set to 0.5 mol/L, 0.3 mol/L, and 0.2 mol/L, respectively, at a specific temperature of 60 °C. A total of 25.0 g of sandstone particles was placed into a polypropylene copolymer container. Prior to the mixing of sandstone and the NaOH solution, they were separately preheated to the reaction temperature. Following this, 25 mL of NaOH solution was added into the container. The containers were sealed and stored at the preset temperature conditions.

ASR products are present in two forms: (1) free silica ions in alkali solution, and (2) precipitated silica to form sodium silica hydrates (N-S-H, ASR gels). At the end of the exposure periods, the container was cooled immediately, and the liquid was filtered. The filtrates were collected, and the content of dissolved silica and the reduction in alkalinity were detected according to the method described in ASTM C289. The residues were dried at 105 °C for 24 h, then ground into powders by passing them through a 0.045 mm sieve. A total of 5.0 g of the powders were immersed in 500 mL of 1:1 dilute hydrochloric acid (HCl) and stirred for 0.5 h [19]. This acid treatment was intended to dissolve the ASR gels in the solids and dilute the silicate ions for silica content detection. The dissolution solution was filtered, and the content of dissolved silica was measured, according to ASTM C289.

The reaction degree of the silica was calculated at different ages. The reaction degree α is defined as:

$$\alpha = \frac{Si_{dissolved}}{Si^{\infty}} = \frac{Si_{solution} + Si_{gel}}{Si^{\infty}} \tag{4}$$

where $Si_{solution}$ is the amount of free silicate ions in alkali solution at time t, Si_{gel} is the amount of ASR gels, Si^{∞} is the asymptotic value for $t = \infty$, of which the value was obtained through curve-fitting, described by Gao [33]. The Si^{∞} is 2.4 g in the reaction system with 25.0 g of sandstone.

2.2.2. Expansion of Rock Prism

The test method for potential alkali reactivity of carbonate rocks was used for reference [34], but some parameters were altered in this experiment: (1) Sandstone instead of carbonate rock was used in the rock prism test method [35]; (2) Rock prisms of 10 mm × 10 mm × 35 mm and 20 mm × 20 mm × 60 mm instead of cylinder sample Φ9 mm × 35 mm were prepared. Each batch of rock prisms contains three mutually perpendicular specimens. Stainless steel reference studs were fixed on the mid-points at the end faces of the rock prisms. Specimens were immersed in deionized water at a room temperature (22 ± 1 °C), and the length was measured at different intervals until the change in length in a 24 h water immersion period did not exceed 0.02%. The stable datum is used as the initial length of rock prism. Following this, the specimens were immersed in 1 mol/L NaOH solution and cured at 60 °C and 80 °C, respectively. The ratio of the NaOH solution to the solid (by mass) is 3. When the concentration of OH^- ions of the alkali solution was less than 0.7 mol/L, an NaOH tablet was added to improve the concentration of OH^- ions to 1.0 mol/L [35]. Prior to the length measurement, the specimens were removed and cooled down to the room temperature (22 ± 1 °C). The length change was calculated as follows:

$$\Delta l = \frac{l_i - l_0}{l_0} \times 100\% \tag{5}$$

where Δl is the length change at the test age, l_i is the length in mm at the test age, and l_0 is the initial length after equilibrium in water. In the three mutually perpendicular specimens, the max expansion specimen was used for plotting.

3. Results and Discussion

3.1. Kinetics of ASR in Sandstone

3.1.1. Reaction Degree of Silica in Sandstone

As mentioned previously, the total dissolved silica consisted of (1) free silicate ions in solution, and (2) precipitated silica as ASR gels. A specific reaction degree α_1 is defined as follows:

$$\alpha_1 = \frac{Si_{solution}}{Si^{\infty}} \tag{6}$$

α_2 is determined by the difference between total dissolved silica and free silica in solution α_1:

$$\alpha_2 = \alpha - \alpha_1 \tag{7}$$

The reaction degree of sandstone immersed in 25 mL 1 mol/L NaOH solution, cured at 40 °C, 60 °C, and 80 °C, respectively, is given in Figure 3.

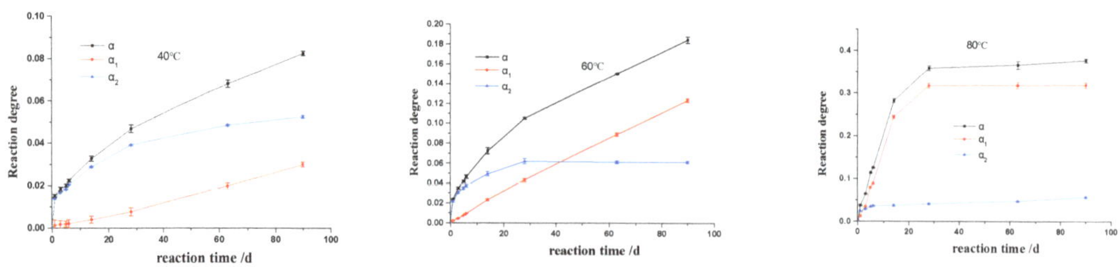

(a) (b) (c)

Figure 3. Reaction degree α as a function of time at different temperatures. (**a**) 40 °C, (**b**) 60 °C, (**c**) 80 °C.

At 40 °C, the reaction degree α was increased continuously in the experimental period. In the early stages, α increased rapidly, which was mainly due to the rapid increase in α_2. After a short plateau at 6 days, α_1 increased steadily. When sandstone was immersed in

1 mol/L NaOH solution for 90 d, the α, α_1, and α_2 were 0.08, 0.03, and 0.05, respectively. At 60 °C, α increased continuously, which is similar to the reaction at 40 °C. The difference is that α_2 was increased to reach an asymptotic value of 0.06. As the temperature increased to 80 °C, the silica dissolved in solution increased rapidly in 28 d, with the highest reaction degree of 0.38, and thereafter, it reached a plateau. The precipitated silica was increased rapidly in the early stages, and then reached a steady value of around 0.04. During the reaction period from 28 d to 90 d, the precipitated silica increased slowly, with the value of α_2 increased from 0.04 to 0.06.

3.1.2. Dissolution Rate Constant

The dissolution of silica is influenced by many factors, such as mineral type of silica, particle size, surface area, pH value, and composition of the solution [36]. In this model, it is assumed that the formation and dissolution of ASR gel is a dynamic equilibrium process, which simplifies the diffusion process of ions in the alkali silica reaction system. The rate of a solid–liquid reaction is usually controlled by chemical dynamic or diffusional dynamic. The chemical control model is usually described as follows [37]:

$$1-(1-\alpha)^{1/3}=k_r t \tag{8}$$

And the diffusional control model is described as [37]:

$$1+2(1-\alpha)-3(1-\alpha)^{2/3}=k_d t \tag{9}$$

where α is the reaction degree, k is the dissolution rate constant, and t is the reaction time. To determine the experimental dissolution rate constants (k), the relationships between the reaction degree (α) and the time (t) were explored according to Equations (8) and (9) for the reaction systems at 40 °C, 60 °C, and 80 °C, respectively. The results are listed in Table 1.

Table 1. Values of dissolution rate constant (k) at 40 °C, 60 °C, and 80 °C, respectively.

Time/Days	40 °C			60 °C			80 °C		
	α	k_r	k_d	α	k_r	k_d	α	k_r	k_d
1	0.015	5.071×10^{-3}	7.842×10^{-5}	0.024	7.932×10^{-3}	1.902×10^{-4}	0.050	1.690×10^{-2}	8.526×10^{-4}
3	0.018	2.054×10^{-3}	3.844×10^{-5}	0.035	3.907×10^{-3}	1.375×10^{-4}	0.085	9.724×10^{-3}	8.375×10^{-4}
5	0.020	1.347×10^{-3}	2.749×10^{-5}	0.042	2.827×10^{-3}	1.196×10^{-4}	0.114	7.887×10^{-3}	9.110×10^{-4}
6	0.022	1.257×10^{-3}	2.869×10^{-5}	0.046	2.620×10^{-3}	1.230×10^{-4}	0.125	7.268×10^{-3}	9.253×10^{-4}
14	0.033	7.882×10^{-4}	2.614×10^{-5}	0.072	1.768×10^{-3}	1.296×10^{-4}	0.282	7.462×10^{-3}	2.178×10^{-3}
28	0.047	5.657×10^{-4}	2.676×10^{-5}	0.105	1.298×10^{-3}	1.384×10^{-4}	0.359	4.918×10^{-3}	1.847×10^{-3}
63	0.068	3.693×10^{-4}	2.549×10^{-5}	0.150	8.384×10^{-4}	1.284×10^{-4}	0.366	2.240×10^{-3}	8.598×10^{-4}
90	0.082	3.141×10^{-4}	2.623×10^{-5}	0.184	7.291×10^{-4}	1.375×10^{-4}	0.377	1.623×10^{-3}	6.421×10^{-4}

As shown in Table 1, the values of $k_d=\frac{1+2(1-\alpha)-3(1-\alpha)^{2/3}}{t}$ at a certain temperature are roughly equal, while the values of $k_r=\frac{1-(1-\alpha)^{1/3}}{t}$ are discrete. The standard deviation values (SD) via ANOVA are listed in Table 2.

Table 2. SD of dissolution rate constant (k) at 40 °C, 60 °C, and 80 °C, respectively.

	SD		
	40 °C	60 °C	80 °C
k_r	0.00157	0.00236	0.00481
k_d	1.8×10^{-5}	2.2×10^{-5}	5.5×10^{-4}

Obviously, the SD of k_r is much greater than that of k_d at each temperature. This means that the rate of silica dissolved in alkali solution is controlled by the diffusional dynamic. The average value of k_d (only the roughly equal values) represents the dissolution rate constants (k) for the reaction system at each temperature. The influence of temperature on the dissolution rate of the silica in sandstone was examined using the Arrhenius equation, and the results are shown in Figure 4. The activation energy (Ea) for the dissolution of silica in sandstone is 80.0 kJ/mol. The value is a little higher than that for flint (78 kJ/mol) [15] and α-cristobalite (73.05 kJ/mol) [19].

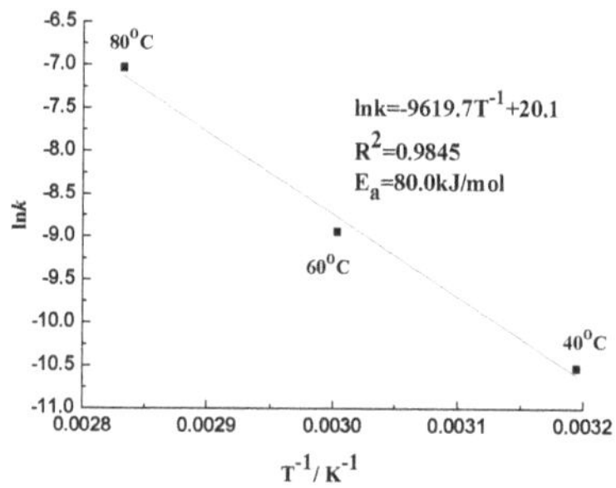

Figure 4. Arrhenius plot for the reacting systems with sandstone immersed in 1 mol/L NaOH.

3.1.3. Reaction Order

To determine the order of the alkali–silica reaction, only the $[OH^-]$ was changed for all of the reaction systems. Assuming the reaction order is n, the kinetics of the reaction could be described as:

$$r \propto k[OH^-]^n \quad (10)$$

where r is the rate of reaction, k is the reaction rate constant depending on the reaction temperature, and n is the reaction order. The following relationship can be deduced from Equation (10) when the temperature is fixed:

$$\lg r \propto n\lg[OH^-] \quad (11)$$

The influence of the concentration of NaOH on silica dissolution at 60 °C is shown in Figure 5. The amount of dissolved silica is the sum of silica in solution and the precipitated silica. The dissolved silica increases rapidly in the first 3 days, and it is mainly precipitated in inner part of sandstone particles. After that, at each concentration of NaOH, the amounts of dissolved silica are approximately proportional to the reaction time. Figure 6 shows the relationship between the magnitude of dissolved silica and the reaction time when different NaOH concentrations were used, and the slope of the fitted line to the different concentration of NaOH represents the reaction rate. The relationship between r and $[OH^-]$ was plotted according to Equation (11), shown in Figure 6. The slope represents the reaction order, n, which is 1.3, indicating that ASR is a first-order reaction. Obviously, the reaction rate is proportional to the $[OH^-]$.

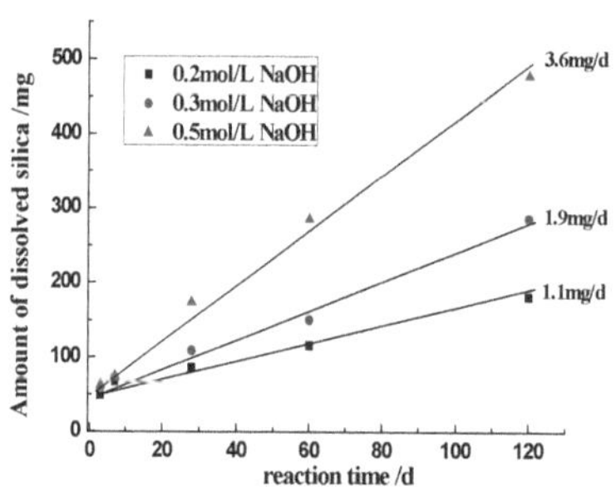

Figure 5. The reaction rate of silica in NaOH solution.

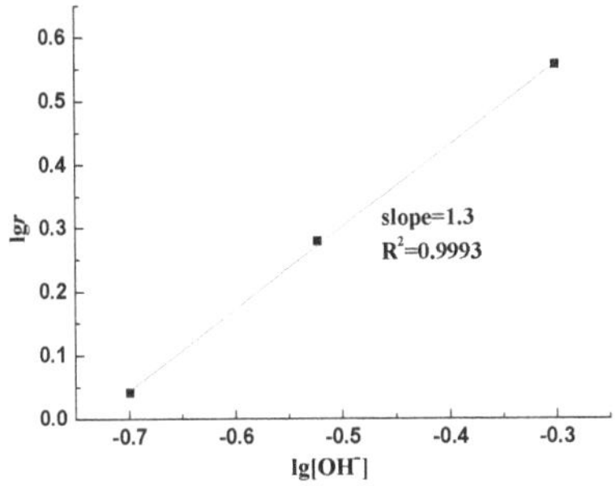

Figure 6. The order of alkali–silica reaction.

3.2. Expansion of Rock Prism

The expansion of rock prisms with the size of 10 mm × 10 mm × 35 mm and 20 mm × 20 mm × 60 mm, cured in 1 mol/L NaOH at 60 °C and 80 °C, respectively, is shown in Figure 7. In the early ages, prisms of 10 mm × 10 mm × 35 mm showed a higher expansion at either 60 °C or 80 °C, while at the later stage, prisms of 20 mm × 20 mm × 60 mm showed a higher expansion. For the prisms of 10 mm × 10 mm × 35 mm, the expansions obtained in 330 days at 60 °C and 80 °C are 0.20% and 0.36%, respectively. For the prisms of 20 mm × 20 mm × 60 mm, the expansions obtained in 228 days at 60 °C and 80 °C are 0.28% and 0.53%, respectively. With increasing temperature, the expansion of the rock prism increased.

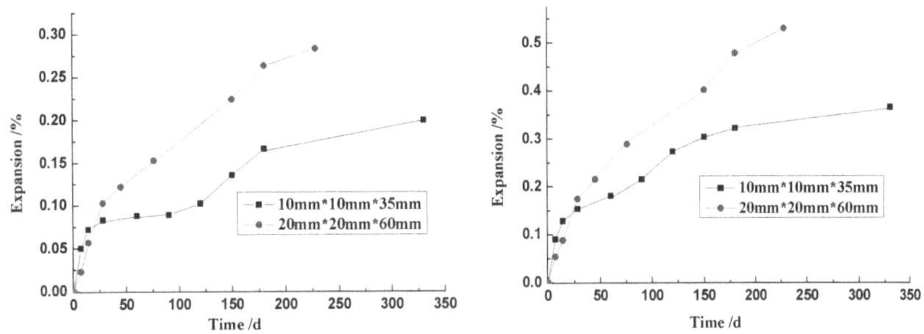

(**a**) 60 °C (**b**) 80 °C

Figure 7. ASR expansions of rock prism cured in 1 mol/L NaOH solution at 60 °C and 80 °C.

3.3. Discussion

The deterioration of concrete structures due to ASR expansion is usually initiated by the chemical reaction accompanying the formation of swelling gels. In sandstone, the microcrystalline quartz particles (reactive SiO_2) are surrounded by other minerals (non-alkali-reactive matrix). The alkali solution diffuses into the sandstone through pores and other channels, and then the chemical reaction takes place at the surface of the microcrystalline quartz to form ASR gels. As the reaction progresses, the amount of ASR gel around the reactive silica sites increases. On the one hand, the ASR gels will diffuse and fill the diffusion channels, leading to a decrease in the diffusion coefficient. On the other hand, cracks generated by ASR expansion will enhance the diffusion coefficient. With the evolution of the reaction, gels increased within the sandstone, and they would create pressure if the gels formed in a confined and limited space, but the gels would exude from the confined space into the alkali solution to release the pressure, accompanied by an increase in free silica ions in the alkali solution. In addition, the reactant NaOH is continuously consumed, and the concentration of OH^- constantly decreases. Therefore, the diffusion of OH^- is not steady. In this model, it is assumed that the formation and dissolution of ASR gel is a dynamic equilibrium process, which simplifies the diffusion process of ions in the alkali silica reaction system.

3.3.1. Effect of the Temperature

The temperature plays an important role in the alkali–silica chemical reaction and ASR gel behavior in the sandstone aggregate. As a chemical reaction, the kinetics of ASR in sandstone follow Arrhenius' law, which means that increasing the temperature may accelerate the process of ASR, forming more products. In many laboratory studies on assessing the performance of concrete due to ASR, it is common to use temperature elevation to accelerate the rate of reaction and to explore the correlation between the expansion of ASR at high and low temperatures for predicting the performance of concrete in service. It is worth noting that the ASR expansion is related to the reacted fraction of aggregates, but it is also affected by the amount of ASR gel retained in the aggregate and confined in a restricted space. As shown in Figure 3, in the early stage, α_2 is greater than α_1, which means that the amount of gel is always higher than that of the free silicate ions. As the reaction proceeds, the amount of silicate ions in the alkali solution increases gradually. Elevated temperature boosts the exudation of ASR gels from the sandstone into the alkali solution. Figure 8 shows the content of ASR gels inside the particle at different temperature, in which the content of gels is defined as α_2/α.

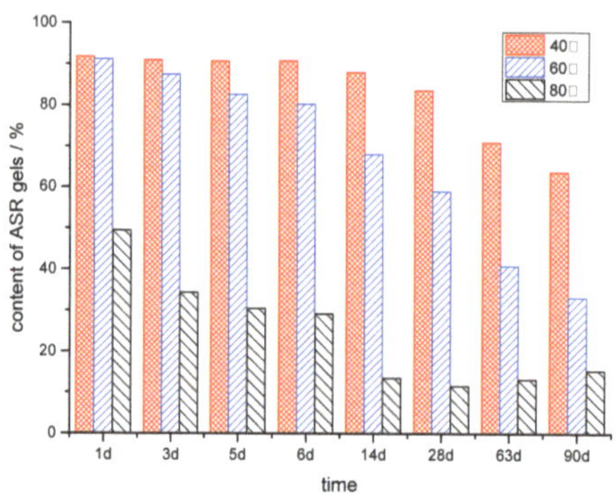

Figure 8. Content of ASR gels inside the particle at different temperatures.

With increasing temperature, the content of ASR gels decreased dramatically. At 40 °C, the reacted silica is mainly in the form of ASR gels, with an amount of 5% at 90 d, and the amount of gel were increased continuously. At 60 °C or 80 °C, the amounts of ASR gel retained in the sandstone came to a plateau, at 6% and 4%, respectively. Although more of the reactive silica reacted at a higher temperature, the products mostly exist in the form of silicate ions in the alkali solution. Research results show that high temperature decreases the viscosity of ASR gel [13,38], which means that the higher temperature enhances the movement of ASR gel, making it more likely to flow out through the pores or cracks. The amount of ASR gels retained inside the aggregate decreased with the increase in temperature, which will reduce the ASR expansion [39].

3.3.2. Effect of Hydroxyl Ions

According to the chemical reaction described in Equations (2) and (3), the hydroxyl ions play a vital role in the dissolution rate of silica, since it acts as a reactant attacking the reactive silica. The results show that the dissolution of reactive silica within sandstone attacked by OH^- ions can be expressed as a first-order reaction, which means that the dissolution rate of reactive silica is only proportional to the concentration of OH^- to the power of one. As ASR proceeds, the hydroxide ions consume continuously, and the $[OH^-]$ reduces gradually. The drop in alkalinity is shown in Figure 9.

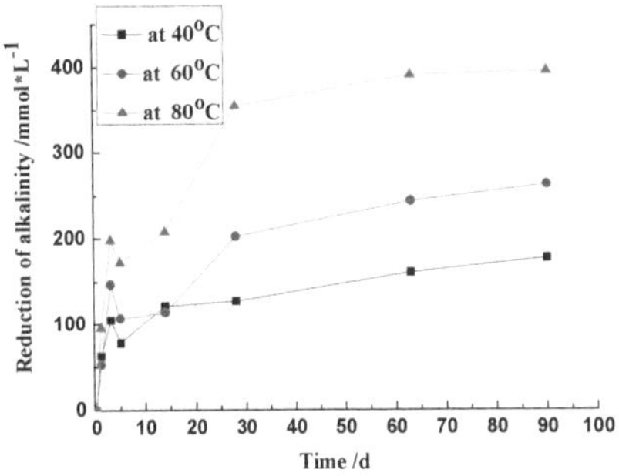

Figure 9. Reduction of alkalinity of 1 mol/L NaOH reacted with sandstone.

During the first 3 days, the concentration of OH^- ions dropped rapidly at every temperature, which might be attributed to the rapid alkali–silica reaction. At the later stages, the decline in the concentration of OH^- ions became slow. It should be noted that the $[OH^-]$ in the solution at 80 °C remained around 0.6 mol/L, even though the reaction degree of silica reached a plateau. It can thus be inferred that the reaction in the system with 25.0 g of sandstone and 25 mL of 1 mol/L NaOH did not come to end, but to an equilibrium of dissolution.

Therefore, the effect of OH^- ions could be described as follows: at the early stage, with a high concentration, OH^- ions rapidly attack the reactive silica, producing abundant gels, leading to a rapid ASR expansion; at the late stage, with a declining concentration, the reaction rate slows down, with a slow growth of ASR gels.

$Ca(OH)_2$ presented in the alkali–silica reaction plays an important role in ASR expansion [40]. Struble et al. [41,42] reported that for a homogenous amorphous silica aggregate, the silica minerals merely dissolved and remained in the solution in the absence of calcium (Ca), while with the presence of Ca, the reaction rim of the aggregates was formed in a confined space and therefore, produced the expansion of concrete. Unlike the homogenous silica aggregate, sandstone itself provides restricted spaces for accommodating the expansive ASR gels. In the absence of Ca in this experiment, the rock prisms expand continuously (Figure 7), but it can be inferred that the presence of $Ca(OH)_2$ accelerates the rate of ASR expansion. High-alkali and low-calcium silica hydrates [C-(Na/K)-S-H) with higher viscosity form on the surface of the aggregate, which can reduce the exudation of gels from the particles [10,13].

3.3.3. Effect of Particle Size

The amount of gel retained inside the aggregate is also influenced by the prism size. Figure 7 shows the ASR expansion of the sandstone rock prisms with different cross-sections. In the experimental system of the rock prism and 1 mol/L NaOH solution, with the increasing size of the rock prism, the expansion of the rock prism increased. Figure 10 shows the micrographs of rock prisms cured at 80 °C. Obviously, ASR damage to the 20 mm × 20 mm × 60 mm prism seems more serious. Ke [43] performed an experiment using the same sandstone as that used in this study. In Ke's experiment [43], the aggregates with sizes of 0.63–1.25 mm and 2.5–5.0 mm were used, respectively, and mortar bars with 20 mm × 20 mm × 80 mm were cast. The effect of the curing temperature on the ASR expansion of mortar bar was explored, and the results are shown in Figure 11. At the same temperature, the expansion of mortar bar with 2.5–5.0 mm aggregates is usually greater than that of the mortar bar with the 0.63–1.25 mm aggregates. Larger aggregates tends to retain a higher amount of gels inside for the longer distance of diffusion, leading to a higher expansion and more serious damage.

by stereomicroscope, 30×

(**a**) Prism measuring 20 mm × 20 mm × 60 mm, 228 days

by stereomicroscope, 30×

(**b**) Prism measuring 10 mm × 10 mm × 35 mm, 330 days

Figure 10. Cracks in the rock prisms cured in 1 mol/L NaOH at 80 °C.

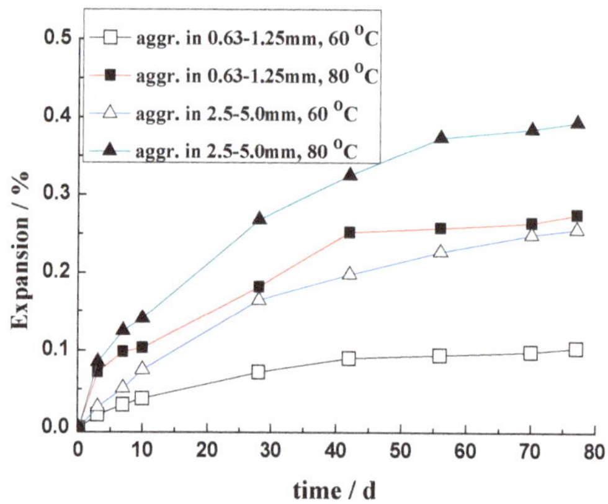

Figure 11. Expansion of mortar bar cured in 1 mol/L NaOH solution [43].

4. Conclusions

The study of the kinetics of the alkali–silica reaction (ASR) in sandstone has provided significant insights into the dissolution behavior of reactive silica under varying conditions, and the following conclusions can be drawn:

(1) The ASR of sandstone is a first-order reaction. The alkali–silica reaction in sandstone conforms to a first-order kinetic model. The reaction rate is directly proportional to the concentration of hydroxide ions ($[OH^-]$) and decreases exponentially over time as the OH^- ions are consumed. This indicates that without the introduction of new hydroxide ions, the silica dissolution rate will progressively diminish. The

reaction kinetics adhere to the principles outlined by Arrhenius' law, demonstrating that temperature is a critical factor in the reaction rate.

(2) There is a temperature influence on ASR dynamics. Elevated temperatures significantly accelerate the ASR process, increasing the rate of silica dissolution and subsequent gel formation. However, higher temperatures also reduce the retention of ASR gels within the aggregate particles, which mitigates the overall expansion caused by the reaction. This dual effect underscores the complexity of predicting ASR behavior solely based on temperature. Larger aggregates tend to retain more ASR gels, leading to greater expansion, which is a crucial consideration for understanding and modeling ASR in real-world scenarios.

(3) The implications for predictive modeling are as follows: The study highlights the necessity of incorporating various parameters, such as temperature and aggregate size, into predictive models for ASR. These factors must be meticulously controlled in experimental setups to ensure the accuracy and applicability of the models. The relationship between ASR expansion at different temperatures and the structural properties of the aggregate provides a basis for developing more reliable predictive tools for concrete deterioration due to ASR.

This study primarily focuses on the behavior of sandstone aggregates under controlled laboratory conditions. While the insights are valuable, real-world conditions often present additional variables, such as fluctuating environmental temperatures and varying chemical compositions of concrete. Future studies should aim to replicate these conditions to further validate the findings. By addressing these aspects, the understanding and management of ASR in concrete can be significantly improved, leading to a more durable and resilient infrastructure.

Author Contributions: Conceptualization, Y.Y. and M.D.; methodology, Y.Y.; software, Y.Y.; validation, Y.Y., L.M., W.L. and M.D.; formal analysis, Y.Y.; investigation, Y.Y.; resources, Y.Y. and W.L.; data curation, Y.Y.; writing—original draft preparation, Y.Y.; writing—review and editing, Y.Y. and W.L.; visualization, Y.Y.; supervision, M.D.; project administration, M.D.; funding acquisition, M.D. All authors have read and agreed to the published version of the manuscript.

Funding: This research was financially supported by the National Key Research and Development Plan of China (2017YFB0309903-01).

Institutional Review Board Statement: Not applicable.

Informed Consent Statement: Not applicable.

Data Availability Statement: Data are contained within the article.

Conflicts of Interest: The authors declare no conflicts of interest.

References

1. Sims, I.; Poole, A. *Alkali-Aggregate Reaction in Concrete: A World Review*; CRC Press: London, UK, 2017.
2. Chatterji, S. Chemistry of alkali-silica reaction and testing of aggregates. *Cem. Concr. Res.* **2005**, *27*, 788–795. [CrossRef]
3. Takahashi, Y.; Ogawa, S.; Tanaka, Y.; Maekawa, K. Scale-dependent ASR expansion of concrete and its prediction coupled with silica gel generation and migration. *J. Adv. Concr. Technol.* **2016**, *14*, 444–463. [CrossRef]
4. Allahyari, H.; Heidarpour, A.; Shayan, A.; Nguyen, V.P. A robust time-dependent model of alkali-silica reaction at different temperatures. *Cem. Concr. Compos.* **2020**, *106*, 103460. [CrossRef]
5. Shi, Z.; Park, S.; Lothenbach, B.; Leemann, A. Formation of shlykovite and ASR-P1 in concrete under accelerated alkali–silica reaction at 60 and 80 °C. *Cem. Concr. Res.* **2020**, *137*, 106213. [CrossRef]
6. Leemann, A.; Saout, G.; Winnefeld, F.; Rentsch, D.; Lothenbach, B. Alkali–silica reaction: The influence of calcium on silica dissolution and the formation of reaction products. *J. Am. Ceram. Soc.* **2011**, *94*, 1243–1249. [CrossRef]
7. Strack, C.; Barnes, E.; Ramsey, M.; Williams, R.K.; Klaus, K.L.; Moser, R.D. Impact of aggregate mineralogy and exposure solution on alkali-silica reaction product composition and structure within accelerated test conditions. *Constr. Build. Mater.* **2020**, *240*, 117929. [CrossRef]
8. Broekmans, M. The Alkali–Silica Reaction: Mineralogical and Geochemical Aspects of Some Dutch Concretes and Norwegian Mylonites. Ph.D. Thesis, Utrecht University, Utrecht, The Netherlands, 2002.

9. Bleszynski, R.; Thomas, M. Microstructural studies of alkali-silica reaction in fly ash concrete immersed in alkaline solutions. *Adv. Cem. Base Mater.* **1998**, *7*, 66–78. [CrossRef]
10. Giorla, A.; Scrivener, K.; Dunant, C. Influence of visco-elasticity on the stress development induced by alkali-silica reaction. *Cem. Concr. Res.* **2015**, *70*, 1–8. [CrossRef]
11. Giorla, A.; Dunant, C. Microstructural effects in the simulation of creep of concrete. *Cem. Concr. Res.* **2018**, *105*, 44–53. [CrossRef]
12. Miura, T.; Multon, M.; Kawabata, Y. Influence of the distribution of expansive sites in aggregates on microscopic damage caused by alkali-silica reaction: Insights into the mechanical origin of expansion. *Cem. Concr. Res.* **2021**, *142*, 106355. [CrossRef]
13. Vayghan, A.; Rajabipour, F.; Rosenberger, J. Composition–rheology relationships in alkali–silica reaction gels and the impact on the Gel's deleterious behavior. *Cem. Concr. Res.* **2016**, *83*, 45–56. [CrossRef]
14. Glasser, L.; Kataoka, N. The chemistry of alkali-aggregate reaction. *Cem. Concr. Res.* **1981**, *11*, 1–9. [CrossRef]
15. Bulteel, D.; Garcia-Diaz, E.; Vernet, C.; Zanni, H. Alkali–silica reaction: A method to quantify the reaction degree. *Cem. Concr. Res.* **2002**, *32*, 1199–1206. [CrossRef]
16. Dron, R.; Brivot, F.; Chaussadent, T. The mechanism of the alkali-silica reaction. *Bull. Lab. Ponts Chaussées* **1998**, *214*, 61–68.
17. Balachandran, C.; Munoz, J.; Peethamparan, S.; Arnold, T.S. Alkali -silica reaction and its dynamic relationship with cement pore solution in highly reactive systems. *Constr. Build. Mater.* **2023**, *362*, 129702. [CrossRef]
18. Broekmans, M.; Jansen, J. Silica dissolution in impure sandstone: Application to concrete. *J. Geochem. Explor.* **1998**, *62*, 311–318. [CrossRef]
19. Kim, T.; Olek, J. Chemical sequence and kinetics of alkali-silica reaction Part I. experiments. *J. Am. Ceram. Soc.* **2014**, *97*, 2195–2203. [CrossRef]
20. Dron, R.; Brivot, F. Thermodynamic and kinetic approach to the alkali-silica reaction. Part 2: Experiment. *Cem. Concr. Res.* **1993**, *23*, 93–103. [CrossRef]
21. *ASTM C289*; Standard Test Method for Potential Alkali-Silica Reactivity of Aggregates (Chemical Method). ASTM: West Conshohocken, PA, USA, 2003.
22. Verstraete, J.; Khouchaf, L.; Bulteel, D.; Garcia-Diaz, E.; Flank, A.M.; Tuilier, M.H. Amorphisation mechanism of a flint aggregate during the alkali–silica reaction: X-ray diffraction and X-ray absorption XANES contributions. *Cem. Concr. Res.* **2004**, *34*, 581–586. [CrossRef]
23. Maraghechi, H. *Development and Assessment of Alkali-Activated Recycled Glass Based Concretes for Civil Infrastructure*; The Pennsylvania State University: State College, PA, USA, 2014.
24. Kim, T.; Olek, J.; Jeong, H. Alkali-silica reaction: Kinetics of chemistry of pore solution and calcium hydroxide content in cementitious system. *Cem. Concr. Res.* **2015**, *71*, 36–45. [CrossRef]
25. *ASTM C1260*; Standard Test Method for Potential Alkali Reactivity of Aggregates (Mortar-Bar Method). ASTM: West Conshohocken, PA, USA, 1994.
26. *ASTM C1293*; Standard Test Method for Determination of Length Change of Concrete due to Alkali Silica Reaction. ASTM: West Conshohocken, PA, USA, 2009.
27. Ichikawa, T.; Miura, M. Modified model of alkali-silica reaction. *Cem. Concr. Res.* **2007**, *37*, 1291–1297. [CrossRef]
28. Kawabata, Y.; Dunant, C.; Yamada, K.; Scrivener, K. Impact of temperature on expansive behavior of concrete with a highly reactive andesite due to the alkali–silica reaction. *Cem. Concr. Res.* **2019**, *125*, 105888. [CrossRef]
29. Ponce, J.; Batic, O. Different manifestations of the alkali-silica reaction in concrete according to the reaction kinetics of the reactive aggregate. *Cem. Concr. Res.* **2006**, *36*, 1148–1156. [CrossRef]
30. Ben, H.; Gallucci, E.; Guidoum, A.; Scrivener, K.L. Relation of expansion due to alkali silica reaction to the degree of reaction measured by SEM image analysis. *Cem. Concr. Res.* **2007**, *37*, 1206–1214.
31. Dunant, C.; Scrivener, K. Micro-mechanical modeling of alkali-silica-reaction-induced degradation using the AMIE framework. *Cem. Concr. Res.* **2010**, *40*, 517–525. [CrossRef]
32. Joo, H.; Takahashi, Y. Analytical and experimental studies on alkali–silica reaction mechanism: Aggregate cracking and chemical composition change of gel. *Cem. Concr. Compos.* **2023**, *139*, 105003. [CrossRef]
33. Gao, X. Contribution to the Requalification of Alkali Silica Reaction (ASR) Damaged Structures: Assessment of the ASR Advancement in Aggregates. Ph.D. Thesis, Toulouse University, Toulouse, France, 2010.
34. *DL/T 5151*; Test Code for Aggregates of Hydraulic Concrete. Standardization Administration of China: Beijing, China, 2014. (In Chinese).
35. Yang, W.; Deng, M.; Huang, B.; Chen, W.; Wang, Z. Effects of NaOH solution concentration on expansion of rock-prisms of siliceous and dolomitic aggregates. *J. Nanjing Tech. Univ.* **2018**, *3*, 18–23. (In Chinese)
36. Iler, R. *The Chemistry of Silica: Solubility, Polymerization, Colloid and Surface Properties and Biochemistry*; Wiley-Interscience: Hoboken, NJ, USA, 1979.
37. Anthony, R. *Basic Solid State Chemistry*; John Wiley & Sons: New York, NY, USA, 2000.
38. Vail, J. *Soluble Silicates—Their Properties and Uses*; Chemistry; Reinhold: New York, NY, USA, 1952.
39. Mbetmi, G.; Multon, S.; Larrard, T.; Duprat, F.; Tieudjo, D. Sensitivity of an alkali-silica reaction kinetics model to diffusion and reactive mechanisms parameters. *Constr. Buld. Mater.* **2021**, *299*, 123913. [CrossRef]
40. Chatterji, S. The role of $Ca(OH)_2$ in the breakdown of Portland cement concrete due to alkali–silica reaction. *Cem. Concr. Res.* **1979**, *9*, 185–188. [CrossRef]

41. Struble, L.; Diamond, S. Swelling properties of synthetic alkali silica gels. *J. Am. Ceram. Soc.* **1981**, *64*, 652–655. [CrossRef]
42. Struble, L.; Diamond, S. Unstable swelling behaviour of alkali silica gels. *Cem. Concr. Res.* **1981**, *11*, 611–617. [CrossRef]
43. Ke, J. *Kinetics of Expansion due to Alkali-Aggregate Reaction and Expansion Prediction*; Nanjing Tech University: Nanjing, China, 2008. (In Chinese).

materials

Article

Influence of Different Types of Fillers on the Performance of PMMA-Based Low-Temperature Rapid Repair Mortar

Zhipeng Zhu, Lingling Xu *, Min Deng *, Shijian Lu, Zemeng Guo, Luchao Yan and Yang Wang

College of Materials Science and Engineering, Nanjing Tech University, Nanjing 211816, China; 202161103006@njtech.edu.cn (Z.Z.); 202161103065@njtech.edu.cn (S.L.); 202061103028@njtech.edu.cn (Z.G.); 202161203204@njtech.edu.cn (L.Y.); 202261103012@njtech.edu.cn (Y.W.)

* Correspondence: xll@njtech.edu.cn (L.X.); dengmin@njtech.edu.cn (M.D.); Tel.: +86-139-5184-6173 (L.X.); +86-136-0518-4865 (M.D.)

Abstract: In order to further optimize the performance of PMMA (Polymethyl Methacrylate) repair mortar. In this paper, fly ash, talcum powder and wollastonite powder are used as fillers to modify the PMMA repair mortar. The effects of these three fillers on the working performance, mechanical performance and durability of PMMA repair mortar were explored. The study shows that the three fillers have good effect on the bond strength of the repair mortar, in which the fly ash has the best effect on the mechanical performance. The mechanical properties of PMMA repair mortar were best when the amount of fly ash was 60 phr (parts per hundred, representing the amount of the material added per hundred parts of PMMA). At this time, the 28 d compressive strength was 71.26 MPa and the 28 d flexural strength was 28.09 MPa, which increased by 13.31% and 15.33%, respectively. Wollastonite powder had the least negative effect on the setting time of the PMMA repair mortar. When the dosage of wollastonite powder was increased to 100 phr, the setting time was only extended from 65 min to 94 min. When the talc dosage was 60 phr, the best improvement in salt freezing resistance was achieved. After 100 cycles of salt freezing, the mass loss rate and strength loss rate decreased to 0.159% and 4.97%, respectively, which were 75.1% and 37.7% higher than that of the control group. The addition of all three fillers reduced the porosity and the proportion of harmful pores in the mortar. This study contributes to a comprehensive understanding how different types of fillers affect PMMA repair mortars, and it also provides theoretical support for the further development of low-temperature rapid repair mortars.

Keywords: PMMA repair mortar; filler; working performance; mechanical performance; durability performance

Citation: Zhu, Z.; Xu, L.; Deng, M.; Lu, S.; Guo, Z.; Yan, L.; Wang, Y. Influence of Different Types of Fillers on the Performance of PMMA-Based Low-Temperature Rapid Repair Mortar. *Materials* **2024**, *17*, 2871. https://doi.org/10.3390/ma17122871

Academic Editor: Weiting Xu

Received: 14 May 2024
Revised: 2 June 2024
Accepted: 7 June 2024
Published: 12 June 2024

1. Introduction

With the increase in vehicle driving speeds, heavy transportation and the impact of extreme weather, the number of concrete pavement damages and the degree of damage are increasing year by year, which leads to an increasingly severe situation of road maintenance [1]. The common damage of cement concrete pavement can be divided into structural damage and non-structural damage. The latter is often referred to as thin-layer disease, and its main forms of damage are categorized into cracks (cracks on the road surface), exposed bones (exposure of aggregates inside the concrete), pitted surfaces (the loss of bonding material on pavement surface) and surface potholes (deteriorated road surfaces with potholes) [2–4]. The method used for repairing thin-layer damage is to pave a new thin layer of repair material in the damaged area, so that the old and new pavements are bonded to each other to form a unified whole. Ultimately, the effects of repeated loads and changes in the external environment on the pavement are shared [5].

The most direct method of repair is to place a layer of cement mortar or concrete with the same material as the old concrete pavement over the damaged area. However, this method has a long maintenance time. Ordinary silicate cement mortar is prone to brittle fracture and high shrinkage, which is not conducive to the compatibility between the repair mortar and the concrete pavement. In order to solve this problem, some scholars added various types of fibers or used early-strength or fast-hardening special cement for modification and optimization [6–13], which greatly improved the toughness of the repair mortar and shortened the maintenance time. Another part of scholars choosed to use epoxy resin, polyurethane, methyl methacrylate and other polymers as repair materials to replace cement mortar [14–17]. Compared with inorganic repair materials, polymer repair materials have higher bond strength and durability, and can be cured quickly, but its cost is also several times higher than inorganic repair materials [15]. Therefore, some scholars proposed the method of composite organic repair materials with inorganic repair materials. The composite repair material they developed fully combines the advantages of organic and inorganic materials. This composite repair material not only significantly improves the mechanical performance and durability of the repair material, but also reduces the cost significantly compared to pure organic repair materials. It is currently a popular material used to repair surface defects in concrete structures [18–20].

Composite repair materials are subdivided into three categories: polymer impregnated mortar/concrete (PIM/PIC), polymer-modified mortar/concrete (PMM/PMC) and polymer mortar/concrete (PM/PC). PM/PC consists only of aggregates and polymers, and it uses polymers as the binder, which drastically reduces the curing time compared to ordinary cement concrete. PM/PC not only has excellent strength and durability, but also has good bonding ability with the substrate, so it is widely used in pavement repair projects. The commonly used polymers in PM/PC are epoxy resin, polyurethane and unsaturated polyester, etc., and the repair materials made from them have good mechanical performance and durability [21–24]. Methyl methacrylate (MMA) has the advantages of good fluidity and strong adhesion. Moreover, it has a wider temperature range [25–27] that can be used in temperatures ranging from -30 °C to 45 °C. Using MMA as the binder material for the repair mortar is expected to provide a composite repair mortar that cures quickly at low temperatures and has high strength for thin-layer lesions on pavements in winter or cold regions.

However, it was found during the study that PM/PC undergoes an exothermic reaction of monomer polymerization during the setting process. The high temperatures resulting from the exothermic reaction cause a large difference in the coefficient of thermal expansion between PM/PC and the concrete substrate. The bonding interface between PM/PC and the substrate can also be subject to large stresses at the interface due to temperature changes, resulting in poor thermal compatibility between the repair material and the substrate [28,29]. In order to solve this problem, researchers have found that the addition of fillers can reduce the coefficient of thermal expansion and modulus of elasticity of polymer mortar through extensive experiments [30]. The smaller particle size of the filler can uniformly fill the internal pores of the repair mortar, thus preventing the expansion of micro-cracks and increasing cohesion. Besides, the addition of filler can also reduce the cost of the polymer mortar.

Rebeiz and Gorninski et al. [31,32] investigated the effect of fly ash filler incorporation on unsaturated polyester mortar. The results showed that the incorporation of fly ash not only significantly improved the mechanical properties and bond strength of the polymer mortar, but also reduced the coefficient of thermal expansion. Pham [33] studied the effect of cement on the mechanical strength of polymer concrete when used as a filler. The findings showed that the filler significantly increased the flexural and compressive strength of polymer concrete. They used a tensile testing machine to test the bond strength of a 300 mm thick PC layer to a substrate. It was found that at low polymer dosages, fillers enhanced the bonding strength. However, at higher polymer levels, the bonded specimens fractured on the substrate with or without fillers. This indicates that the bonding is effective. Guanjie Li et al. [34] analyzed the changes in the performance of epoxy repair mortars filled with different forms of fillers. It was found that any form of filler could improve the densification of the epoxy mortar, thus increasing its compressive and tensile bond strengths and reducing its shrinkage. Among them, the spherical particles of coal gasifier slag can transfer the internal stress of the mortar more uniformly, so its improvement effect on the densification of epoxy mortar is the most significant. Chandrasekaran [35] studied the effect of heavy calcium carbonate and fly ash on the performance of unsaturated polyester mortar. It was finding that spherical fly ash enhanced the PM strength more significantly with almost zero water absorption as compared to heavy calcium carbonate. Yemam [36] investigated the effect of sand washing waste on epoxy resin mortar. The results showed that a certain amount of sand washing waste could significantly improve the mechanical properties and bond strength of epoxy mortar. They concluded that sand washing waste can be used as a potential filler for epoxy resin mortar, but specific issues such as dimensional compatibility need to be further explored.

In this study, PMMA repair mortar for quick repair at low temperature of −10 °C is used as the base proportion. Fly ash in micro-spherical form, talcum powder in flake form and wollastonite powder in pin-rod form were used as fillers. By changing the amount of fillers, the effects of each form of fillers on the working performance, mechanical properties and durability of PMMA repair mortar were investigated. The structural changes of PMMA repair mortar after the addition of fillers were also analyzed by microscopic morphology. It is expected to obtain the best choice of filler type and dosage to optimize the performance of PMMA repair mortar and reduce the cost, so as to lay the foundation for its wider application.

2. Materials and Experiments

2.1. Raw Materials

MMA prepolymer was prepared by using methyl methacrylate as polymerization monomer, benzoyl peroxide (BOD) as initiator, dioctyl phthalate (DOP) as the plasticizer, N, N-dimethylaniline (DMA) as curing agent, and polyvinyl acetate (PVAc) + styrene as low shrinkage additives. Bagged ISOA standard sand from Xiamen Aceo Standard Sand Co., Ltd. was used as fine aggregate. Spherical fly ash, lamellar talc and needle-and-rod wollastonite powder were used as fillers, and the particle size distribution map, micro-morphology map and particle size distribution data are shown in Figure 1 and Table 1. The average particle size of fly ash, talcum powder and wollastonite powder are 11.55 μm, 8.82 μm and 18.58 μm. The γ-methacryloxypropyltrimethoxysilane (KH570) from Shanghai Yuanye Biotechnology Co. Ltd. (Shanghai, China) was used as the silane coupling agent.

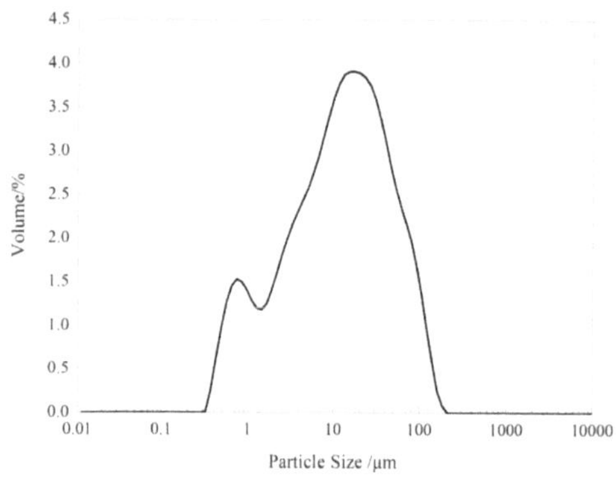

(a)

Figure 1. *Cont.*

(**b**)

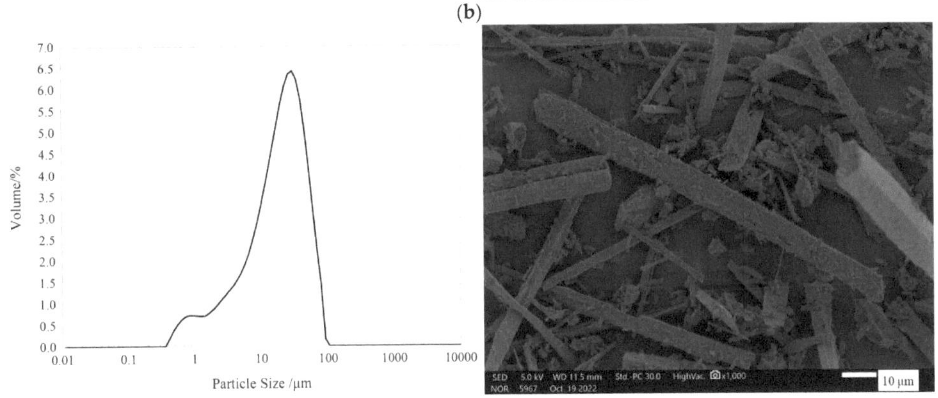

(**c**)

Figure 1. Particle size distribution and microscopic morphology of fillers. (**a**) fly ash (**b**) talc (**c**) wollastonite.

Table 1. Particle size distribution of fillers.

Materials	D10/μm	D50/μm	D90/μm	D (mean)/μm
fly ash	1.06	11.55	47.21	21.44
talc	1.75	8.82	31.82	13.92
wollastonite	2.62	18.58	47.39	22.26

2.2. Specimen Preparation

2.2.1. Preparation of MMA Prepolymers

Through the preliminary study [37], 12% of (PVAc+ Styrene) was selected as the low shrinkage additive. The dosage is the percentage of the mass of the solution formed by PVAc dissolved in styrene to the total mass of the whole MMA prepolymer. The distribution of each component of the prepared MMA prepolymer is shown in Table 2. Finally, the MMA prepolymer is a light yellow transparent and homogeneous fluid, free of impurities, without precipitation and delamination.

Table 2. Raw material ratio of MMA prepolymer/wt%.

MMA	BPO	DOP	PVAc	Styrene
67.38	0.41	20.21	8.40	3.60

2.2.2. Preparation of PMMA Repair Mortar

The basic ratio of PMMA repair mortar is PMMA:DMA:standard sand:silane coupling agent = 100:4:550:0.75 (mass ratio),its performance is shown in Table 3. The difference is the addition of different types of fillers in different dosages. Different dosages of fly ash, talc and wollastonite powder were used as fillers to modify the PMMA repair mortar. The amount of filler is defined on the basis of 100 parts of PMMA, i.e., when the amount of filler is 40 parts, the mass ratio of filler:PMMA is 40:100. Phr is parts per hundred, which stands for the amount of the substance added per hundred parts of PMMA. F1, F2, F3, F4, and F5 represent fly ash in dosages of 20, 40, 60, 80, and 100 parts, and similarly T1–5 and W1–5 represent talc and wollastonite powders in dosages of 20–100 parts.

Table 3. Results of the performance tests of the benchmark PMMA repair mortar.

Fluidity (mm)	**Setting Time (min)**	**Flxural Strength (MPa)**			**Compressive Strength (MPa)**			**28 d Bonding Condition**	
		1 d	**7 d**	**28 d**	**1 d**	**7 d**	**28 d**	**Bonding Strength (MPa)**	**Fracture Surface**
220.00	65.00	21.65	23.59	24.79	50.88	60.13	61.79	>6.25	Reference mortar specimen

In order to simulate the real construction process. Prior to the preparation of the PMMA repair mortar, the MMA prepolymer, curing agent and silane coupling agent KH570 prepared in Section 2.2.1 were left at −10 °C for 24 h and taken out before preparing the mortar. When preparing the repair mortar, the MMA prepolymer is first added to the corresponding proportion of hardener and silane coupling agent and stirred well. Then, the sand, filler and mixed solution are poured into the mixing pot and stirred into shape. After molding, put the specimen together with the mold into the curing box at −10 °C, take off the mold after 24 h and put it into the curing box at −10 °C again to curing to the corresponding age.

2.3. *Testing Method*

2.3.1. Mechanical Strength of Repair Mortars

After the resulting PMMA repair mortar was maintained to the corresponding test age, the flexural strength and compressive strength were tested with reference to GB/T 17671-2021 [38]. The compressive strength and flexural strength of the specimens were determined with WDW-20E microcomputer-controlled electronic universal testing machine under the loading speeds of 2.4 kN/s and 0.05 kN/s. The test results were as follows.

2.3.2. Fluidity

Referring to GB/T 2419-2005 [39], NLD-3 cementitious sand fluidity tester (Wuxi Jianyi Building Material Instrument Factory, Wuxi, China) was used to conduct the test. Firstly, the PMMA repair mortar was divided into two layers and quickly put into the test mold and pounded. After pounding, the test mold was gently lifted upward and immediately rotated on the beating table, completing 25 beatings in 25 ± 1 s at a frequency of once per second. After jumping, use a steel ruler to measure the diameter of the bottom of the PMMA repair mortar perpendicular to each other in both directions, and the average of the two diameters obtained is the flow of the PMMA repair mortar.

2.3.3. Setting Time

(1) The setting time of the polymer PMMA was determined by filling the test tube with the polymer solution. Start timing from the MMA prepolymer add curing agent mixing, pour the solution into the test tube, tilt the test tube from time to time. When the liquid no longer flows while the test tube is placed flat, which means that the

liquid in the test tube loses its fluidity, the timing is stopped. And this time is the setting time of PMMA.

(2) Determination of setting time of PMMA repair mortar with reference to JGJ/T 70-2009 [40]. Fill and smooth the fresh repair mortar mix into a slurry container with an inner diameter of 140 mm and a height of 75 mm, and place it in a maintenance box at −10 °C for maintenance. It was taken out every 10 min and tested using ZKS-100 mortar setting time tester (Tianjin Beichen Testing Instrument Factory, Tianjin, China). The penetration resistance method was used to determine the setting time of the PMMA repair mortar by the resistance generated by the mortar to the penetrating test needle, which was measured every 2 min as it approached the setting time.

2.3.4. Bonding Strength

The baseline mortar specimens were broken in the center after being cured for 7 d under standard curing conditions. Put half of the broken specimen into a mold with the size of 40 mm × 40 mm × 160 mm, and pour PMMA repair mortar into the other side of the mold. Finally, the resulting specimens were cured at −10 °C until 28 d. After the maintenance was completed, the flexural bond strength of the bonded specimen was tested according to the test method of flexural strength introduced in Section 2.3.1 as shown in Figure 2. The two fracture surfaces of the specimen after fracture are shown in Figure 3. If the fracture surface occurs at the bond interface between the repair mortar and the reference mortar, the resulting flexural strength is the bond strength of the repair mortar. But if the bond specimen fractures on the reference mortar specimen, the bond strength of the repair mortar is greater than the value of the resulting flexural strength.

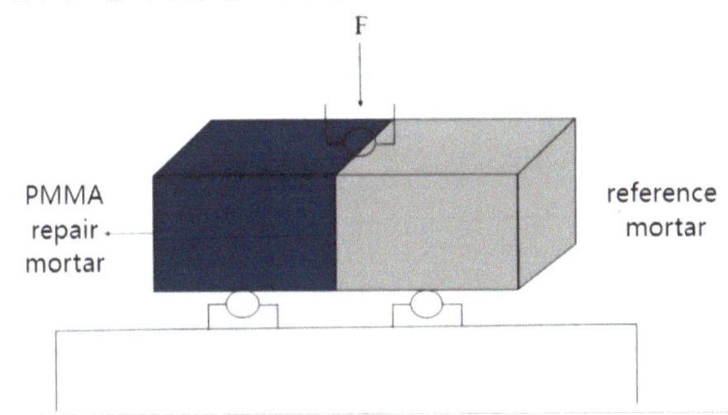

Figure 2. Schematic diagram of flexural bonding experiment.

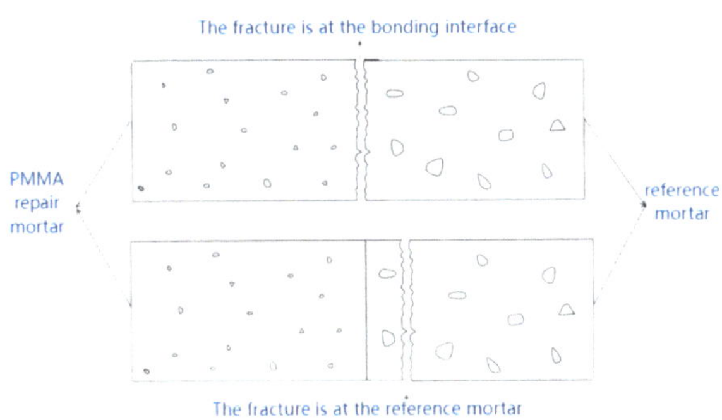

Figure 3. Schematic diagram of the two types of fracture interfaces.

2.3.5. Abrasion Resistance

The abrasion resistance test was carried out with reference to GB/T 3810.6-2016 [41], and the abrasion volume and abrasion mass were used to characterize the abrasion resistance of the material. The abrasion mass was determined by the mass difference of the specimen before and after the abrasion resistance test, and the abrasion pit volume was calculated from the measured chord length of the abrasion pit through Equations (1) and (2):

$$V = \left(\frac{\pi \cdot \alpha}{180} - sin\alpha\right) \times \left(\frac{h \times d^2}{8}\right) \quad (1)$$

$$\sin\frac{\alpha}{2} = \frac{L}{d} \quad (2)$$

where the V is the volume of the grinding pit (mm^3), the L is the chord length of the grinding pit (mm), the α is the angle formed between the chord length of the grinding pit and the center of the friction wheel (°), the h is the thickness of the friction wheel, which is 10 mm in the present test, and d is the diameter of the friction wheel, which is 200 mm in the test.

2.3.6. Water Absorption Capability

Referring to the test method in DL/T5126-2021 [42] for determination. The specimen was cured in the curing box at −10 °C until 28 d, and then put into the oven at 80 ± 2 °C for 48 h and taken out, then put in the desiccator to cool to room temperature and quickly weigh the weight of the specimen at this time, G_0, and then put the specimen into the water at 20 ± 2 °C for 48 h. During the immersion process, the upper surface of the specimen should be more than 50 mm away from the water surface. After the end of the immersion with a wet cloth to wipe off the surface water and quickly weigh the weight of the test piece G_1. Finally, the water absorption rate is calculated according to the Equation (3):

$$W = \frac{G_1 - G_0}{G_0} \times 100\% \quad (3)$$

where W is the water absorption rate (%) of the specimen; G_0 is the mass of the specimen after drying (g); G_1 is the mass of the specimen after water absorption (g).

2.3.7. Salt Scaling Resistance

The PMMA repair mortar specimens were cured in a curing box at −10 °C for up to 24 d. Subsequently, the specimens were immersed in a salt solution configured with a mass ratio of 97% distilled water and 3% NaCl at 20 °C for 4 d. Remove the specimens immediately after the end of immersion, dry the surface with a damp cloth and weigh it. Each salt freezing cycle should be frozen in a freezer at −20 °C for more than 4 h and then immediately thawed in salt solution at 20 °C for more than 4 h. The specimens were weighed and tested for compressive strength at the end of every 25 cycles. The rate of loss of compressive strength and the rate of mass loss of the PMMA repair mortar were used to express its resistance to salt freezing. The rate of loss of compressive strength is calculated in Equation (4):

$$\Delta f_c = \frac{f_{c0} - f_{cn}}{f_{c0}} \times 100 \quad (4)$$

where Δf_c is the compressive strength loss rate (%) of the repair mortar after N salt freezing cycles, f_{c0} is the compressive strength measurement value (MPa) of the repair mortar in the blank group without salt freezing cycles, and f_{cn} is the compressive strength measurement value (MPa) of the repair mortar after N salt freezing cycles.

The rate of mass loss is calculated according to Equation (5):

$$\Delta W_{ni} = \frac{W_{0i} - W_{ni}}{W_{0i}} \times 100 \quad (5)$$

where ΔW_{ni} is the mass loss rate (%) of the i-th mortar specimen after N salt freezing cycle, W_{0i} is the mass (g) of the i-th mortar mortar before salt freezing cycle, and W_{ni} is the mass (g) of the i-th mortar specimen after N salt freezing cycle.

2.3.8. Volumetric Deformation

Referring to JC/T 603-2004 [43], PMMA repair mortar was filled into 25 mm × 25 mm × 280 mm molds that had been pre-assembled with measuring nail heads. After 24 h of curing in the curing box at −10 °C and demolding, the initial length L_0 of the mortar should be determined by a specific length meter within 30 min after demolding, and then continue to be placed at −10 °C for curing to different ages to measure its length. The formula for shrinkage at different ages is:

$$S_n = \frac{L_0 - L_n}{L_e} \times 100 \quad (6)$$

where S_n is the shrinkage (%) of the specimen at different age n, L_0 is the initial length of the specimen (mm), L_n is the length of the specimen at different age n (mm), and L_e is the effective length of the specimen with the nail head portion removed, which is 250 mm in the present experiment.

2.3.9. Pore Size Distribution

The internal pore size distribution of PMMA repair mortar was tested by using mercuric pressure method. The operation steps are shown below:

(1) Use a firm hammer to knock the PMMA repair mortar specimen into 3~5 mm size granular sample;
(2) Put it into a constant temperature blower drying oven to dry the moisture;
(3) Put the sample into the Quantachrome PoreMaster 60 mercury pore analyzer for testing.

2.3.10. Scanning Electron Microscopy

A scanning electron microscope (SEM) model JSM-5900 was used to observe and analyze the internal structure of the PMMA repair mortar and the bonding interface between the PMMA repair mortar and the reference mortar in terms of microscopic morphology. The test procedure is shown below:

(1) Crack the PMMA repair mortar and select the particles of which the diameter is about 10 mm;
(2) Place the selected mortar particles in a constant temperature blast drying oven to dry the moisture;
(3) Use conductive adhesive to fix the sample to be tested in the sample stage, to be coated and processed into the scanning electron microscope observation.

3. Results and Discussions

3.1. Working Performance

Figure 4 shows the effect of filler on the working performance of PMMA repair mortar. It can be seen that the increase in the amount of filler significantly affects the working performance of the mortar system. Unlike talc and wollastonite powder, the flowability of the mortar shows a tendency to increase and then decrease as the amount of fly ash increases. his is due to the spherical fly ash has a rolling ball effect, which enhances the wetting ability of PMMA, so with the increase of fly ash dosage, the ease of PMMA repair mortar can be improved. However, when the dosage of fly ash is too much, the fly ash particles will appear localized agglomeration phenomenon, so the fluidity is reduced again. Talc and wollastonite powder, due to the special morphology of lamellar and needle and rod on their surfaces, respectively, make them adsorb more PMMA while filling the internal voids of the PMMA repair mortar, so the proportion of PMMA used in the bonding aggregate decreases, thus reducing the fluidity of the repair mortar.

In addition, all three fillers dosages extended the setting time of the PMMA repair mortar at −10 °C environment. When adding more fillers, the lower the proportion of PMMA components in the mortar system that play a cementing role, the smaller the exothermic effect during the reaction, the lower the speed of the final formation of a continuous polymer network, which leads to prolonging the setting time of the mortar. The setting time of PMMA repair mortar with fly ash, talc and wollastonite powder increased from 65 min to 586 min, 112 min and 94 min, respectively, when the dosage of the three fillers was increased to 100 phr. In addition, at the same dosage, the effect of fly ash in retarding the setting was much stronger than that of the other two fillers. This is also related to the unique spherical particles of fly ash, which have a smooth surface and play a certain dispersing effect on PMMA, making the intermolecular distance of PMMA larger, thus substantially delaying the setting time of PMMA repair mortar. Wollastonite powder has the least effect on the setting time of PMMA repair mortar, which is more suitable to be used in the project of rapid repair of pavement.

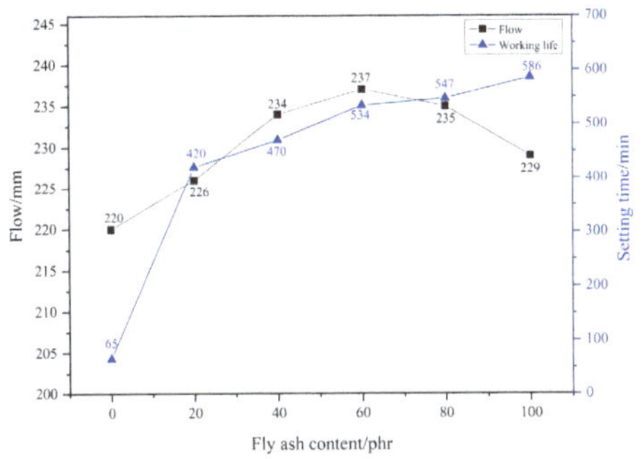

(a)

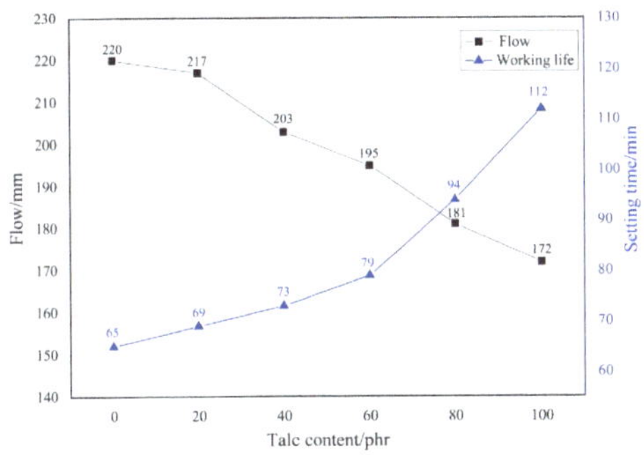

(b)

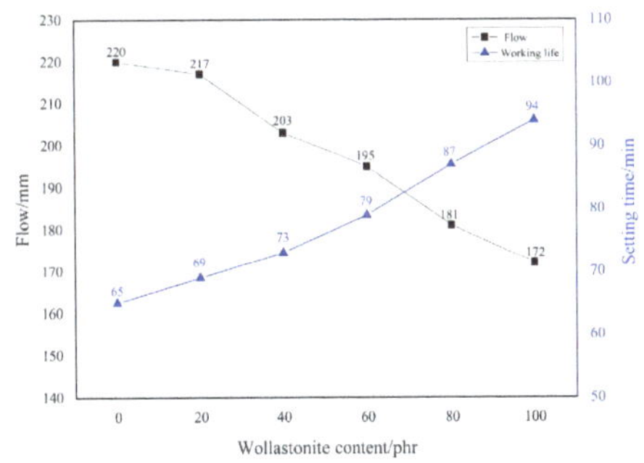

(c)

Figure 4. Influence of fillers contents on the workability of PMMA repair mortar. (**a**) fly ash (**b**) talc (**c**) wollastonite.

3.2. Mechanical Performance

3.2.1. Flexural and Compressive Strength

From Figure 5, it can be seen that the strength of PMMA repair mortar increases and then decreases with the increase of filler dosage. The addition of all three fillers increased the flexural and compressive strength of PMMA repair mortar at 28 d to different degrees. When the dosage of fly ash was 60 phr, it had the best reinforcing effect, and the flexural and compressive strengths at 28 d reached 28.09 MPa and 71.26 MPa, which were 13.31% and 15.33% higher than that of the reference group. When the dosage of talc was 40 parts, there was an increase in flexural and compressive strength at 28 d by 7.94% and 8.32%.

At 60 phr of wollastonite powder, the increased flexural and compressive strength of PMMA repair mortar at 28 d compared to the reference group were 2.38% and 5.13%. In summary, Dosing of fly ash at 60 phr had the most significant effect on the improvement of mechanical performance of PMMA repair mortar. Compared with lamellar and needle-and-rod particles, microspherical particles can more evenly transfer the external force to the resin matrix, reducing the stress concentration inside the PMMA repair mortar, thus reducing or delaying the generation and expansion of microcracks.

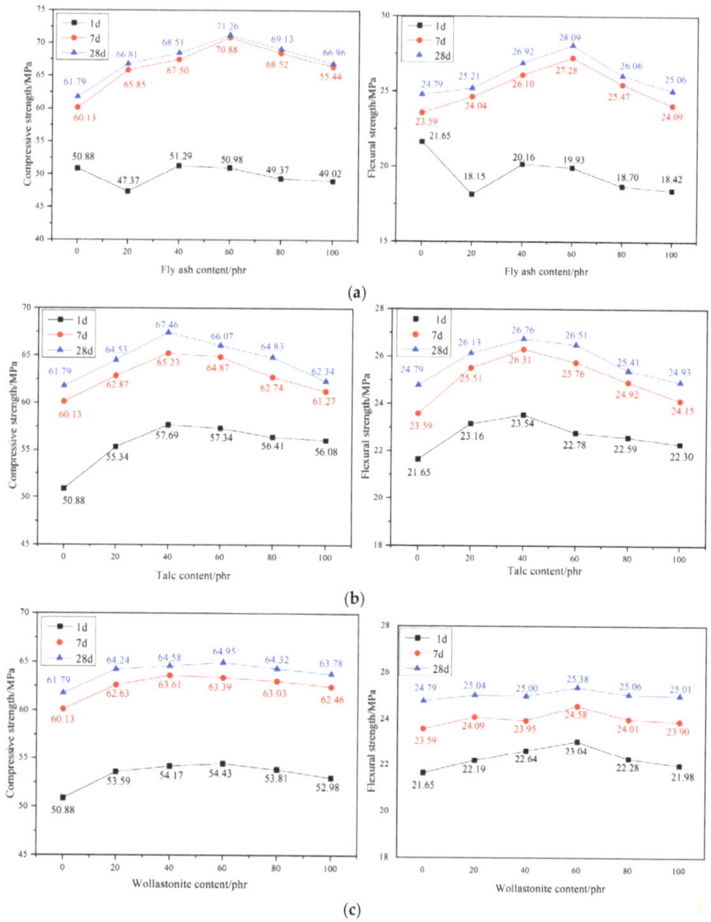

Figure 5. Influence of fillers contents on the mechanical property of PMMA repair mortar. (**a**) fly ash (**b**) talc (**c**) wollastonite.

Since fly ash substantially prolongs the setting time of mortar, its 1 d early strength is reduced when used as a filler compared to no filler. However, its enhancement is best in the later stages of strength development. Talc and wollastonite powders were more effective in enhancing the 1 d early strength than the 28 d late strength. This may be due to the fact that in the early stage of the maintenance of PMMA repair mortar, there is still some MMA that has not reacted with the curing agent, so the filler contributes more to the strength of PMMA repair mortar. At the late stage of maintenance, the PMMA has been cured completely. At this time, although the addition of filler can still increase the strength of PMMA repair mortar, but the enhancement effect is not as obvious as in the early stage.

3.2.2. Bond Strength

In addition to the mechanical performance of PMMA repair mortar itself, the bond strength between the repair mortar and the reference mortar is also an important factor affecting the actual use of PMMA repair mortar. From Table 1, it is known that the base proportion of PMMA repair mortar has good bonding performance with the reference mortar, so it is also necessary to study the effect of the addition of filler on the bonding performance of PMMA repair mortar under the temperatures at −10 °C, the results are shown in Table 4.

Table 4. Influence of fillers contents on the flexural bonding strength of PMMA repair mortar.

Dosage/phr	Fly Ash		Talc		Wollastonite	
	Flexural Bonding Strength/MPa	Fracture Surface	Flexural Bonding Strength/MPa	Fracture Surface	Flexural Bonding Strength/MPa	Fracture Surface
20	>5.97	M	>6.32	M	>5.99	M
40	>5.88	M	>5.92	M	>5.81	M
60	>6.13	M	>6.04	M	>6.04	M
80	>5.98	M	>6.19	M	>6.17	M
100	>6.01	M	>6.23	M	>5.92	M

Note: "M" in the table represents the fracture on the reference mortar specimen.

As can be seen in Table 4, the addition of filler did not reduce the flexural bond strength between the PMMA repair mortar and the reference mortar. The fracture surfaces of all the filler-added bond specimens were still located on the reference mortar specimen. This indicates that the flexural bond strength of the filled PMMA repair mortar is 5.5 MPa higher than that of the reference mortar.

3.2.3. Abrasion Performance

The results of the effect of different filler dosages on the abrasion performancer of PMMA repair mortar maintained at −10 °C for 28 d are shown in Table 5.

As the filler fills the voids of PMMA repair mortar with tiny particle size, it increases the denseness of the mortar, which improves the mechanical performance of the mortar, so the abrasion resistance of PMMA repair mortar is also greatly improved. The abrasive volume and mass of PMMA repair mortar were found to decrease and then increase with the increase of filler dosage. It can be learned from Table 5 that the optimum dosage of the three fillers were 60 phr of fly ash, 40 phr of talc, and 60 phr of wollastonite powder, separately. Compared to the reference group, the abraded volumes were 83 mm^3, 95 mm^3 and 107 mm^3, which were 28.45%, 18.10% and 7.76% lower, and the abraded masses were 0.25 g, 0.27 g and 0.31 g, reduced by 21.88%, 15.63% and 3.13%, respectively. Compactness is the main factor affecting the abrasion performance of PMMA repair mortar, and the PMMA repair mortar under these ratios corresponds to the best mechanical performance, so it has the highest compactness and the best abrasion resistance. The 28 d compressive strength increased at any amount of fly ash, so the abrasion resistance of PMMA repair mortar with fly ash as filler was superior to that without filler.

Table 5. Influence of fillers contents on the wear resistance of PMMA repair mortar.

Sample	Dosage/phr	Chord Length of Abrasion Pit/mm	Abrasion Volume/mm^3	Abrasion Mass/g
A0	0	24.0	116	0.32
F1	20	22.3	92	0.27
F2	40	22.0	89	0.27
F3	60	21.5	83	0.25
F4	80	21.9	88	0.25
F5	100	23.5	109	0.30
T1	20	23.0	102	0.31
T2	40	22.5	95	0.27
T3	60	22.7	98	0.29
T4	80	23.2	105	0.31
T5	100	23.5	109	0.31
W1	20	23.8	112	0.32
W2	40	23.5	109	0.31
W3	60	23.3	107	0.31
W4	80	24.0	116	0.33
W5	100	24.3	123	0.33

3.3. Anti Salt Freezing Performance

In cold regions or in winter, freeze-thaw cycles can lead to microcracks in concrete pavements that have been hardened. Especially when the environment temperature is −10 °C or below, deicing salt is usually sprinkled and spreaded on the pavement. The salt solution causes much more concrete damage than water, and the presence of salt solution on the surface during freezing can cause more serious erosion damage. Since this study is about PMMA repair mortar with a maintenance temperature of −10 °C, it is necessary to investigate the salt freezing performance of PMMA repair mortar. Adopting cyclic conditions of air freezing and 3% NaCl solution thawing to investigate the effect of filler dosage on the rate of mass loss and compressive strength loss of PMMA repair mortar, and the results are shown in Figure 6.

As can be seen in Figure 6, the addition of filler can effectively reduce the rate of mass loss and compressive strength loss to PMMA repair mortar in a number of salt freezing cycles. After 100 cycles of salt freezing, there was a mass loss of 0.638% and a compressive strength loss of 7.98% in the reference group. The lowest mass loss rate of 0.332%, 0.159% and 0.334% were achieved when fly ash was used with a dosage of 20 phr, Talc with a dosage of 80 phr, and wollastonite powder with a dosage of 20 phr, respectively. The loss of compressive strength reached its lowest rate at 4.86%, 4.97% and 6.22% when the dosage of fly ash was 60 phr, talc was 40 phr and wollastonite powder was 60 phr, respectively. Among the three fillers, talc had the best effect on the improvement of salt freezing performance of PMMA repair mortar.

According to the analysis of the mechanism of salt freezing damage [44,45], the overall good resistance of PMMA repair mortar to salt freezing is due to the dense polymer film formed during the molding process that can block the intrusion of foreign media. In addition, the filler's refinement of the pore size makes the PMMA repair mortar matrix more compact, so it has a stronger blocking capacity and better reduce the expansion stress caused by salt freezing. Talc has an overlapping arrangement of the palace structure, which makes the invasion path of foreign media more tortuous, more difficult to penetrate. Therefore, the PMMA repair mortar with talc as filler has higher anti-salt freezing performance.

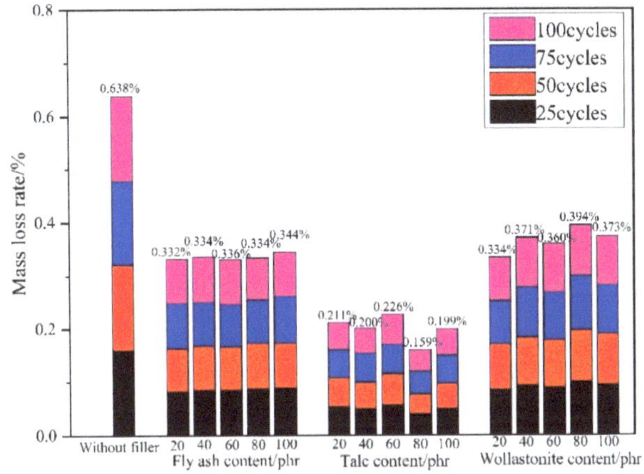

(**a**)

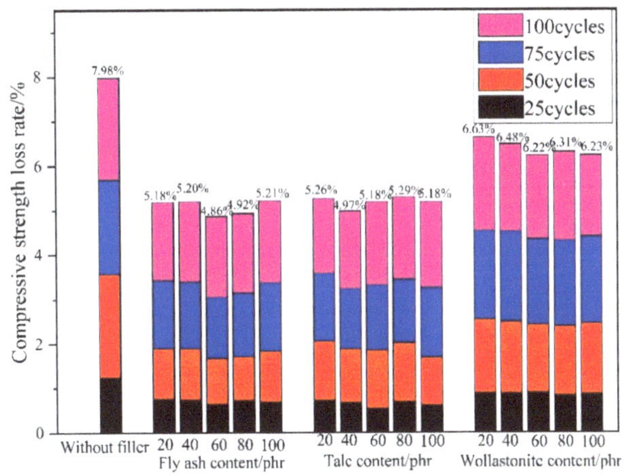

(**b**)

Figure 6. Influence of fillers contents on salt freeze resistance of PMMA repair mortar. (**a**) mass loss rate (**b**) compressive strength loss rate.

3.4. Shrinkage

Shrinkage is an important performance indicator of repair mortar, which is related to the volume compatibility of repair mortar and reference mortar after repair. Shrinkage of the newly laid repair mortar will generate internal stresses at the repair interface, and if the shrinkage rate is too large, the repair interface will be prone to cracking and lead to repair failure. Therefore, the effect of filler dosage on the shrinkage of PMMA repair mortar was investigated under the condition of maintenance at −10 °C. The results are shown in Figure 7.

As can be seen from Figure 7, the shrinkage of PMMA repair mortar with added filler is lower than that of PMMA repair mortar without filler, and it can reduce the shrinkage at all ages. The final shrinkage of PMMA repair mortar without filler is about 0.071%. With the increase of filler dosage, the shrinkage of PMMA repair mortar are first reduced and then increased. When the dosage of fly ash is 60 phr, its final shrinkage is about 0.021%, which is 70.42% lower than that of the unfilled mortar. When the dosage of talc is 40 phr, its final shrinkage is about 0.038%, which is 46.48% lower than when no filler is added. When the dosage of wollastonite powder is 60 phr, its final shrinkage is about 0.023%, which is 67.61% lower than when no filler is added.

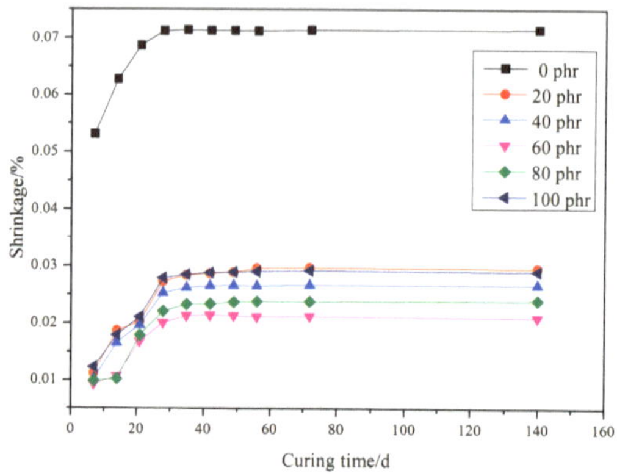

(a)

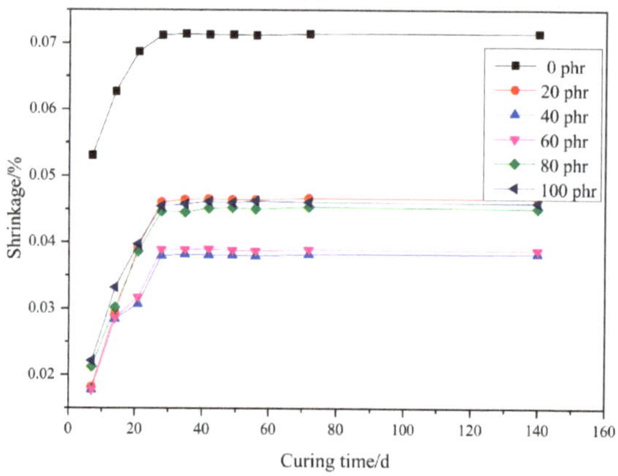

(b)

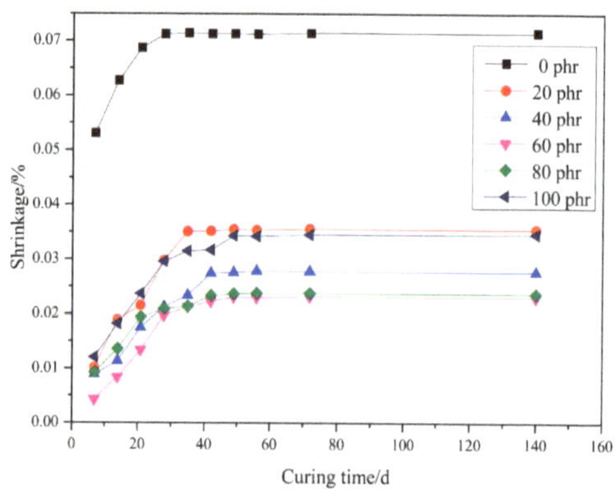

(c)

Figure 7. Influence of fillers contents on shrinkage of PMMA repair mortar, (**a**) fly ash (**b**) talc (**c**) wollastonite.

Among the three kinds of fillers, fly ash has the strongest effect in reducing shrinkage, followed by wollastonite powder and finally talcum powder. The role of filler to reduce shrinkage is, on the one hand, filler particles dispersed in the PMMA resin, the PMMA resin shrinkage played a role in hindering; on the other hand, small particle size filler can be filled in the gap between the PMMA resin and aggregates to improve the denseness of the PMMA repair mortar, the two aspects of the joint effect of the reduction of the shrinkage of the PMMA repair mortar.

3.5. Discussion

Mortar is a porous material and pore structure is inseparable and closely related to the macroscopic properties [46–48]. For analyzing the influence mechanism of filler on PMMA repair mortar from a microscopic point of view, the porosity of PMMA repair mortar was tested by using piezomercury method. Through the experimental results of the mechanical and durability performance, it was found that for the fillers, 60 phr fly ash, 40 phr talc and 60 phr wollastonite powder had the best comprehensive performance. Therefore, these three mortars and the filler-free PMMA repair mortar were selected for mercury compression testing, and the obtained pore structure parameters and the distribution ratio of each pore size are shown in Table 6 and Figure 8, respectively.

Table 6. Pore structure parameters of several PMMA repair mortars.

	Total Pore Volume/mL/g	Average Pore Size/nm	Porosity/%
A0	0.05	18.41	9.39
F3	0.02	10.67	6.03
T2	0.02	10.15	5.75
W3	0.03	12.07	6.84

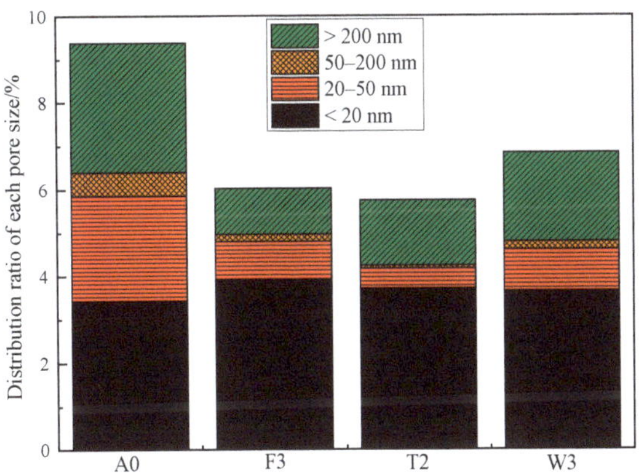

Figure 8. Distribution rate of each pore size of PMMA repair mortar with different ratios.

From Table 6, it can be seen that the addition of all three fillers significantly reduced the porosity and average pore size of the mortar. Since the porosity and average pore size of the mortar after mixing the filler are mainly related to the particle size of the filler itself, the effects of the three fillers on the porosity of the PMMA repair mortar are talcu > fly ash > wollastonite powder in descending order, which coincides with the average particle size data of the three fillers in Table 1. However, according to the mechanical property test results, it can be found that although the porosity of PMMA repair mortar mixed with talc is the lowest, but its strength is not the highest. Therefore, although the porosity can reflect the strength, the porosity is not directly proportional to the strength.

Micropores in concrete can be categorized into four types: harmless pores (pore size less than 20 nm), less harmful pores (pore size 20–50 nm), harmful pores (pore size 50–200 nm), and multi-hazardous pores (pore size greater than 200 nm) [49]. Figure 8 shows the histogram of the pore size distribution ratio of each repair mortar, from which it can be seen that the filler can effectively improve the proportion of harmless pores in PMMA repair mortar. It can be observed that the proportion of harmless pores is 65.01%, 64.52% and 53.36% when 60 phr fly ash, 40 phr talc and 60 phr wollastonite powder are incorporated, respectively. Comparing the strength data obtained in 3.2.1, the strength of PMMA repair mortar was the highest when fly ash was used as filler, which corresponded to the largest proportion of harmless pores to the total porosity. Therefore, we can learn that the proportion of harmless pores in the pore structure of mortar is the main factor affecting the strength of mortar.

Figure 9 represents the effect of filler dosage on the water absorption of PMMA repair mortar after 28 d of curing at -10 °C. It is known that with the increase of filler dosage, the water absorption of PMMA repair mortar all showed a trend of decreasing first and then increasing. Same as the porosity test results, the water absorption of PMMA repair mortar was lowest when the amount of fly ash was 60 phr, talc was 40 phr, and wollastonite powder was 60 phr. The water absorption was reduced by 38.23%, 47.21% and 40.61%, respectively, compared with that when no filler was added. Because of the lamellar structure of the talc, in addition to filling the pore effect, it can also block the pore channel, so the water absorption of PMMA repair mortar with talc is reduced to the greatest extent. Compared to wollastonite powder, fly ash has a smaller average particle size and a higher ability to fill pores. Therefore, the PMMA repair mortar with fly ash has the second highest water absorption, while the PMMA repair mortar with wollastonite powder has the relatively highest water absorption among the three fillers.

As can be seen from the Figure 10, the fracture surfaces of the three mortars are obviously rough and uneven. Fly ash and talcum powder are embedded in the repair mortar matrix in the form of microspheres and lamellar layers, respectively. And the wollastonite powder in the repair mortar disappears, leaving only its pin-rod-like pits. The filling of filler increases the compactness of PMMA repair mortar. When subjected to external forces, the filler will play an obstructive role in the development of microcracks in the path of the need for additional de-attachment of the filler work and pull-out work to increase the development of microcracks consumed by the energy to impede the propagation of cracks. When subjected to the same torque, fly ash particles and talc particles are still embedded in the mortar, wollastonite powder is "pulled out", proving that the degree of adhesion between wollastonite powder and PMMA repair mortar is not as high as that of fly ash and talc. From the particle size data of the fillers in 2.1, it can be seen that the average particle size of wollastonite powder is significantly larger than that of fly ash and talc. This is one of the reasons why wollastonite powder has a weaker effect on the strength enhancement of PMMA repair mortar than fly ash and talcum powder.

Figure 11 shows the effect of filler amount on the coefficient of thermal expansion of PMMA mortar. The coefficient of thermal expansion of ordinary cement concrete pavement is $8 - 12 \times 10^{-6}/°C$, and the coefficient of thermal expansion of PMMA repair mortar without filler is $50.03 \times 10^{-6}/°C$, which is 4–5 times of the coefficient of thermal expansion of ordinary cement concrete pavement. And when the filler dosage was 100 phr, the coefficients of thermal expansion of mortar doped with fly ash, talc, and wollastonite powder were $29.8 \times 10^{-6}/°C$, $34.1 \times 10^{-6}/°C$ and $34.1 \times 10^{-6}/°C$, respectively, which were 40.44%, 31.84%, and 39.24% lower than that of the mortar with no filler added, respectively. When the external temperature changes, if the material in contact with each other can be free deformation, it will not produce internal stress inside the material. But the material is the interaction force and mutual binding force between the two, if the difference in the coefficient of thermal expansion between the two is too large, it will lead to the production of internal stresses in the material, thus affecting the bonding situation between the two [50]. After adding 100 phr filler, the coefficient of thermal expansion of PMMA repair mortar decreased by 30–40% compared with that without filler. The thermal compatibility between PMMA repair mortar and ordinary concrete pavement was significantly improved by adding filler.

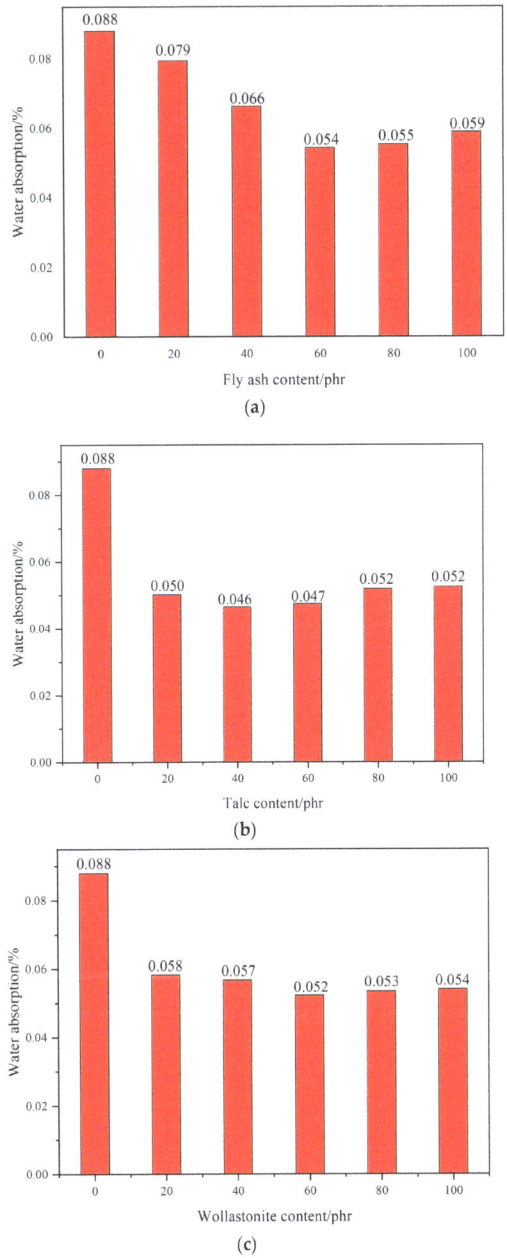

Figure 9. Influence of fillers contents on water absorption of PMMA repair mortar. (**a**) fly ash (**b**) talc (**c**) wollastonite.

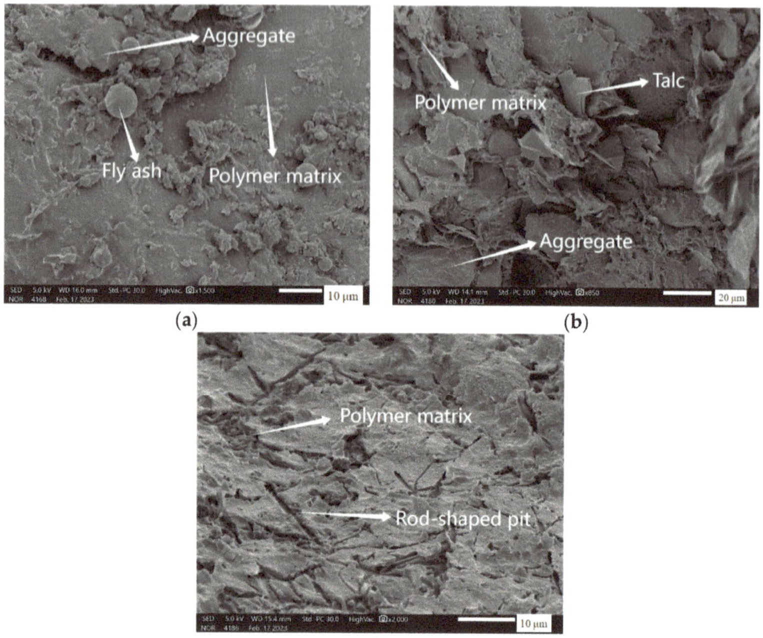

Figure 10. SEM microsgraphs of PMMA repair mortars containing different fillers. (**a**) fly ash (**b**) talc (**c**) wollastonite.

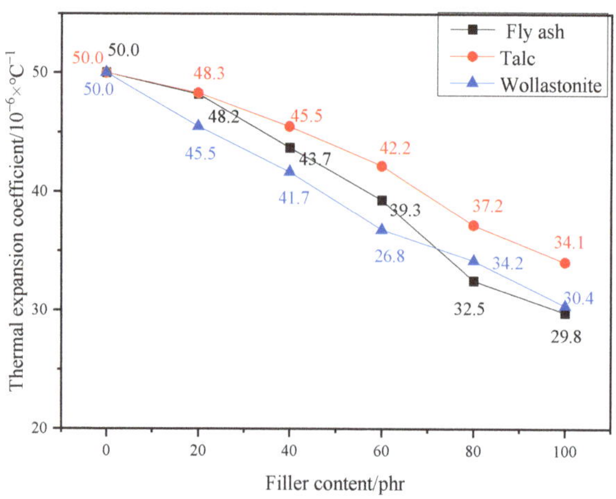

Figure 11. Influence of fillers contents on thermal expansion coefficient of PMMA mortar.

Pore structure, water absorption and SEM analysis revealed the role of different types of fillers in repair mortars. On the one hand, fillers filled the internal pores of the mortar, reduced the pore size, decreased the proportion of harmful pores, and provided better densification. On the other hand, the uniformly distributed fillers increased the energy required for the diffusion of microcracks, thus achieving the effect of increasing the strength of the PMMA repair mortar. Finally, the results of the coefficient of thermal expansion test showed that the filler, in addition to improving the performance of the repair mortar itself, can greatly improve the thermal compatibility between the PMMA repair mortar and the cement concrete pavement, and reduce the damage due to temperature stress fatigue.

4. Conclusions

In this paper, the effects of fly ash, talc and wollastonite powder on the working performance, mechanical performance and durability of PMMA repair mortar were investigated, and the main conclusions are as follows:

(1) 60 phr of fly ash has the greatest effect on the overall performance of PMMA repair mortar.
(2) All three fillers prolonged the setting time of the PMMA repair mortar. Wollastonite powder had the least negative effect on the setting time of the repair mortar. The larger the dosage of talc and fly ash, the lower the fluidity of the mortar. With the increase in the amount of fly ash, the fluidity of the mortar tended to increase and then decrease.
(3) All fillers improved the mechanical properties of the repair mortar itself to different degrees. Among them, fly ash had the most significant enhancement effect on the mechanical performance. The addition of fillers had no negative effect on the bond strength of the PMMA repair mortar to the reference mortar.
(4) The filler effectively improved the anti-salt-freezing performance of PMMA repair mortar. Thanks to the unique overlapping layer structure of talcum powder, its comprehensive anti-salt freezing performance was the most excellent. Since the filler itself does not shrink, so the shrinkage of the PMMA repair mortar after the addition of the filler was also significantly reduced.
(5) Fillers improve the densification of PMMA repair mortars. The inclusion of fillers greatly improves the thermal compatibility between the PMMA repair mortar and the cement concrete pavement, reducing damage at the bond interface due to temperature stress fatigue.

Author Contributions: Conceptualization, Z.Z. and L.X.; methodology, Z.Z., M.D. and L.X.; data curation, Z.Z., S.L., L.Y., Z.G. and Y.W.; testing and providing help in calculation, Z.Z., S.L., L.Y., Z.G. and Y.W.; writing—original draft preparation, Z.Z.; writing—review and editing, Z.Z., M.D. and L.X. All authors have read and agreed to the published version of the manuscript.

Funding: The work was supported by the Priority Academic Program Development of Jiangsu Higher Education Institutions (PAPD).

Institutional Review Board Statement: Not applicable.

Informed Consent Statement: Not applicable.

Data Availability Statement: All data in this article are listed in this paper. The data presented in this study are available on request from the corresponding author.

Conflicts of Interest: The authors declare no conflict of interest.

References

1. Yin, F.; Mei, S. Statistical Bulletin of Transportation Industry Development in 2022. *Waterw. Ports* **2023**, *44*, 1002–1006.
2. Shi, J.; Liu, B.; Qin, J.; Jiang, J.; Wu, X.; Tan, J. Experimental study of performance of repair mortar: Evaluation of in-situ tests and correlation analysis. *J. Build. Eng.* **2020**, *31*, 101325. [CrossRef]
3. Yu, H.; Lu, L.; Qiao, P. Localization and size quantification of surface crack of concrete based on Rayleigh wave attenuation model. *Constr. Build. Mater.* **2021**, *280*, 122437. [CrossRef]
4. Zailani, W.W.A.; Abdullah, M.M.A.B.; Arshad, M.F.; Razak, R.A.; Tahir, M.F.M.; Zainol, R.R.M.A.; Nabialek, M.; Sandu, A.V.; Wysłocki, J.J.; Błoch, K. Characterisation at the Bonding Zone between Fly Ash Based Geopolymer Repair Materials (GRM) and Ordinary Portland Cement Concrete (OPCC). *Materials* **2020**, *14*, 56. [CrossRef] [PubMed]
5. Li, Y.R. Discussion on the Construction Technology of Cement Concrete Pavement Repaired by Thin Layer of Highway. *Value Eng.* **2016**, *35*, 151–153.
6. Song, X.; Song, X.; Liu, H.; Huang, H.; Anvarovna, K.G.; Ugli, N.A.D.; Huang, Y.; Hu, J.; Wei, J.; Yu, Q. Cement-Based Repair Materials and the Interface with Concrete Substrates: Characterization, Evaluation and Improvement. *Polymers* **2022**, *14*, 1485. [CrossRef]

7. Xu, S.; Mu, F.; Wang, J.; Li, W. Experimental study on the interfacial bonding behaviors between sprayed UHTCC and concrete substrate. *Constr. Build. Mater.* **2018**, *195*, 638–649. [CrossRef]
8. Javed, A.; Gillani, S.A.A.; Abbass, W.; Riaz, M.R.; Hameed, R.; Abbas, S.; Salmi, A.; Deifalla, A.F. Mechanical Performance of Amorphous Metallic Fiber-Reinforced and Rubberized Thin Bonded Cement-Based Overlays. *Sustainability* **2022**, *14*, 8226. [CrossRef]
9. Yucel, H.E.; Jashami, H.; Sahmaran, M.; Guler, M.; Yaman, I.O. Thin ECC overlay systems for rehabilitation of rigid concrete pavements. *Mag. Concr. Res.* **2013**, *65*, 108–120. [CrossRef]
10. Gillani, S.A.A.; Shahzad, S.; Abbass, W.; Abbas, S.; Toumi, A.; Turatsinze, A.; Mohamed, A.M.; Sayed, M.M. Debonding of Thin Bonded Rubberised Fibre-Reinforced Cement-Based Repairs under Monotonic Loading: Experimental and Numerical Investigation. *Materials* **2022**, *15*, 3886. [CrossRef]
11. Beßling, M.; Groh, M.; Koch, V.; Auras, M.; Orlowsky, J.; Middendorf, B. Repair and Protection of Existing Steel-Reinforced Concrete Structures with High-Strength, Textile-Reinforced Mortars. *Buildings* **2022**, *12*, 1615. [CrossRef]
12. Liu, F.; Pan, B.; Zhou, C. Experimental Study on a Novel Modified Magnesium Phosphate Cement Mortar Used for Rapid Repair of Portland Cement Concrete Pavement in Seasonally Frozen Areas. *J. Mater. Civ. Eng.* **2022**, *34*, 04021483. [CrossRef]
13. Liu, Y.; Jia, M.; Song, C.; Lu, S.; Wang, H.; Zhang, G.; Yang, Y. Enhancing ultra-early strength of sulphoaluminate cement-based materials by incorporating graphene oxide. *Nanotechnol. Rev.* **2020**, *9*, 17–27. [CrossRef]
14. Xu, L.; Li, X.; Jiang, F.; Yu, X.; Wang, J.; Xiao, F. Thermosetting characteristics and performances of polyurethane material on airport thin-overlay. *Constr. Build. Mater.* **2022**, *344*, 128252. [CrossRef]
15. Knight, M.L.; Wilson, G.S.; Seger, W.J.; Mahadevan, S. Overlay Types Used as Preventive Maintenance on Tennessee Bridge Decks. *Transp. Res. Rec. J. Transp. Res. Board* **2004**, *1866*, 79–84. [CrossRef]
16. Genedy, M.; Matteo, E.N.; Stenko, M.; Stormont, J.C.; Taha, M.R. Nanomodified Methyl Methacrylate Polymer for Sealing of Microscale Defects in Wellbore Systems. *J. Mater. Civ. Eng.* **2019**, *31*, 04019118. [CrossRef]
17. Tomić, N.Z.; Marinković, A.D.; Veljović, D.; Trifković, K.; Lević, S.; Radojević, V.; Heinemann, R.J. A new approach to compatibilization study of EVA/PMMA polymer blend used as an optical fibers adhesive: Mechanical, morphological and thermal properties. *Int. J. Adhes. Adhes.* **2018**, *81*, 11–20. [CrossRef]
18. Kim, T.-K.; Park, J.-S. Comparative Analysis of Domestic and International Test Guidelines for Various Concrete Repair Materials. *Materials* **2022**, *15*, 3267. [CrossRef] [PubMed]
19. Lakhiar, M.T.; Bai, Y.; Wong, L.S.; Paul, S.C.; Anggraini, V.; Kong, S.Y. Mechanical and durability properties of epoxy mortar incorporating coal bottom ash as filler. *Constr. Build. Mater.* **2022**, *315*, 125677. [CrossRef]
20. Shi, C.; Wang, P.; Ma, C.; Zou, X.; Yang, L. Effects of SAE and SBR on properties of rapid hardening repair mortar. *J. Build. Eng.* **2020**, *35*, 102000. [CrossRef]
21. Li, Z. Application of modified epoxy resin in concrete pavement repair works. *China Highw.* **2021**, *12*, 108–110. [CrossRef]
22. Lu, G.; Renken, L.; Li, T.; Wang, D.; Li, H.; Oeser, M. Experimental study on the polyurethane-bound pervious mixtures in the application of permeable pavements. *Constr. Build. Mater.* **2019**, *202*, 838–850. [CrossRef]
23. Leng, C.; Lu, G.; Gao, J.; Liu, P.; Xie, X.; Wang, D. Sustainable Green Pavement Using Bio-Based Polyurethane Binder in Tunnel. *Materials* **2019**, *12*, 1990. [CrossRef] [PubMed]
24. Rebeiz, K.S.; Fowler, D.W.; Paul, D.R. Polymer concrete and polymer mortar using resins based on recycled poly(ethylene terephthalate). *J. Appl. Polym. Sci.* **1992**, *44*, 1649–1655. [CrossRef]
25. Yeon, K.-S.; Cha, J.-Y.; Yeon, J.H. Effects of DMT and TMPTMA on Working Life of Acrylic Polymer Concrete Exposed to Low Curing Temperatures. *Polymers* **2015**, *7*, 1587–1598. [CrossRef]
26. Yeon, K.-S.; Cha, J.-Y.; Yeon, J.H.; Seung, I.B. Effects of TMPTMA and silane on the compressive strength of low-temperature cured acrylic polymer concrete. *J. Appl. Polym. Sci.* **2014**, *131*, 40939. [CrossRef]
27. Son, S.-W.; Yeon, J.H. Mechanical properties of acrylic polymer concrete containing methacrylic acid as an additive. *Constr. Build. Mater.* **2012**, *37*, 669–679. [CrossRef]
28. Mirza, J.; Durand, B.; Bhutta, A.R.; Tahir, M.M. Preferred test methods to select suitable surface repair materials in severe climates. *Constr. Build. Mater.* **2014**, *50*, 692–698. [CrossRef]
29. Liang, P.J. Application of phosphate cement thin-layer repair process in the disposal of surface-type lesions on airport pavements. *Transp. Manag. World* **2022**, *18*, 28–30.
30. Yeon, J. Deformability of Bisphenol A-Type Epoxy Resin-Based Polymer Concrete with Different Hardeners and Fillers. *Appl. Sci.* **2020**, *10*, 1336. [CrossRef]
31. Rebeiz, K.S.; Serhal, S.P.; Craft, A.P. Properties of Polymer Concrete Using Fly Ash. *J. Mater. Civ. Eng.* **2004**, *16*, 15–19. [CrossRef]
32. Gorninski, J.; Molin, D.D.; Kazmierczak, C. Comparative assessment of isophtalic and orthophtalic polyester polymer concrete: Different costs, similar mechanical properties and durability. *Constr. Build. Mater.* **2007**, *21*, 546–555. [CrossRef]
33. Khang, P.H.; Hiep, N.T. Experimental Research on performance of Polymer Concrete Used for Repairing Portland Cement Concrete Pavement of Airport. *J. Mater. Eng. Struct.* **2020**, *7*, 583–589.
34. Li, G.J.; Luo, F. Influence of filler morphology on the performance of epoxy resin road repair mortar. *Nonmet. Min.* **2019**, *42*, 35–38.
35. Chandrasekaran, K.; Srinivasan, L.P.; Meyappan, N. Characterisation of Fibre Reinforced Resin Concrete. *Mater. Plast.* **2022**, *58*, 158–170. [CrossRef]

36. Yemam, D.M.; Kim, B.-J.; Moon, J.-Y.; Yi, C. Mechanical Properties of Epoxy Resin Mortar with Sand Washing Waste as Filler. *Materials* **2017**, *10*, 246. [CrossRef] [PubMed]
37. Guo, Z.; Xu, L.; Lu, S.; Yan, L.; Zhu, Z.; Wang, Y. Study on the Effect of PVAc and Styrene on the Properties and Microstructure of MMA-Based Repair Material for Concrete. *Materials* **2023**, *16*, 3984. [CrossRef]
38. *GBT 17671-2021;* Test Method for Strength of Cement Mastic Sand (IOS Method). China National Standards: Beijing, China, 2021.
39. *GBT 2419-2005;* Test Method for Flowability of Cementitious Sand. Chinese National Standards: Beijing, China, 2005.
40. *JGJT 70-2009;* Building Mortar Basic Performance Test Method Standard. China Industry Standards: Beijing, China, 2009.
41. *GBT 3810.6-2016;* Test Method for Ceramic Tiles Part 6: Determination of Abrasion Depth of Unglazed Tiles. Chinese National Standards: Beijing, China, 2016.
42. *DLT5126-2021;* Test Procedure for Polymer Modified Cement Mortar. China Electric Power Industry Standards: Beijing, China, 2021.
43. *JCT 603-2004;* Test Method for Drying Shrinkage of Cementitious Sand. China Building Materials Industry Standard: Beijing, China, 2004.
44. Powers, T.C. A working hypothesis for further studies of frost resistance of concrete. *J. Proc.* **1945**, *41*, 245–272.
45. Powers, T.C.; Helmuth, R.A. Theory of volume changes in hardened Portland cement paste during freezing. *Highw. Res. Board Proc.* **1953**, *32*, 285–297.
46. Yang, K.; Tang, Z.; Cheng, Z.; Zhao, H.; Feng, R.; Long, G. Mechanical properties of ultra-high strength cement-based materials (UHSC) incorporating metal powders and steel fibers. *Constr. Build. Mater.* **2021**, *318*, 125926. [CrossRef]
47. Wang, Y.; Ge, Y.; Wang, X.; Chen, X.; Li, Q. The effect of powder activated carbon on mechanical properties and pore structures of cement-based mortars. *Constr. Build. Mater.* **2022**, *316*, 125798. [CrossRef]
48. Cai, R.; Qi, H.; Mao, J.; Lv, J.; Jin, D. Improved Crack Resistance and Pore Structure of Cement-Based Materials by Adding EVA Powder. *J. Mater. Civ. Eng.* **2022**, *34*, 04022012. [CrossRef]
49. Wu, Z.W.; Lian, H.Z. *High Performance Concrete;* China Railway Press: Beijing, China, 1999.
50. Perkins, P.H. *Concrete Structures: Repair Waterproofing and Protection;* Applied Science Publishers: Basel, Switzerland, 1997. [CrossRef]

materials MDPI

Article

Influence of Ultrafine Fly Ash and Slag Powder on Microstructure and Properties of Magnesium Potassium Phosphate Cement Paste

Zheng Jia [1], Yuhui Zhang [1] and Liwu Mo [1,2,*]

[1] College of Materials Science and Engineering, Nanjing Tech University, Nanjing 210009, China; jz1473436935@163.com (Z.J.); serenobody@outlook.com (Y.Z.)
[2] State Key Laboratory of Materials-Orientated Chemical Engineering, Nanjing 210009, China
* Correspondence: andymoliwu@njtech.edu.cn

Abstract: This study investigated the influences of ultrafine fly ash (UFA) and ultrafine slag powder (USL) on the compressive strengths, autogenous shrinkage, phase assemblage, and microstructure of magnesium potassium phosphate cement (MKPC). The findings indicate that the aluminosilicate fractions present in both ultrafine fly ash and ultrafine slag participate in the acid–base reaction of the MKPC system, resulting in a preferential formation of irregularly crystalline struvite-K incorporating Al and Si elements or amorphous aluminosilicate phosphate products. UFA addition mitigates early age autogenous shrinkage in MKPC-based materials, whereas USL exacerbates this shrinkage. In terms of the sustained mechanical strength development of the MKPC system, ultrafine fly ash is preferred over ultrafine slag powder. MKPC pastes with ultrafine fly ash show greater compressive strength compared to those with ultrafine slag powder at 180 days due to denser interfaces between the ultrafine fly ash particles and hydration products like struvite-K. The incorporation of 30 wt% ultrafine fly ash enhances compressive strengths across all testing ages.

Keywords: magnesium potassium phosphate cement; ultrafine fly ash; ultrafine slag powder; compressive strength; microstructure; chemically bonded ceramics

check for updates

Citation: Jia, Z.; Zhang, Y.; Mo, L. Influence of Ultrafine Fly Ash and Slag Powder on Microstructure and Properties of Magnesium Potassium Phosphate Cement Paste. *Materials* **2024**, *17*, 2556. https://doi.org/10.3390/ma17112556

Academic Editor: Alexandre Bogas

Received: 19 April 2024
Revised: 17 May 2024
Accepted: 23 May 2024
Published: 25 May 2024

1. Introduction

Magnesium phosphate cement (MPC) is a ceramic type formed via hydration solidification between dead-burned magnesium oxide and phosphate, such as $NH_4H_2PO_4$ and KH_2PO_4 [1]. In the case of MKPC, this hydration process is represented by Equation (1) [2,3]:

$$MgO(s) + KH_2PO_4(s) + H_2O(l) \rightarrow MgKPO_4 \cdot 6H_2O(s) \quad (1)$$

Compared to conventional silicate cement, magnesium phosphate cement offers several advantages, including rapid condensation, great early-age strength, environmental friendliness, and good bonding strength [4,5]. Its applications range widely, including effective fixation of metal contaminants [6], repair grouting [4,7–10], storage of radioactive waste [11], and biomedical applications [12]. However, MPC faces limitations as a repair material, including the fact that it is non-waterproof, excessive condensation time, and reduced validity under high environmental temperatures or when used in large volumes. In order to enhance MPC properties, mineral admixtures like blast furnace slag [13], fly ash [13,14], calcined kaolin [15], and silica powder [16] are commonly used in MPC paste. However, the impact of mineral admixtures on MPC properties varies based on their variety, phase composition, and additive amount.

In the traditional sense, fly ash was typically viewed as a mere inert material or retarder in the MPC reaction [16]. Its special micromorphology enhances the operability of MPC because it is spherical [8,11]. When combined with silica powder in MKPC, fly ash was reported to enhance water resistance by refining pore structures [16]. Nevertheless, some studies have suggested that adding fly ash decreases MPC's tensile and residual strength under high temperature curing [17,18]. Recent studies have begun to show that fly ash is more than just a filler in MPC systems [13,19]. Xu et al. [19] used a two-component formula to prepare MKPC mortar mixed with fly ash. When the water–cement ratio was the same, fly ash was regarded as inert or active. In the first recipe, FA was applied as a substitute for two MKPC raw materials (dead-burned magnesium oxide and phosphate) so that the magnesium–phosphorus ratio in the fixed formulation remained constant at 8:1. In another recipe, FA was considered as a substitute only for dead-burned magnesium oxide so that the magnesium–phosphorus ratio in the MKPC changed to 5:1. Due to fly ash's chemical reaction, replacing MgO alone with fly ash improved the setting time, operability at the end of the stirring, and mechanical properties of MKPC more than replacing both MgO and KH_2PO_4 with fly ash [19]. Nicholson and Wilson [20] pointed out that the aluminum–silicon glass phase of FA may hydrate with phosphate after dissolution to produce a gelling material that is conducive to the development of mechanical properties and relatively stable in volume.

The current research indicates that MPC represents a novel gel material characterized by a rapid setting time, high strength, and environmental friendliness. This innovative gel material has found extensive applications in areas such as quick-drying concrete repairs, artificial bone manufacturing, and the treatment of solid waste [21]. Nevertheless, significant challenges persist regarding the practical application of magnesium phosphate cement (MPC), particularly its high preparation cost attributed to the expensive nature of a key ingredient, magnesia. Identifying low-cost alternatives to magnesia beyond conventional magnesite sources is a potential solution to this economic constraint. To mitigate the practical issues associated with MPC, Yang et al. [22] and Li et al. [23] studied the effect of FA on the durability of MKPC in water and salt solutions (sulfate and chloride) and found that the incorporation of fly ash can improve the mechanical properties of MKPC and its pore structure, leading to an improved durability of MKPC with fly ash. Xu et al. [24] found that the addition of FA with a high CaO content, up to 50 wt%, can significantly improve the compressive strength of MKPC. When the fly ash content exceeded 50 wt%, the observed decrease in MKPC strength could be attributed to the formation of $CaK_3H(PO_4)_2$ and $Mg_3(PO_4)_2{\cdot}22H_2O$. Mo et al. [25] demonstrated that the addition of FA can improve the pore structure of MKPC. Gardner et al. [26] found that FA (or slag) additions at 50 wt% (based on the sum of MgO, KH_2PO_4, and H_2O) resulted in a compact microstructure in the MKPC hardened matrix with an improved mechanical strength at a water-to-solids ratio of 0.24. In summary, incorporating FA can markedly enhance several properties of magnesium potassium phosphate cement (MKPC). These improvements include enhanced workability, mechanical strengths, and durability, alongside a reduction in the heat of hydration and an extension of the setting time. Moreover, employing FA, an industrial by-product, can significantly lower MKPC production costs without adversely affecting its mechanical performance.

With the development of modern civil engineering and environmental engineering, higher requirements have been put forward for the mechanical properties and durability of cementitious materials. Based on this, researchers have developed ultrafine fly ash (UFA) and ultrafine slag powder (USL) using innovative processing techniques on fly ash and slag. UFA is the second superfine grinding product of fly ash, a byproduct from thermal power plants. The ultrafine grinding of fly ash releases the smaller microspheres contained in the larger/hollow fly ash particles [27]. Ultrafine slag powder is the abbreviation of ultrafine granulated blast furnace slag powder, which is a kind of high-quality cement admixture. The industrial solid waste slag obtained from the smelting of pig iron in the blast furnaces of iron plants is mainly composed of calcium silicaluminate as the molten substance. Most of the industrial solid waste slag obtained after water quenching and grinding is glassy and has potential water hard gelling properties. At present, UFA and USL are widely used as mineral admixtures in high performance concrete [28,29]. UFA is an ideal material to utilize due to its smaller particles, which, when mixed with Portland cement, can lead to the formation of ultrahigh-performance concretes [29] and accelerated hydration [30]. These systems were demonstrated to have an improved filling effect associated with the fine particle size of UFA [31], leading to detectable improvements in physical properties, microstructures, and durability [30,32,33]. Similarly, USL, as a common auxiliary hydraulic cementing material, can be incorporated well in the hydration reaction when mixed with Portland cement, thereby improving its performance. In geopolymers, the inclusion of UFA also leads to a reduced porosity [34]. The above studies indicate that UFA could better improve the properties of cementitious materials compared to FA. The performance of solid waste-based MPC is mainly manifested in two aspects: mechanical properties and stability. Current research results have proven that the heavy metal elements in solid waste will result in a decrease in the mechanical properties and stability of MPC in most cases, while the silicon and aluminum elements are able to improve the performance of MPC. Therefore, solid wastes with high silicon and aluminum contents, such as UFA, are appropriate to be solidified and stabilized by MPC. In addition, solid wastes with a low silica–aluminum content and a high content of other elements should be used for a comparative study of specific reaction effects in MKPC.

Therefore, this study investigated the reaction behavior and mechanism in MKPC pastes containing ultrafine fly ash/ultrafine slag powder. The inclusion of ultrafine mineral additives, especially at higher levels, can alter the reaction conditions (such as acidity and alkalinity, degree of hydration, ion mobility, etc.) within the MKPC system, potentially affecting its reaction behavior, microstructure formation, and subsequent strength development. This study seeks to explore how ultrafine fly ash and ultrafine slag powder impact the microstructure, compressive strength, and volume stability of MKPC pastes at five different substitution levels: 0 wt%, 20 wt%, 30 wt%, 40 wt%, and 50 wt% of the total MgO weight. Additionally, we examined the potential chemical reactions of ultrafine fly ash and ultrafine slag powder within the same MKPC paste system using XRD, TG/DSC, and SEM/EDS. Moreover, we measured the early volumetric self-shrinkage properties of MKPC mortar for the first time using a German Micro-Epsilon displacement sensor (MICRO-EPSILON MESSTECHNIK GmbH & Co. KG, Ortenburg, Germany), which has not been widely used in this field before. This also opens up new research perspectives and provides a more direct and reliable method for future researchers. This research aims to offer foundational insights into the properties of MKPC altered by ultrafine fly ash or ultrafine slag powder, along with their associated mechanisms, which is valuable for advancing the properties of MKPC.

2. Experimental

2.1. MKPC Paste Preparation

Yancheng Hua Nai Magnesium Industry Co., Ltd., Yancheng, China, provided the dead-burned MgO, which was obtained by calcining magnesite at 1500 °C. Shi fang Dingli Co., Ltd. in Sichuan, China supplied industrial-grade KH_2PO_4. Sodium tetraborate was added as the solidification retarder in a constant amount of 10.0 wt% relative to the weight of MgO. Ultrafine fly ash and ultrafine slag powder were used to partially substitute MgO. The chemical compositions of MgO, ultrafine fly ash, and ultrafine slag powder are outlined in Table 1. Specifically, the ultrafine slag powder contains 11.89 wt% Al_2O_3, 33.39 wt% SiO_2, and 41.51 wt% CaO. In contrast, the ultrafine fly ash has higher levels of Al_2O_3 and SiO_2, at 31.65 wt% and 50.53 wt%, respectively. Al_2O_3 and SiO_2 in the ultrafine slag powder are primarily present as amorphous aluminosilicate, while in the ultrafine fly ash, they exist as mullite crystalline phases and a glassy aluminosilicate phase. Additionally, a portion of quartz is present in the ultrafine fly ash. The ultrafine fly ash and ultrafine slag powder used in this paper are in accordance with T/CBMF194-2022 [35] China Building Materials Association standard industry document specifications. Particle size distributions of dead-burned magnesium oxide, ultrafine fly ash, and ultrafine slag powder were analyzed using a powder laser particle size analyzer, as displayed in Figure 1.

Table 1. Chemical compositions of raw materials/wt%.

Raw Material	MgO	CaO	Fe_2O_3	Al_2O_3	SiO_2	LOI
Magnesia	95.39	1.57	0.73	0.17	1.54	0.32
Ultrafine fly ash	0.92	4.07	4.48	31.65	50.53	2.77
Ultrafine slag powder	8.82	41.51	0.63	11.89	33.39	−0.28

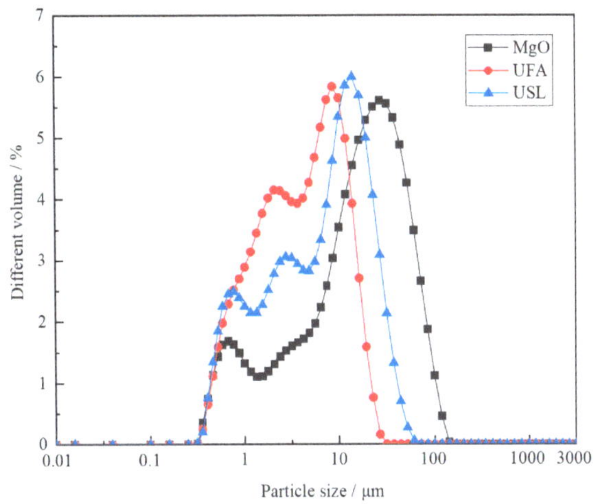

Figure 1. Particle size distributions of MgO, ultrafine fly ash, and ultrafine slag powder.

The mechanical properties of MPC are significantly affected by the M/P ratio [36,37]. Previous research has shown that the best mechanical performance of MKPC is achieved with an M/P molar ratio of 3~4:1 [38,39]. In this study, the control mix had an initial magnesium-to-phosphorus ratio of 3:1. Nine sets of MKPC were prepped, with ultrafine fly ash and ultrafine slag powder utilized as partial substitutes for MgO alongside the control. Table 2 details the composition of alkaline components in MKPC. Various levels of substitution of MgO with ultrafine fly ash and ultrafine slag powder were tested, ranging from 20 wt% to 50 wt%. The water–cement ratio remained at 0.18. Before adding water, the powdered raw materials (including magnesium oxide, KH_2PO_4, sodium tetraborate, UFA, and USL) were pre-blended to ensure uniformity.

Table 2. The proportions of alkaline components in MKPC/wt%.

Mix ID	MgO	UFA/USL
Control	100	0
UF20/US20	80	20
UF30/US30	70	30
UF40/US40	60	40
UF50/US50	50	50

UFA: ultrafine fly ash, USL: ultrafine slag powder.

Then, water was added to the MKPC mixture and mixed continuously until a homogeneous mixture was achieved. The resulting mixture was poured into a 40 mm × 40 mm × 40 mm mold to form a cubic mortar at a temperature of 20 ± 2 °C and a relative humidity (RH) of 50 ± 5% in a laboratory environment. The mold was covered with plastic film and left to stand for 2 h. The specimens were then removed from the molds and cured at room temperature for different times: 2 h, 24 h, 28 days, and 180 days.

2.2. Compressive Strength Test

The compressive strengths of MKPC-bonded specimens were evaluated in multiple stages, specifically at 2 h, 24 h, 28 days, and 180 days. For each examination, an average compressive strength value was derived from five individual paste cubes.

2.3. Autogenous Shrinkage Test

The German Micro-Epsilon displacement sensor was used to test the early volume shrinkage of the mortar. The length of the threaded pipe was 42 cm, the outer diameter was 28 mm, and the inner diameter was 20 mm. One end of the threaded pipe was fixed and the sensor probe was adjusted so that the shrinkage or expansion of the mortar was within the measuring range of the instrument. During the test, the threaded tube device and laser equipment were kept stable, and the ambient temperature was controlled at 20 ± 3 °C.

2.4. Mineral Phase Examination

The required samples were examined using an X-ray diffractometer. The samples were prepared using the powder diffraction method by first drying the specimens at 50 °C, then grinding them to less than 80 μm using an agate mortar and pestle, and then placing the fine powder into a 0.5 mm-deep glass sample stage and slightly compacting the powder surface using a smooth glass sheet. For XRD measurements, the device was set to work with a target of Cu Kα, a tube current of 100 mA, and a tube voltage of 40 kV. The scanning was carried out with a speed setting of 10°/min and covered an angular range of 5° to 60°.

2.5. Pore Structure Test

The pore composition within set MKPC was meticulously examined utilizing a Pore Master GT-60 Mercury Intrusion Porosimeter (MIP). To prepare for this analysis, the solidified MKPC paste was fragmented into diminutive pieces ranging from 2 to 5 mm. These fragments were then submerged in pure ethyl alcohol to cease hydration processes, followed by a vacuum drying process at a temperature of 45 °C to ensure complete dryness prior to the pore size evaluation.

2.6. Microstructure Examination Using SEM

A selected sample from a newly damaged area of the cured MKPC specimen was soaked in pure ethanol to stop continued reaction. After halting hydration, the sample was dried at 50 °C and thinly coated with gold. The microstructural characteristics and composition of emerging compounds within the MKPC pastes were examined using an LEO 1530VP Scanning Electron Microscope (SEM) (Zeiss AG, Baden-Württemberg, Germany) equipped with an Oxford Energy Dispersive Spectroscopy (EDS) system.

Moreover, for further examination, a cubic piece of the set MKPC paste was sectioned into thin slices, underwent a similar cessation of hydration in absolute ethyl alcohol, and was then vacuum-dried. The arid specimen was infused with an epoxy resin, polished to achieve a smooth surface, coated with carbon to make it conductive, and analyzed with the SEM operated in the backscattered electron (BSE) mode for high-contrast imaging. Concurrently, the EDS was used to carry out a precise elemental analysis of selected areas of interest.

3. Results and Discussion

3.1. Phase Assemblage in MKPC Pastes

Figure 2 presents XRD spectra of MKPC pastes at 28 d with different doses of ultrafine fly ash and ultrafine slag powder. Regardless of the presence of ultrafine fly ash or ultrafine slag powder, struvite-K was still the primary crystalline hydration product in all MKPC specimens, as shown in the figure. No new crystalline phases were detected as the curing period extended to 28 d, indicating the complete reaction of KH_2PO_4 upon the addition of ultrafine fly ash or ultrafine slag powder. Across all MKPC pastes, abundant unreacted MgO was evident in all test samples, as indicated by XRD spectra. Similarly, the inclusion of ultrafine fly ash and ultrafine slag powder did not lead to the formation of novel crystalline products compared to the control sample.

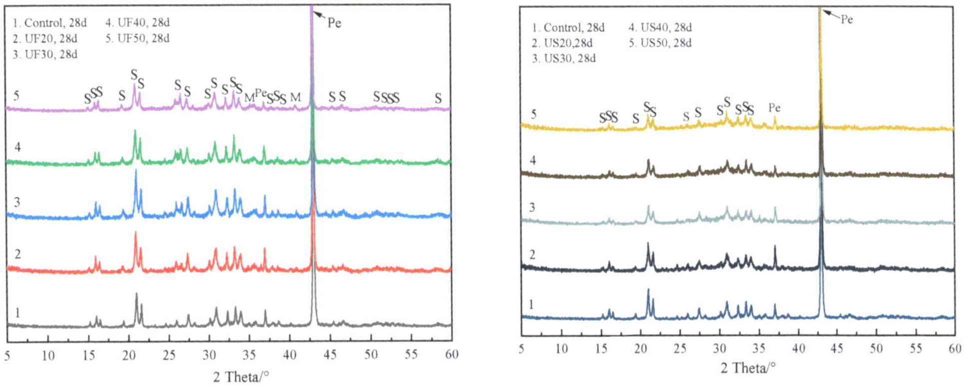

(a) (b)

Figure 2. Powder diffraction patterns for MKPC pastes at 28 days of curing: (**a**) mixed with ultrafine fly ash and (**b**) mixed with ultrafine slag powder. Pe: periclase, S: struvite-K, Q: quartz, M: mullite.

Figure 3 illustrates TG/DSC curves at 28 d for MKPC pastes with additions of ultrafine fly ash or ultrafine slag powder. Endothermic peaks can be observed around 100 °C in all MKPC pastes, attributed to dehydration of hydration products like struvite-K. Within the temperature range of 60–150 °C, MKPC paste mass losses were 14.4 wt%, 14.9 wt%, 13.6 wt%, 12.6 wt%, and 12.2 wt% for UF20, UF30, UF50, US20, and US40, respectively, primarily due to struvite-K dehydration. Consequently, adding 30 wt% ultrafine fly ash slightly increased the struvite-K quantity compared to UF20 paste, but UF50 showed lower mass loss than UF30. Substituting MgO with 30 wt% UFA increased the water for cement reaction owing to dilution, facilitating the reaction somewhat. However, it reduced MgO and KH_2PO_4 quantities, potentially decreasing the overall struvite-K quantity. For UF50, decreased total MgO and KH_2PO_4 quantities might mainly reduce the struvite-K quantity. Unlike in MKPC pastes with ultrafine slag powder, the struvite-K quantity gradually decreased with the increasing dosage. Exothermic peaks around 420 °C can be observed in MKPC pastes with ultrafine fly ash or ultrafine slag powder, albeit with no obvious mass losses (Figure 3a). Further investigation is needed to understand this phenomenon. No endothermic peak indicating KH_2PO_4 dehydration, typically occurring within 200–300 °C, was observed, indicating the absence of KH_2PO_4 in the pastes, consistent with XRD analysis.

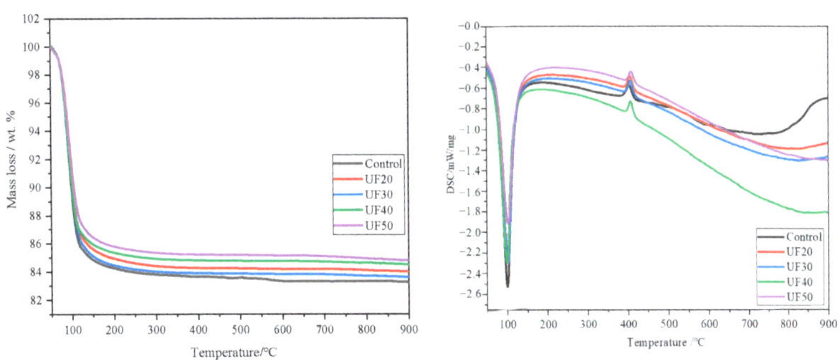

Figure 3. *Cont.*

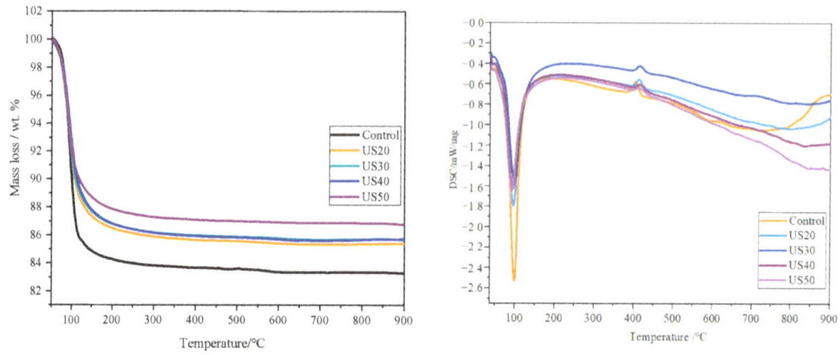

(**a**) (**b**)

Figure 3. TG/DSC curves for MKPC pastes with ultrafine fly ash or ultrafine slag powder (28 d). (**a**) TG curves and (**b**) DSC curves.

3.2. Compressive Strength

Figure 4 illustrates the compressive strengths of MKPC mortars at various time points: 2 h, 24 h, 28 days, and 180 days. The compressive strength of the control mortar was 34.5 MPa at 2 h, rising to 82.1 MPa and 99.6 MPa at 28 days and 180 days, respectively. UF30 displayed notably higher compressive strengths compared to the control mixture at the same age. As shown in Figure 4a, at 180 days, the compressive strengths of the control, UF20, UF30, UF40, and UF50 were 99.6 MPa, 108.7 MPa, 110.9 MPa, 107.4 MPa, and 99.3 MPa, respectively. However, ultrafine slag powder exhibited varied effects on the compressive strengths of MKPC. For example, US20, US30, and US40 showed higher compressive strengths than the control at the early stage of 2 h (Figure 4b), but they did not progress as well as the control at later stages (180 d). The increased addition of ultrafine fly ash in MKPC mortar from 30 wt% to 50 wt% led to a slight decline in compressive strength in the early stages of curing at 2 h, 24 h, and 28 days. Nevertheless, at 180 days, UF50 achieved a compressive strength proximate to the control. Compared to MKPC containing ultrafine slag powder, MKPC containing ultrafine fly ash demonstrated superior strength development overall. For example, the compressive strengths of UF30 at 2 h, 28 days, and 180 days were 39.5 MPa, 96.1 MPa, and 110.9 MPa, respectively, higher than the corresponding 37.6 MPa, 76.5 MPa, and 90.0 MPa of US30. Furthermore, at the higher incorporation rate of 50 wt%, more significant variations were observed between the compressive strengths of MKPC mortar with ultrafine fly ash and that mixed with ultrafine slag powder at 180 days. This indicates that ultrafine fly ash contributes more to the long-term compressive strength development of MKPC mortar compared to ultrafine slag powder.

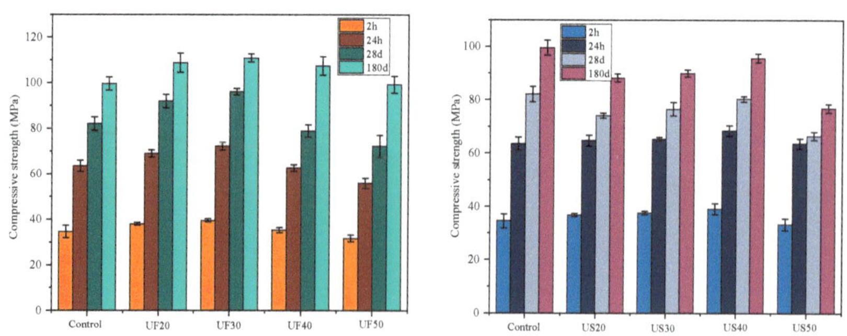

(**a**) (**b**)

Figure 4. Compressive strengths of the MKPC pastes containing different doses of (**a**) ultrafine fly ash and (**b**) ultrafine slag powder.

3.3. Autogenous Shrinkage

In this study, the early volume autogenous shrinkage of MKPC mortar was examined using a Micro-Epsilon displacement sensor from Germany. Figure 5 illustrates the impact of ultrafine fly ash and ultrafine slag powder on the autogenous shrinkage of MKPC mortar. It is evident from Figure 5 that the incorporation of ultrafine fly ash has a positive effect on the early volume autogenous shrinkage of magnesium phosphate cement-based materials. The final volume autogenous shrinkage of the UF30 mortar was 251 με, which represents a 26% reduction compared to the control group mortar at 339 με. More notably, the MKPC mortar with 50 wt% ultrafine fly ash (UF50) exhibited a final volume autogenous shrinkage of 65 με, achieving an 80.8% decrease compared to the final volume autogenous shrinkage of the control group mortar. In contrast, the addition of ultrafine slag powder demonstrated a different effect by increasing the early autogenous shrinkage of the magnesium phosphate cement mortar. For instance, compared to the control group mortar, the final volume autogenous shrinkage of the US20 mortar increased by 348 με. This phenomenon may be associated with the consumption of retarders due to the addition of ultrafine slag powder. In summary, the incorporation of ultrafine fly ash is more beneficial for improving the early volume stability of MKPC cement mortar.

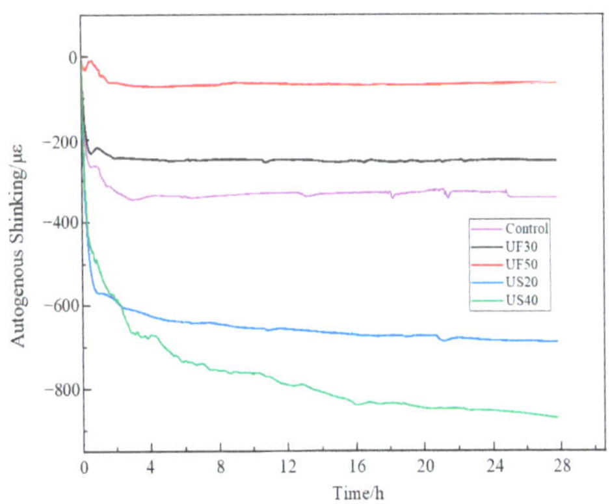

Figure 5. Autogenous shrinkage of MKPC mortar mixed with ultrafine fly ash or ultrafine slag powder.

3.4. Microstructure

3.4.1. Pore Structure Analysis

Figure 6 depicts the pore structures of MKPC pastes with and without the inclusion of ultrafine fly ash/ultrafine slag powder. The addition of 30 wt% ultrafine fly ash (UF30) resulted in a decrease in the volume of pores with diameters ranging from 0.1 to 10 μm at 28 days compared to the control (Figure 6b). However, similar pore structures between UF50 and US40 were observed within the 0.1 to 10 μm range at 28 days. With an increase in the addition of ultrafine fly ash to 50 wt% (UF50), there was a notable increase in the volume of pores with diameters larger than 0.1 μm, leading to a higher total porosity compared to the control and UF30.

The addition of ultrafine slag powder in moderate amounts had a similar effect on the pore microstructure of MKPC as the incorporation of ultrafine fly ash. US20 showed fewer pores with diameters from 0.1 μm to 10 μm and a smaller porosity of 10.7% at 28 days compared to the control. In comparison to MKPC containing ultrafine slag powder, MKPC supplemented with ultrafine fly ash had a visibly denser structure with less porosity, especially at the 30 wt% addition. For example, the total porosity of UF30 was 4.2%, which is less than the total porosity of 10.7% for US20 at 28 days (Figure 6a).

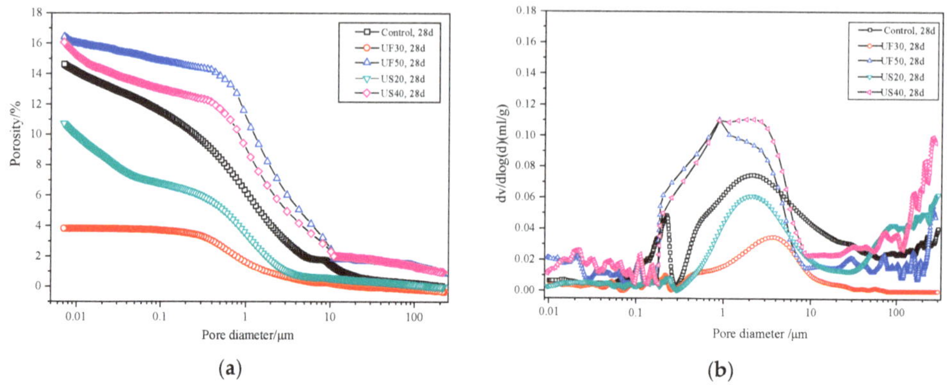

Figure 6. Pore structure analysis plots of MKPC pastes with or without the addition of ultrafine fly ash/ultrafine slag powder: (**a**) porosity and (**b**) pore size distribution.

3.4.2. Morphology

Figure 7 displays the typical struvite-K morphology observed in the control paste. The formation location heavily influences the morphology of struvite-K. In porous and loose regions, struvite-K generally forms sizable tissues (Figure 7a). Large lamellar structures (Figure 7a) and laminated ribbed particles (Figure 7b) are visible within pores, as depicted in Figure 7, consistent with previous findings indicating struvite-K's propensity to grow larger in open spaces like pores [40]. In contrast, in closed regions, irregular but tightly packed crystals emerge (Figure 7c). Ding et al. [34] identified two common forms of struvite-K in solidified paste: crystalline and amorphous phases. MKPC paste reaction products exhibit visible cracks. Previous studies attributed cracks to potential product dehydration under a vacuum [41]. However, Ma et al. [42] proposed that cracks represent gaps between portions of attached struvite-K particles because vacuum drying at room temperature does not dehydrate struvite-K. Sample processing, such as fracturing and vacuum drying, may induce micro-cracks. Furthermore, our research indicates that product gaps appear in large crystals with regular shapes (e.g., as depicted in Figure 7a,b), resembling cracks under SEM observations, especially in BSE mode.

Figure 8 showcases scanning electron microscope pictures of MKPC containing either ultrafine fly ash or ultrafine slag powder. In a manner akin to the control, a number of large lamellar structures formed within a pore in the UF30 sample (Figure 8a). Far from pores, distinct ultrafine fly ash pellets with a spherical shape were embedded in the solidified UF30 paste (Figure 8b). Careful observation of the region delineated by a white square revealed irregular struvite-K formation, contributing to a denser microstructure compared to the control paste. Trace amounts of Si and Al elements were discernible in the EDS spectra at spot P8-1 (Figure 8c), suggesting the integration of Si and Al elements into products. There were also some cracks in the reaction product of the MKPC paste containing ultrafine fly ash, but the number of cracks was significantly lower than in the control sample.

In the MKPC incorporating ultrafine slag powder, depicted in Figure 8d, lumpy reaction products of variable shapes developed surrounding particles of ultrafine slag powder. Based on the EDS energy spectrum at spot P8-2, these reaction phases comprised struvite-K with certain levels of Si and Al (Figure 8d). Likewise, as illustrated in Figure 8e, layered mass crystals formed around particles of ultrafine slag powder, also recognized as struvite-K, yet showing more pronounced peak intensities of Si, Al, and even Ca in the EDS spectrum at hydration spot P8-3. It is worth noting that with the addition of ultrafine fly ash or ultrafine slag powder, the resultant reaction product, struvite-K, exhibited irregular phase shapes, a rough interface, and bulges in comparison with the control. This occurrence could be attributed to the integration of Al and Si elements into the struvite-K crystals or the presence of fresh aluminum silicate potassium phosphate crystals [13].

Figure 9 displays a representative backscatter picture of the reference. The solidified MKPC was composed of struvite-K and numerous unreacted magnesia particles. Large prismatic crystals, demonstrating a large prismatic structure, were commonly located in close proximity to pores or within porous areas (Figure 9b). In the central portion of Figure 9a, larger aggregates of crystals with fissures are noticeable. Struvite-K crystals developed around the columnar magnesium oxide fractions. However, it appears that not all magnesium oxide fractions were effectively bound to reaction products due to noticeable gaps between the original blended cement and the MgO pellets. Several fissures were evident in the paste, aligning with the observations from SEM analyses.

Figure 10 showcases standard BSE images of MKPC pastes incorporating ultrafine slag powder. As depicted in Figure 10a, particles of ultrafine slag powder were distributed across the MKPC paste, leading to a porous microstructure with many cracks, in contrast to UF30. A detailed inspection of the area indicated by a white square unveiled the creation of compact hydration products encasing the ultrafine slag powder particles (Figure 10b). EDS examination at point P10-1 confirmed the existence of Al, Si, and Ca elements in the product, aligning with the SEM analysis.

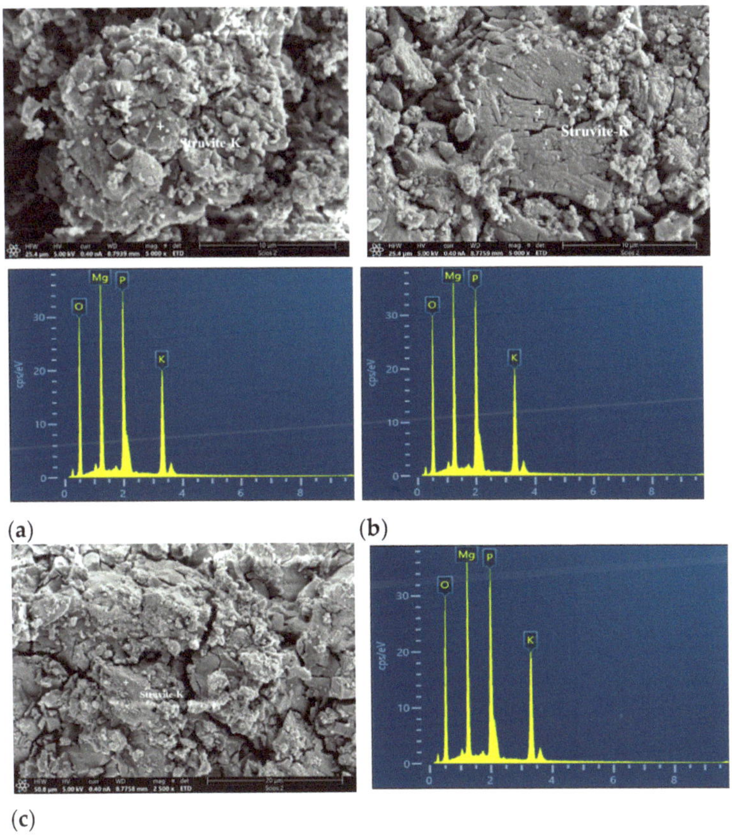

Figure 7. SEM microscopy morphology of the control group at 28 d: (**a**) loosely agglomerated sheeted struvite-K crystals, (**b**) compact columnar struvite-K crystals, (**c**) irregularly shaped struvite-K. Note: "+" is the location of the DES scanning point.

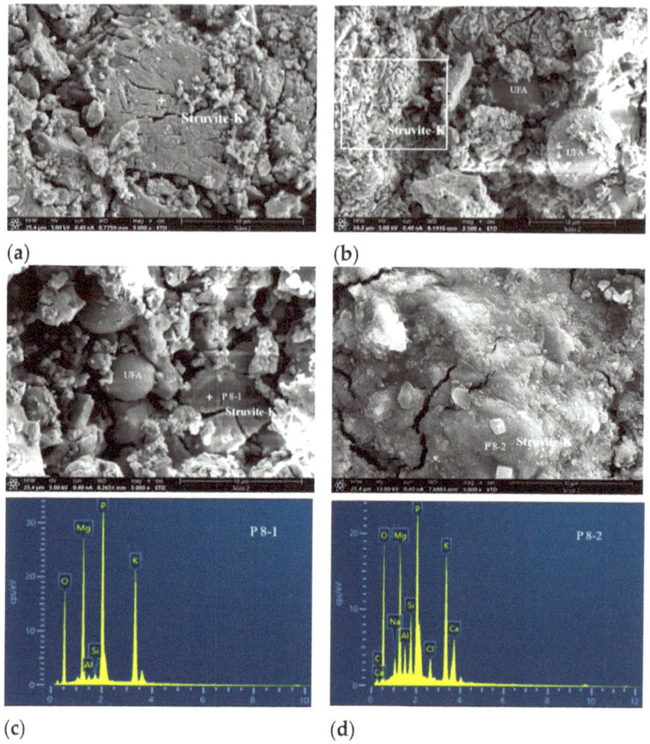

Figure 8. SEM images of MKPC pastes with and without ultrafine fly ash or ultrafine slag powder: (**a**) control, (**b**,**c**) UF30, (**d**) US40. Note: "+" is the location of the DES scanning point, and the white area is the significant accumulation area of struvite-K.

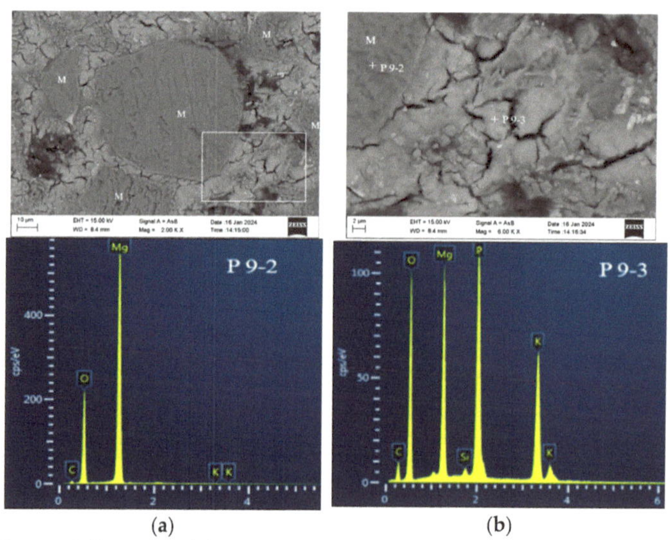

Figure 9. BSE images of the control paste: (**a**) low magnification, (**b**) high magnification, and close observation of the white square-marked area in (**a**). Note: "+" is the position of the DES scanning point, and the white area is the enlarged observation area.

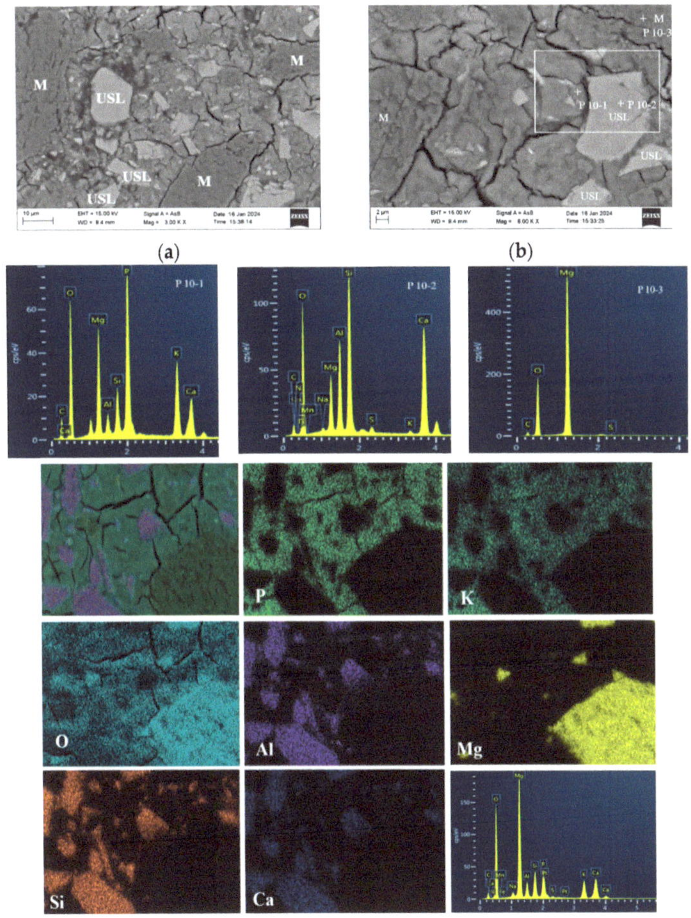

(**a**) (**b**) (**c**)

Figure 10. BSE images of the MKPC with ultrafine slag powder: (**a**) paste morphology, (**b**) paste with higher magnification, and (**c**) backscatter surface scanning images and elemental maps. Note: "+" is the position of DES scanning point, and the white area is the interface area between USL and paste.

Figure 11b exhibits an MKPC paste with 30 wt% ultrafine fly ash at 28 days. Ultrafine fly ash particles were embedded in the UF30 paste matrix, resulting in a denser paste with fewer cracks compared to the control and US20 pastes (Figures 9 and 10). Moreover, distinct hydration product rims encased the ultrafine fly ash particles, establishing a significantly denser surface between the ultrafine fly ash fraction and the original cement (Figure 11a). EDS analysis (Figure 11a) indicated that the hydration products were composed of the elements Mg, O, P, K, Si, and Al, suggesting that silicon and aluminum were integrated in the structure, consistent with the observations in Figure 10. Gardner et al. [13] also noted the integration of Al and Si into the combined substrate of MKPC incorporating FA based on BSEM images and elemental plots. This integration was attributed to the solubilization of the glassy portion of aluminosilicates in the FA, leading to the creation of an aluminosilicate potassium phosphate-binding phase [13]. Secondary electron microscopy demonstrated that struvite-K crystals effectively encapsulated the ultrafine fly ash pellets, indicating favorable tissue compatibility between them [19]. Nonetheless, a number of ultrafine fly ash particles exhibited noticeable gaps at the boundary with struvite-K. Within the UF30 paste, certain MgO particles remained unreacted. As can be observed in Figure 11b, needle-like or bulk phases were interspersed with amorphous struvite-K, creating an intensive blend microstructure. The hybrid phase may represent an aluminosilicate-rich product. This blend structure is presumed to enhance strength.

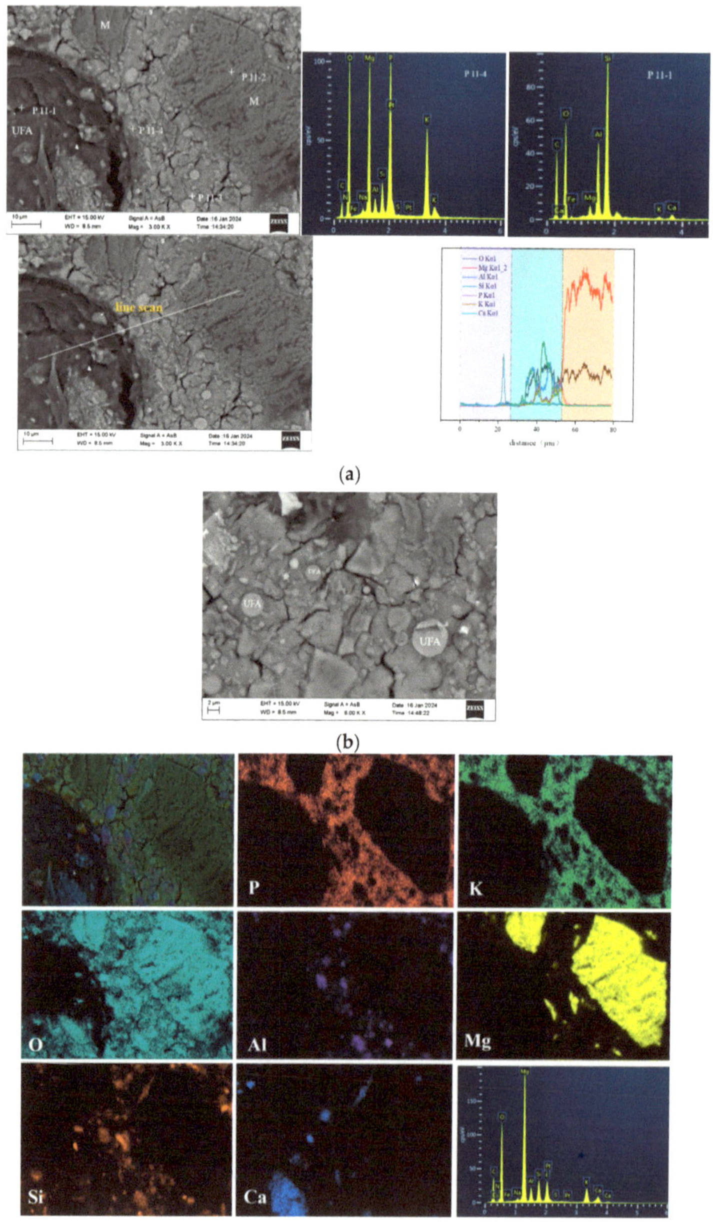

Figure 11. BSE images of the MKPC paste with ultrafine fly ash: (**a**) UF30, (**b**) UF30 with higher magnification, and (**c**) backscatter surface scanning images and elemental maps of UF30. Note: "+" is the position of DES scanning point.

3.5. Discussion

3.5.1. Physicochemical Effects of Ultrafine Fly Ash and Ultrafine Slag Powder in MKPC System

The reactions of ultrafine fly ash and ultrafine slag powder in MKPC have both physical and chemical effects. Physically, the incorporation of ultrafine fly ash or ultrafine slag powder has a diluting impact on the MKPC paste. In this study, the addition of a high volume of ultrafine admixture actually increased the amount of water reacting between MgO and KH_2PO_4 due to the fixed water–cement ratio. Despite incorporating ultrafine fly ash instead of raw materials, the quantities of reaction products, particularly struvite-K, appeared similar or showed only a slight difference (as indicated by XRD and DSC/TG) compared to those of the control. This suggests that MKPC pastes with UFA achieved a great degree of reaction. Given the constant water–cement ratio, the inclusion of UFA or USL may impact pH levels and consequently affect the hydration. According to Rouzic et al. [29], the kind of hydration product in MKPC systems heavily relies on the acidity and alkalinity. For example, newberyite is dominant in a pH range of four to six, a combination of newberyite and struvite-K typically forms at pH six to seven, and struvite-K becomes the primary product when the pH exceeds seven. Furthermore, the dispersion of ultrafine fly ash or slag powder particles within MKPC pastes provides nucleation sites for struvite-K precipitation, thereby influencing the position distribution of reaction products.

Regarding the chemical points, both UFA and USL participate in the reaction process, albeit to varying degrees. When either UFA or USL was added, as observed in X-ray diffraction (XRD) patterns, no other product was formed in MKPC apart from struvite-K. However, scanning electron microscopy (SEM) coupled with energy-dispersive X-ray spectroscopy (EDS) reveals that in the presence of either UFA or USL, a number of Si and Al elements are present in the hydration product, struvite-K. Gardner et al. [26] discovered through microstructure characterization and multi-nuclear NMR spectra that the aluminosilicate glassy fractions of both fly ash and GBFS dissolve within the MKPC binders under near-neutral pH conditions and subsequently form second-reactive species rich in silicon and aluminum. This has the potential to form an aluminosilicate potassium phosphate phase.

In our study, binding products containing Si and Al were also observed in the MKPC pastes with either UFA or USL (Figures 10 and 11). This phenomenon suggests that Si and Al elements are bound to main hydration products. Additionally, amorphous hydration products, such as the potassium aluminosilicate phosphate phase, might form due to the geopolymer synthesis of soluble aluminum silicate species. However, owing to their amorphous character, they cannot be analyzed via X-ray diffraction. Apart from aluminum, it has been reported that other components of FA, particularly magnesium and calcium, compete against magnesium oxide and react with KH_2PO_4, forming phosphate-containing colloids, thereby altering the initially designed M/P molar ratios and potentially enhancing mechanical properties [19]. A fraction of magnesium oxide in FA is anticipated to contribute to the formation of struvite-K [19]. The calcium component of FA also hydrates with KH_2PO_4 to form an amorphous calcium phosphate product [43]. Ma et al. [42] proposed that synergistic effects between GBFS and MKPC might largely be due to the fact that CaO in GBFS reacts with KH_2PO_4 to create calcium phosphate, given that GBFS contains a higher CaO content and exhibits better mechanical properties. Nevertheless, our study appears to contradict this assumption, since ultrafine fly ash has a lower content of calcium oxide but still achieved better mechanical properties.

Alumina oxide demonstrates reactivity and can potentially result in the formation of magnesium alumina phosphate in MPC [44]. As elaborated by Wagh [44], in a phosphate solution, alumina dissolves and reacts with phosphate ions in solution to produce $AlH_3(PO_4)_2{\cdot}H_2O$, which then further reacts with the remaining alumina to yield the ultimate product of $AlPO_4$. It is recognized as an amorphous gel-like tissue that can form under ambient temperatures [44–46] or at elevated temperatures ranging from 50 to 100 °C [47]. From the structural analysis of the products, substituting a fraction of the magnesium atoms with aluminum possibly results in a material resembling a geopolymer within the MPC matrix [48,49]. Ultrafine fly ash demonstrates higher chemical instability compared to alumina, primarily attributable to the presence of its glassy aluminosilicate phase and the structural disruption induced by milling. The envisaged reaction mechanism suggests that, within the acidic or nearly neutral conditions of the MKPC environment, Al could leach from ultrafine fly ash, likely in an aluminosilicate phase. Subsequently, the Al-rich entities could serve as attachment sites or furnish crystallization sites for the deposition of K^+, Mg^{2+}, and PO_4^{3-} ions, culminating in the formation of struvite-K.

The factors affecting the solubilization rate of aluminum silicate in slag, fly ash, and other admixtures containing silica–alumina phases have been reported, including the temperature, alkalinity, crystallinity, dissolution type, and structural defects [50,51]. Due to the high content of the aluminosilicate glass phase in ultrafine fly ash, it exhibits an elevated rate of dissolution in the acidic or nearly neutral milieu of the magnesium potassium phosphate cement (MKPC) system. Within this system, ultrafine fly ash demonstrates enhanced reactivity, potentially resulting in reaction products such as struvite-K containing increased levels of aluminum and silicon. Given the significant role of pH levels, our forthcoming research endeavors will delve into the dissolution behaviors of ultrafine fly ash and slag powder within MKPC systems.

3.5.2. Influence of Ultrafine Fly Ash or Ultrafine Slag Powder on the Microstructure of MKPC

The integration of ultrafine fly ash and ultrafine slag powder into the MKPC matrix forms a ternary composite structure. This structure comprises the unreacted magnesium oxide (MgO), the ultrafine constituents (fly ash/slag powder), and the hydration products, notably amorphous aluminosilicate phosphate and struvite-K. The incorporation of these ultrafine materials modifies the positional distribution of other amorphous reaction products and struvite-K. These products precipitate around the MgO particles and the ultrafine components, fostering a cohesive matrix by connecting these particles. Incorporating 30 wt% of ultrafine fly ash or 20 wt% ultrafine slag powder into the MKPC paste leads to a reduction in pore size and overall porosity, enhancing the material's density. Conversely, increasing the content of these ultrafine materials beyond these proportions to 40 wt% or 50 wt% inversely affects the paste's total porosity, making it more porous. Specifically, comparing compositions with 30 wt% ultrafine fly ash (UFA) and 20 wt% ultrafine slag powder (USL), the paste containing 30 wt% UFA showcases a more compact microstructure compared to the 20 wt% USL variant. This densification correlates with a notable decrease in total porosity, especially in the reduction in pores within the 0.1 μm to 10 μm range, likely due to the enhanced reactivity of ultrafine fly ash within the MKPC pastes. An interesting observation is the decreased incidence of cracks within the MKPC paste matrix following the addition of ultrafine fly ash. This phenomenon could be attributed to the augmented strength and improved interconnectivity among the amorphous reaction products, which contribute to the structural integrity of the paste.

The microstructural characteristics of struvite-K are fundamental to the mechanical performance, transport characteristics, and longevity of magnesium potassium phosphate cement (MKPC). The present study observes that struvite-K within MKPC pastes manifests in two distinct forms: crystalline and amorphous. The occurrence of these forms is contingent upon the specific precipitation sites within the matrix. Crystalline struvite-K is typically precipitated within pores or porous areas, while amorphous struvite-K forms in denser or more constrained spaces. The formation of voids within MKPC pastes is predominantly a result of water evaporation during the reaction process, with higher water contents leading to an increased void volume. Consequently, the water content critically influences the crystalline structure and the morphological traits of the hydration products in magnesium phosphate cement (MPC) mortar. Supporting evidence can be found in the literature [52], where an elevation in water content from 5 wt% to 8 wt% was observed to correlate with the augmented formation of crystalline magnesium ammonium phosphate hexahydrate. It is worth noting that the morphological and structural attributes of the struvite-K crystals are key indicators of the MKPC's physical properties. The delicate balance between crystalline and amorphous phases, driven by the water content in the mix, plays a pivotal role in optimizing the material's attributes for enhanced durability and performance.

Previous studies have reported that the morphological characteristics of struvite-K are influenced by the magnesium–phosphorus ratio, as well as the acidity or alkalinity of the solution [36,53,54]. Specifically, at a magnesium–phosphorus ratio of 3, struvite-K was found to crystallize into a mixture of needle-like and plate-like forms. When the ratio was elevated to 5, the crystalline structure transitioned into a distinctly prismatic configuration. However, at a Mg–P ratio of 10, the resulting struvite-K was characterized by an indistinct shape [36]. Chau et al. [55] observed that a pH value of 6 fostered the formation of struvite-K crystals with a plate-like morphology, while at a pH of 7.5, prismatic crystals were predominant. A converse relationship between pH and crystal size was noted, with smaller struvite-K crystals forming as the pH diminished [56]. However, our study discovered that the addition of UFA or USL promoted the genesis of amorphous products. This suggests that the involvement of silicon and aluminum ions during the deposition process significantly alters the conventional crystalline structure of struvite-K.

3.5.3. Effect of Ultrafine Fly Ash or Ultrafine Slag Powder on the Mechanical Properties and Volume Stability of MKPC

In the case of MKPC, the presence of both unreacted dead-burned magnesium oxide and the hydration product forms a sturdy substrate, contributing to its strength [57]. The unreacted MgO plays the role of an adsorption aggregate within the paste, thereby helping to improve its mechanical properties [57,58]. Considering the excellent surface properties of calcined magnesium oxide, the mechanical strength can be comprehensively improved by increasing proportion of MgO, i.e., increasing the magnesium oxide content in the MPC [36]. Nevertheless, there is an optimal ratio of unreacted MgO "hard core" to struvite "binder" to achieve optimal mechanical properties [4,36]. Beyond this optimal ratio, increasing the initial MgO content may actually decrease the compressive strength because of reduced hydraulicities [36]. It has also been argued that the strength of MKPC pastes is largely derived from the bond strength between unreacted dead-burned magnesium oxide and the hydration product [58]. In our study, there is still an observable excess of unreacted MgO in the MKPC pastes compared to KH_2PO_4 due to its stoichiometry. However, there appeared to be insufficient binding between the MgO and the struvite-K, as a clear gap was observed on the SEM. Similar to the unreacted magnesium oxide, the remaining ultrafine fly ash/ultrafine slag powder can also serve as a filler in the MKPC matrix. Moreover, higher dosages of ultrafine fly ash/ultrafine slag powder in MKPC result in decreased residual magnesium oxide. The ultrafine doped particles are softer than the dead-burned MgO particles, leading to reduced bonding between unreacted MgO and struvite-K crystals and enhanced bonding between ultrafine fly ash/ultrafine slag powder and struvite-K. As indicated by the pronounced hydration rims around the ultrafine fly ash particles, a robust bond appears to form between the ultrafine fly ash and the product, which is believed to positively impact the mechanical properties. At the same time, when ultrafine fly ash is used as a filler in MKPC and there is an early increase in the dosage of MgO and phosphate, the fast reaction bond is reduced and the volume stability is gradually improved. In the right amount, ultrafine slag powder can play a role as a filler to reduce porosity; however, if the dosage continues to increase, the ultrafine slag powder calcium ions and retarder reaction also increase, resulting in early rapid contraction of the MKPC hardening reaction and increased volume shrinkage. With the addition of 50 wt% UFA, the amount of reaction products formed in the MKPC pastes was not sufficient to effectively bind all of the UFA particles to the unreacted MgO. Hence, the added 50 wt% UFA slightly reduced the strength of the MKPC pastes, especially at early ages.

Densification of the cement paste has a great influence on the mechanical properties of MKPC. Introducing 30 wt% ultrafine fly ash (UFA) and 20 wt% ultrafine slag (USL) reduces the overall porosity, yet it has varying effects on the strengths of MKPC paste over time. Specifically, adding 20 wt% ultrafine slag powder slightly reduces the compressive strength, while incorporating 30 wt% ultrafine fly ash substantially enhances the MKPC paste's compressive strength. However, further increasing the addition of ultrafine slag powder/ultrafine fly ash to 40 wt%/50 wt% raises the total porosity of the MKPC paste, leading to significant decreases in compressive strengths. This suggests that factors other than microstructure (pore structure) can affect the mechanical properties of MKPC. It is possible that the glass-like aluminosilicate fractions in UFA react in the aqueous phosphoric acid solution of the MKPC–UFA system, leading to the formation of secondary amorphous products with silico-phosphates [20]. These emergent products amalgamate with struvite-K, thereby enhancing the integrity of the entire hardened body and consequently augmenting its mechanical robustness. Additionally, the morphological attributes of struvite-K may contribute to this phenomenon. The development of a cohesive structure, facilitated by the entanglement and linkage of struvite-K particles, is pivotal for the augmentation of mechanical properties [57]. Higher MKPC specimen strengths are also typically associated with improved crack resistance. First, the high-strength cement design results in a denser, more homogeneous paste matrix that provides better crack resistance. This dense matrix reduces the permeability of the cement sample, limiting the entry of water, chemicals, and other harmful substances that can cause cracking over time. In addition, the MKPC hydration process is an important factor in strength development and crack resistance enhancement. During the cement mixing process, a properly designed MKPC specimen promotes repeated hydration of the internal particles, resulting in a more cohesive and durable slurry that is less likely to crack. Tchakouté [59] synthesized phosphate metakaolin-based geopolymer cementitious material in a laboratory setting, achieving a strength of 54 MPa at 28 d post-formation, by initiating a reaction between metakaolin and a phosphoric acid solution. Their findings indicate that the dispersion of berlinite within the matrix, serving to fortify the matrix, was a primary factor behind the observed strength enhancement. Although such berlinite inclusions were not detected in the MKPC paste incorporating UFA, the formation of a fortified hybrid microstructure was nonetheless observed (Figure 11a). Moreover, the intrinsic physical properties of struvite-K may also influence the overall strength of MKPC pastes, although this aspect has not been explored within the current research. In summary, the long-term compressive strength of MKPC mixed with ultrafine fly ash is higher than that of MKPC mixed with ultrafine slag powder observed after a long curing time. This is mainly attributed to the activation degree and structure of ultrafine fly ash in the MKPC system, which causes (I) the aluminum and silicon elements present in the struvite-K to contribute to the formation of adhesive and reinforced structures, (II) the formation of excess secondary amorphous aluminosilicate phosphate products beyond struvite-K, and (III) denser interfacial structures that are formed between ultrafine fly ash particles and crystals.

4. Conclusions

The effects of ultrafine fly ash and ultrafine slag powder on the microstructure and properties of MKPC were studied. From our experimental work, the following main conclusions can be drawn:

1. Whether ultrafine fly ash (UFA) or ultrafine slag powder (USL) is added, the primary hydration phase formed in MKPC paste is still struvite-K. Both ultrafine fly ash (UFA) and ultrafine slag powder (USL) contribute aluminosilicate fractions to the acid–base hydration reactions of MKPC. Consequently, aluminum and silicon are integrated into the struvite-K crystals in the cured MKPC pastes containing either of the two ultrafine mineral admixtures. However, in the MKPC paste containing ultrafine slag powder, Al, Si, and even Ca elements are present in the resulting struvite-K product.

2. The addition of ultrafine slag powder initially increases the compressive strength of the MKPC paste at early ages, but it does not perform as well as the control paste at later ages. In contrast, the positive effect of ultrafine fly ash on the strength of MKPC is clearly observed at later testing ages. At 28 d of curing, the compressive strength of MKPC containing 30 wt% ultrafine fly ash was increased by 17.1% compared to the control group. Notably, at 180 days of curing, the MKPC mortar with 50 wt% ultrafine fly ash achieved a compressive strength similar to that of the control paste. Overall, the MKPC phases with ultrafine fly ash show better compressive strength development compared to those with ultrafine slag powder. The incorporation of 30 wt% ultrafine fly ash enhances compressive strengths across all testing ages. This is due to the formation of a stronger microstructure in MKPC mixed with ultrafine fly ash and denser interfaces between the hydration products and ultrafine fly ash particles.

3. Regardless of the addition of ultrafine fly ash or ultrafine slag powder, the MKPC mortar presents the phenomenon of autogenous shrinkage under the rapid reaction of the paste in the early stage. The addition of ultrafine fly ash effectively improves the autogenous shrinkage of MKPC mortar in the early stage, while the addition of ultrafine slag powder significantly increases the early autogenous shrinkage of MKPC. It is noteworthy that the final volumetric autogenous shrinkage of MKPC mortar with 50 wt% UFA addition was reduced by as much as 80.8% compared to the control mortar. The German Micro-Epsilon displacement sensor technology is not only applicable to the displacement measurement of cement and other cementitious materials but also can be applied to sensing the deformation of products in a fixed space in order to improve the quality and safety of actual products; however, there are some limitations, such as the sensor accuracy, environmental adaptability, and cost constraints. Further research can be devoted to addressing these limitations to further advance the development and application of the technology.

4. The addition of 30 wt% ultrafine fly ash and 20 wt% ultrafine slag powder effectively refines the paste structure of MKPC, reducing the total porosity, particularly in the pore volume within the diameter range of 0.1~10 μm, thereby enhancing internal bonding compactness. For example, the MKPC with the addition of 20 wt% ultrafine slag powder had fewer pores with pore sizes from 0.1~10 μm at 28 d compared to the control, with a porosity of 10.7%. In comparison to MKPC containing ultrafine slag powder, MKPC supplemented with ultrafine fly ash had a visibly denser structure with less porosity, especially in the 30 wt% addition. The total porosity of UF30 is 4.2%, which is less than the total porosity of 10.7% for US20 at 28 days. However, the total porosity of MKPC also increases with the elevated dosage of ultrafine admixture. Moreover, with the addition of an appropriate quantity of ultrafine admixture, particularly ultrafine fly ash, amorphous aluminosilicate phosphate products intercalate with struvite, promoting the development of long-term strength properties in MKPC.

Author Contributions: Z.J.: Methodology, experiment, data analysis, writing—original draft, conceptualization, verification, writing—review and editing. Y.Z.: Experiment, data analysis, writing—review and editing, verification. L.M.: Conceptualization, resources, writing—review and editing, supervision, funding acquisition, investigation. All authors have read and agreed to the published version of the manuscript.

Funding: This research was funded by National Natural Science Foundation of China (52272018), State Key Laboratory of Materials-Oriented Chemical Engineering (No. SKL-MCE-22A07), National Key R&D Program of China (2021YFB3802002), and a project funded by the Priority Academic Program Development (PAPD) of Jiangsu Higher Education Institutions.

Institutional Review Board Statement: Not applicable.

Informed Consent Statement: Not applicable.

Data Availability Statement: The original contributions presented in the study are included in the article, further inquiries can be directed to the corresponding author.

Conflicts of Interest: The authors declare that they have no conflicts of interest in connection with this work. We also declare that we do not have any commercial or associative interest that represents a conflict of interest in relation to the work submitted.

References

1. Wagh, A.S. (Ed.) Chapter 9—Magnesium Phosphate Ceramics. In *Chemically Bonded Phosphate Ceramics*, 2nd ed.; Elsevier: Amsterdam, The Netherlands, 2016; pp. 115–131.
2. Ma, H.; Xu, B.; Li, Z. Magnesium potassium phosphate cement paste: Degree of reaction, porosity and pore structure. *Cem. Concr. Res.* **2014**, *65*, 96–104. [CrossRef]
3. Ma, H.; Xu, B. Potential to design magnesium potassium phosphate cement paste based on an optimal MAGNESIA-to-phosphate ratio. *Mater. Des.* **2017**, *118*, 81–88. [CrossRef]
4. Qiao, F.; Chau, C.K.; Li, Z. Property evaluation of magnesium phosphate cement mortar as patch repair material, Constr. *Build. Mater.* **2010**, *24*, 695–700. [CrossRef]
5. Yang, J.; Tang, Q.; Wu, Q.; Li, X.; Sun, Z. The effect of seawater curing on properties of magnesium potassium phosphate cement. *Constr. Build. Mater.* **2017**, *141*, 470–478. [CrossRef]
6. Su, Y.; Yang, J.; Liu, D.; Zhen, S.; Lin, N.; Zhou, Y. Effects of municipal solid waste incineration fly ash on solidification/stabilization of Cd and Pb by magnesium potassium phosphate cement. *Biochem. Pharmacol.* **2016**, *4*, 259–265. [CrossRef]
7. Yang, Q.; Wu, X. Factors influencing properties of phosphate cement-based binder for rapid repair of concrete. *Cem. Concr. Res.* **1999**, *29*, 389–396. [CrossRef]
8. Yang, Q.; Zhu, B.; Wu, X. Characteristics and durability test of magnesium phosphate cement-based material for rapid repair of concrete. *Mater. Struct.* **2000**, *33*, 229–234. [CrossRef]
9. Yang, Q.; Zhu, B.; Zhang, S.; Wu, X. Properties and applications of magnesia-phosphate cement mortar for rapid repair of concrete. *Cem. Concr. Res.* **2000**, *30*, 1807–1813. [CrossRef]
10. Yang, Q.; Zhang, S.; Wu, X. Deicer-scaling resistance of phosphate cement-based binder for rapid repair of concrete. *Cem. Concr. Res.* **2002**, *32*, 165–168. [CrossRef]
11. Covill, A.; Hyatt, N.; Hill, J. Development of magnesium phosphate cements for encapsulation of radioactive waste. *Adv. Appl. Ceram.* **2011**, *110*, 151–156. [CrossRef]
12. Wu, F.; Wei, J.; Guo, H.; Chen, F.; Hong, H.; Liu, C. Self-setting bioactive calcium magnesium phosphate cement with high strength and degradability for bone regeneration. *Acta Biomater.* **2008**, *4*, 1873–1884. [CrossRef] [PubMed]
13. Hu, X.; Tao, R.; Xin, Y. The effect of aluminum-silicon glass phase in fly ash on the microstructure and properties of magnesium phosphate cement. *Constr. Build. Mater.* **2024**, *416*, 135159. [CrossRef]
14. Liao, W.; Ma, H.; Sun, H.; Huang, Y.; Wang, Y. Potential large-volume beneficial use of low-grade fly ash in magnesia-phosphate cement based materials. *Fuel* **2017**, *209*, 490–497. [CrossRef]
15. Lu, X.; Chen, B. Experimental study of magnesium phosphate cements modified by metakaolin. *Constr. Build. Mater.* **2016**, *123*, 719–726. [CrossRef]
16. Zheng, D.; Ji, T.; Wang, C.; Sun, C.; Lin, X.; Muhammed, K.; Hossain, A. Effect of the combination of fly ash and silica fume on water resistance of magnesium-potassium phosphate cement. *Constr. Build. Mater.* **2016**, *106*, 415–421. [CrossRef]
17. Li, Y.; Shi, T.; Chen, B.; Li, Y. Performance of magnesium phosphate cement at elevated temperatures. *Constr. Build. Mater.* **2015**, *91*, 126–132. [CrossRef]
18. Joshi, S.S.; Kuntal, V.S.; Bolander, J.E.; Nagai, K. Reproducible estimations of internal corrosion distribution from surface cracks using MPC-RBSM. *Eng. Fract. Mech.* **2023**, *292*, 109–113. [CrossRef]
19. Xu, B.; Ma, H.; Shao, H.; Li, Z.; Lothenbach, B. Influence of fly ash on compressive strength and micro-characteristics of magnesium potassium phosphate cement mortars. *Cem. Concr. Res.* **2017**, *99*, 86–94. [CrossRef]
20. Wilson, A.; Nicholson, J. *Acid–Base Cements: Their Biomedical and Industrial Applications*; Cambridge University Press: Cambridge, UK, 1993.
21. Yang, Y.; Liu, J.; Wang, B.; Liu, R. Properties of fly ash blended magnesium potassium phosphate cement cured in presence of sulfuric acid. *Constr. Build. Mater.* **2020**, *244*, 118349. [CrossRef]
22. Yang, J.M.; Wang, L.M.; Cheng, J.; Dong, S. Effect of fly ash on the corrosion resistance of magnesium potassium phosphate cement paste in sulfate solution. *Constr. Build. Mater.* **2020**, *237*, 117639.
23. Li, Y.; Shi, T.; Li, J. Effects of fly ash and quartz sand on water-resistance and salt-resistance of magnesium phosphate cement. *Constr. Build. Mater.* **2016**, *105*, 384–390. [CrossRef]
24. Xu, B.W.; Lothenbach, B.; Ma, H.Y. Properties of fly ash blended magnesium potassium phosphate mortars: Effect of the ratio between fly ash and magnesia. *Cem. Concr. Compos.* **2018**, *90*, 169–177. [CrossRef]
25. Mo, L.; Lv, L.; Deng, M.; Qian, J. Influence of fly ash and metakaolin on the microstructure and compressive strength of magnesium potassium phosphate cement paste. *Cem. Concr. Res.* **2018**, *111*, 116–129. [CrossRef]
26. Gardner, L.J.; Bernal, S.A.; Walling, S.A.; Corkhill, C.L.; Provis, J.L.; Hyatt, N.C. Characterisation of magnesium potassium phosphate cements blended with fly ash and ground granulated blast furnace slag. *Cem. Concr. Res.* **2015**, *74*, 78–87. [CrossRef]

27. Zhou, S.; Li, Y.; Yin, J.; Gao, Y. Research on properties of ultra fine fly ash. *J. Chin. Ceram. Soc.* **2003**, *31*, 513–516.
28. Ferdosian, I.; Camoes, A. Eco-efficient ultra-high performance concrete development by means of response surface methodology. *Cem. Concr. Compos.* **2017**, *84*, 146–156. [CrossRef]
29. Shi, C.; Wu, Z.; Xiao, J.; Wang, D.; Huang, Z.; Fang, Z. A review on ultra high-performance concrete: Part I. Raw materials and mixture design. *Constr. Build. Mater.* **2015**, *101*, 741–751. [CrossRef]
30. Supit, S.W.M.; Shaikh, F.U.A.; Sarker, P.K. Effect of ultrafine fly ash on mechanical properties of high-volume fly ash mortar. *Constr. Build. Mater.* **2014**, *51*, 278–286. [CrossRef]
31. Obla, K.H.; Hill, R.L.; Thomas, M.D.A.; Shashiprakash, S.G.; Perebatova, O. Properties of concrete containing ultra-fine fly ash. *ACI Mater. J.* **2003**, *100*, 426–433.
32. Gao, Y.L.; Ma, B.G.; Zhou, S.Q. Production and engineering application ofC60 high-performance pump pebble concrete containing ultra-fine fly ash. *Can. J. Civ. Eng.* **2008**, *35*, 757–763. [CrossRef]
33. Krishnaraj, L.; Ravichandiran, P.T.; Annadurai, R.; Rajkumar, P.K. Study on micro structural behavior and strength characteristics of ultra-fine fly ash as a secondary cementitious material with Portland cement. *Int. J. ChemTech Res.* **2015**, *7*, 555–563.
34. Deb, P.S.; Sarker, P.K. Effects of ultrafine fly ash on setting, strength, and porosity of geopolymers cured at room temperature. *J. Mater. Civ. Eng.* **2017**, *29*, 0001745. [CrossRef]
35. *T/CBMF194*; Ultrafine Compound Mineral Admixtures. China Building Materials Industry Press: Beijing, China, 2022.
36. Rouzic, M.L.; Chaussadent, T.; Stefan, L.; Saillio, M. On the influence of Mg/P ratio on the properties and durability of magnesium potassium phosphate cement pastes. *Cem. Concr. Res.* **2017**, *96*, 27–41. [CrossRef]
37. Qiao, F. Reaction Mechanisms of Magnesium Potassium Phosphate Cement and Its Application. Ph.D. Thesis, Hong Kong University of Science and Technology, Hong Kong, 2010.
38. Li, Y.; Sun, J.; Chen, B. Experimental study of magnesia and M/P ratio influencing properties of magnesium phosphate cement. *Constr. Build. Mater.* **2014**, *65*, 177–183. [CrossRef]
39. Wang, A.; Yuan, Z.; Zhang, J.; Liu, L.; Li, J.; Liu, Z. Effect of raw material ratios on the compressive strength of magnesium potassium phosphate chemically bonded ceramics. *Mater. Sci. Eng. C* **2013**, *33*, 5058–5063. [CrossRef] [PubMed]
40. Chong, L.; Yang, J.; Shi, C. Effect of curing regime on water resistance of magnesium–potassium phosphate cement. *Constr. Build. Mater.* **2017**, *151*, 43–51. [CrossRef]
41. Gardner, L.J.; Bernal, S.A.; Walling, S.A.; Corkhill, C.L.; Provis, J.L.; Hyatt, N.C. Response to the discussion by Hongyan Ma and Ying Li of the paper 'Characterization of magnesium potassium phosphate cement blended with fly ash and ground granulated blast furnace slag'. *Cem. Concr. Res.* **2018**, *103*, 249–253. [CrossRef]
42. Ma, H.; Li, Y. Discussion of the paper 'Characterisation of magnesium potassium phosphate cement blended with fly ash and ground granulated blast furnace slag' by L.J. Gardner et al. *Cem. Concr. Res.* **2018**, *103*, 245–248. [CrossRef]
43. Mahyar, M. Phosphate-activated High-calcium Fly Ash Acid-base Cements. *Cem. Concr. Compos.* **2015**, *63*, 96–103. [CrossRef]
44. Wagh, A.S. *Chemically Bonded Phosphate Ceramics*, 2nd ed.; Elsevier: Amsterdam, The Netherlands, 2016.
45. Sugama, T.; Carciello, N. Hydrothermally synthesized aluminium phosphate cements. *Adv. Cem. Res.* **1993**, *5*, 31–40. [CrossRef]
46. Douiri, H.; Louati, S.; Baklouti, S.; Arous, M.; Fakhfakh, Z. Structural, thermal and dielectric properties of phosphoric acid-based geopolymers with different amounts of H_3PO_4. *Mater. Lett.* **2014**, *116*, 9–12. [CrossRef]
47. Roncal-Herrero, T.; Rodríguez-Blanco, J.D.; Benning, L.G.; Oelkers, E.H. Precipitation of iron and aluminum phosphates directly from aqueous solution as a function of temperature from 50 to 200 °C. *Cryst. Growth Des.* **2009**, *9*, 5197–5205. [CrossRef]
48. Fan, S.; Chen, B. Experimental research of water stability of magnesium alumina phosphate cements mortar. *Constr. Build. Mater.* **2015**, *94*, 164–171. [CrossRef]
49. Liu, N.; Chen, B. Experimental research on magnesium phosphate cements containing alumina. *Constr. Build. Mater.* **2016**, *121*, 354–360. [CrossRef]
50. Panagiotopoulou, C.; Kontori, E.; Perraki, T.; Kakali, G. Dissolution of aluminosilicate minerals and by-products in alkaline media. *J. Mater. Sci.* **2007**, *42*, 2967–2973. [CrossRef]
51. Provis, J.; Van Deventer, J. Geopolymerisation kinetics: 2. Reaction kinetic modelling. *Chem. Eng. Sci.* **2007**, *62*, 2318–2329. [CrossRef]
52. Hall, D.A.; Stevens, R.; El Jazairi, B. Effect of water content on the structure and mechanical properties of magnesia–phosphate cement mortar. *J. Am. Ceram. Soc.* **1998**, *56*, 1550–1556. [CrossRef]
53. Ronteltap, M.; Maurer, M.; Hausherr, R.; Gujer, W. Struvite precipitation from urine influencing factors on particle size. *Water Res.* **2010**, *44*, 2038–2046. [CrossRef]
54. Abbona, F.; Boistelle, R. Growth morphology and crystal habit of struvite crystals. *J. Cryst. Growth* **1979**, *46*, 339–354. [CrossRef]
55. Chau, C.K.; Vyas, P.M.; Joshi, M.J. Growth and characterization of struvite-K crystals. *Cryst. Res. Technol.* **2011**, *46*, 187–194.
56. Zhang, X.; Hu, J.; Spanjers, H.; van Lier, J.B. Struvite crystallization under a marine/brackish aquaculture condition. *Bioresour. Technol.* **2016**, *218*, 1151–1156. [CrossRef] [PubMed]
57. Ding, Z.; Dong, B.; Xing, F.; Han, N.; Li, Z. Cementing mechanism of potassium phosphate based magnesium phosphate cement. *Ceram. Int.* **2012**, *38*, 6281–6288. [CrossRef]

58. Stratful, I.; Scrimshaw, M.; Lester, J. Conditions influencing the precipitation of magnesium ammonium phosphate. *Water Res.* **2001**, *35*, 4191–4199. [CrossRef] [PubMed]
59. Tchakouté, H.K.; Rüscher, C.H.; Kamseu, E.; Djobo, J.; Leonelli, C. The influence of gibbsite in kaolin and the formation of berlinite on the properties of metakaolin-phosphate-based geopolymer cements. *Mater. Chem. Phys.* **2017**, *199*, 280–288. [CrossRef]

materials

MDPI

Article

Durability of Prestressed Piles in a Leachate Environment

Yu Wang [1], Min Deng [1,*], Rihong Zhang [2,3], Xuming Yu [4], Junzhong Xue [4] and Jing Zhang [5]

[1] College of Materials Science and Engineering, Nanjing Tech University, Nanjing 211800, China; 202262103005@njtech.edu.cn
[2] Faculty of Architectural, Civil Engineering and Environment, Ningbo University, Ningbo 315000, China; zhangrh@zcone.com.cn
[3] Ningbo ZCONE High-Tech Holdings Co. Ltd., Ningbo 315000, China
[4] CGN Solar Energy (Jiaxing) Co., Ltd., Jiaxing 314000, China; xl.xujia@163.com (X.Y.); xue_jz@126.com (J.X.)
[5] SGIDI Engineering Consulting (Group) Co., Ltd., Shanghai 200093, China; zhangjing@sgidi.com
* Correspondence: dengmin@njtech.edu.cn

Abstract: Prestressed pipe piles are common concrete components characterized by dense concrete structures and favorable mechanical properties, and thus, extensively used as coastal soft soil foundations. However, their durability in harsh environments has not been fully clarified. In this study, leachate from an actual landfill site was collected from the east coast of China as the corrosive medium, and the corrosion process was accelerated by electrifying prestressed pipe piles. The results demonstrated that the concentration of chloride ions in the concrete of the prestressed pile increased with the increase in corrosion time. Moreover, the experimental corrosion of these prestressed piles in the drying–wetting cycle proved to be the most severe. However, a protective layer of epoxy resin coating can effectively inhibit the diffusion of chloride ions into the interior of the piles. The final theoretical corrosion amounts of the piles were 1.55 kg, 1.20 kg, and 1.64 kg under immersion, epoxy resin protection, and a drying–wetting cycle environment. The application of epoxy resin reduced chloride penetration by 22.6%, and the drying–wetting cycle increased chloride penetration by 5.8%, respectively, with corresponding corrosion potentials following similar patterns. The actual corrosion depth of the welding seam was 3.20 mm, and there was a large corrosion allowance compared with the requirement (6.53 mm) for the ultimate bending moment. In summary, these prestressed piles exhibited good durability in a leachate environment.

Keywords: prestressed piles; harsh environments; leachate; durability

Citation: Wang, Y.; Deng, M.; Zhang, R.; Yu, X.; Xue, J.; Zhang, J. Durability of Prestressed Piles in a Leachate Environment. *Materials* **2024**, *17*, 2497. https://doi.org/10.3390/ma17112497

Academic Editor: Yuri Ribakov

Received: 27 March 2024
Revised: 24 April 2024
Accepted: 13 May 2024
Published: 22 May 2024

1. Introduction

With the acceleration of urbanization in China, the demand for land resources has increased dramatically, and an increasing number of buildings have to be constructed in harsh landfill environments [1,2]. As a common category of foundations, prestressed concrete pipe piles are inevitably used in landfills. In landfills, domestic and industrial wastes can form leachate under the combined action of rainwater and surface runoff. The leachate accumulates in the soil layer of landfills and is not easily absorbed. Further, the underground soil of landfills fills with leachate in the long run. Landfill leachate constitutes a weak alkaline solution with a complex composition and many hazardous ions, and the corrosive media include Cl^-, SO_4^{2-}, NH_4^+, and organic biomass [3]. Due to the inherent pores and microcracks in the concrete, these corrosive media may penetrate the interior of reinforced concrete materials, which would affect the durability of the concrete, thus reducing the service life of concrete structures [4–6].

Compared with ordinary concrete structures, prestressed concrete piles require a series of special processes, such as prestressing [7], centrifugation [8,9], and steam curing [10–12]. They have a higher density, lower porosity, and greater bending capacity [13,14]. Special production processes and hollow shapes contribute to the superiority of these piles over ordinary reinforced concrete in terms of microstructures and mechanical properties. In particular, prestressed high-performance concrete (PHC) piles are generally considered to have favorable durability.

In recent years, scholars have conducted in-depth research on the durability of prestressed concrete piles. In 1965 and 1970, a Japanese company constructed two factory buildings, A and B, using prestressed concrete (PC) piles as pile foundations. The durability of relevant units in the factory buildings and the PC piles in the foundation was investigated 32–37 years after operation [15]. It was found that the compressive strength of the concrete of these piles was still higher than the design requirement. The elastic modulus was higher than 4×10^4 MPa, the maximum neutralization depth was 1 mm, and the tensile strength of the steel reinforcement inside the pile was not significantly reduced. Japan's Kao Group conducted a field exposure test on the long-term durability of PC piles in 1968 [16–18]. The pile concrete was mixed with calcium chloride equivalent to 0.05%, 0.5%, 1.0%, 2.0%, and 4.0% of the cement mass. After 32 years, the PC steel bar elongation of the test piles with 4.0% of calcium chloride added was less than 1%; after 50 years, the PC steel bar of the test piles with 4.0% of calcium chloride added fractured. No obvious rust was found on the PC steel bars of the rest of the test piles, and the depth of neutralization was less than 2 mm. Zhan et al. [19] investigated the effects of maintenance and admixture on the durability of prestressed piles. They found that the prefabrication of the prestressed piles with a polycarboxylic water reducer and 8 h steam curing provided the highest cost-effectiveness in terms of durability, with an electrical flux of 85 °C and a chloride ion migration coefficient of 0.54×10^{-12} m^2/s. Liu et al. [20] simulated pipe pile joints immersed in 5% NaCl solution for 50 years through energized acceleration. They revealed that the seams of the joints did not crack after deterioration, and that the welding seams of the welded square pile joints rusted mildly after deterioration. Li et al. [21] predicted effects on pipe piles under a chlorine ion environment by solving the diffusion equations of the chlorine ions in the layered structure of the piles. The results confirmed that increasing the thickness of the mortar layer shortened the diffusion and corrosion periods, thereby significantly reducing the service life. Increasing the thickness of protective layers and sealing pile ends were identified as effective measures to extend the service life of prestressed concrete piles.

In underground environments, dry and wet cycles often occur, and leachate environments are no exception. In recent years, scholars have studied the effects of dry and wet cycling on concrete corrosion and concluded that dry and wet coupling cycles and the diffusion of chloride ions increased the development of cracks and pores within the concrete structure and accelerated the transport of the chloride ions through the concrete.

Epoxy resin is one of the polymer materials with the best chemical resistance, the best adhesion with all kinds of concrete, no volatile by-products from curing, small shrinkage from curing, and excellent anti-corrosion performance. Epoxy resin coating can be used for the corrosion protection of prestressed pipe piles.

In summary, prestressed piles exhibit favorable long-term durability performance in ordinary environments or mildly corrosive environments. However, there is a lack of research on the deterioration process of prestressed piles in harsh environments, such as the waste leachate field. The issue of the durability of precast piles in leachate environments is closely related to the longevity of buildings in leachate environments, and the present study is undoubtedly pioneering as leachate environments are unique environments that have emerged in the course of modern urbanization. In this study, the durability of prestressed piles in the landfill leach field was evaluated by combining the energized accelerated test with the actual situation of the project. These findings may provide a theoretical basis for the protection and application of prestressed concrete piles in landfill environments.

2. Materials and Methods

2.1. Raw Materials

In this test, P-II 52.5 silicate cement produced by Onada Cement Corp (Nanjing, China) was used. The specific surface area was 366 m^2/kg, the initial setting time was 163 min, and the final setting time was 220 min. The slag powder was a by-product of granulated blast furnace production from Ningbo Ziheng Building Materials Technology (Ningbo, China). The ratios of cement to slag powder are shown in Table 1. The coarse aggregate used was continuously graded gravel produced by the Ningbo Guoding Mining Industry (Ningbo, China), with the size ranging from 5 mm to 32 mm. The fine aggregate used was mechanism sand produced by the Ningbo Guoding Mining Industry, with a fineness modulus of 2.8. The Point-C300 high-efficiency concrete water-reducing agent produced by Kozijie New Materials (Nanjing, China) was selected as the water-reducing agent.

Table 1. Chemical composition of cement and slag powder (%).

Chemical Composition	SiO_2	Al_2O_3	Fe_2O_3	CaO	MgO	SO_3	R_2O	L.O.I.
Cement	19.40	4.73	2.98	64.04	2.30	2.58	0.68	2.80
Slag Powder	38.27	8.41	1.58	42.65	7.40	0.14	0.45	1.10

The diameter of the prestressed reinforcement was 10.7 mm, and the tensile strength and elongation after the fracture of the prestressed reinforcement were 1490 MPa and 8.8%. The spiral stirrup was a cold-drawn mild steel wire with a diameter of 5.0 mm, and the tensile strength and elongation after the fracture of the spiral stirrup were 620 MPa and 4.5%.

The leachate was extracted from a landfill site on the eastern coast of China and had a brown-black color and a strong, irritating odor. The pH value was 7.9, the ammonium nitrogen content was 2.55 g/L, the sulfate content was 7.43 mg/L, and the chloride content was 3.14 g/L.

2.2. *Prestressed Piles and the Mix Ratio of Concrete*

The prestressed concrete pipe was PHC 400(95) AB, using grade C80 concrete mix, and its strength was 84.1 MPa on day 28. The outer diameter of the pipe pile was 400 mm, the thickness of the protective layer was 41 mm, and seven prestressed reinforcements were uniformly arranged.

The concrete mixing ratio is shown in Table 2, and its quality meets the requirements of the Zhejiang Provincial Atlas G22 (*Pre-tensioned Prestressed Concrete Pipe Pile*) [22]. The reinforcement of the test pile is illustrated in Figure 1, with a pitch spacing of 80 mm for the helical reinforcement and no densified zone at either end.

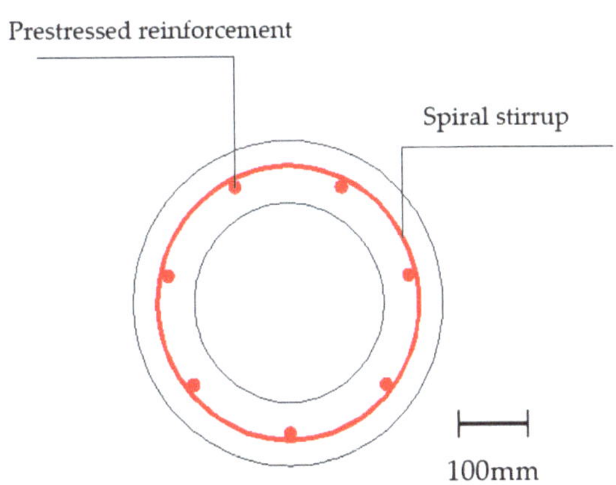

Figure 1. The section of prestressed pipe piles.

Table 2. The mix ratio of concrete (kg/m^3).

Grade	Cement	Water	Fine Aggregate	Coarse Aggregate (5–16 mm)	Coarse Aggregate (16–32 mm)	Slag Powder	Water Reducer
C80	414	160	690	537	537	62	5.7

After the prestressed pile had been prepared, it underwent steam curing. Following 28 days of steam curing, the pile was cut and welded to form samples, as illustrated in Figure 2. The test piece was cut into a 1000 mm test pile.

Figure 2. Test pile cutting diagram.

A total of four prestressed concrete pipe piles were used in the test, named PHC-1, PHC-2, PHC-3, and PHC-4, respectively. PHC-1 was a common pipe pile with both ends blocked, PHC-2 was a pile with an epoxy coal tar protective layer applied to the body, PHC-3 was a pile formed by welding two ordinary piles together with a joint, and PHC-4 was a common pile placed in a drying–wetting cycle leachate environment with a cycle of 8 days/time. After demolding, the test pile was fabricated and cured in a steam environment (85 °C) for 6 h, with the empty holes at both ends of the piles sealed before the test.

An electric field was applied to accelerate the migration rate of the solution ions at both ends of the pile. Electrified acceleration can increase the corrosion degree of reinforced concrete specimens in a shorter period of time compared to reinforced concrete specimens corroded for a longer time in the natural environment. Figure 3 illustrates the accelerated corrosion of a test pile, where the positive pole is the steel bar and the negative pole is the stainless steel mesh surrounding the pile.

Before the test, it was necessary to calculate the natural resistance of the piles in a harsh environment. After the test piles were placed into the maintenance box, they were submerged in the leachate from the landfill, covering the top of the test piles by at least 10 cm. After 10 days, the resistance was measured by a multimeter; the resistance results of PHC-1, PHC-2, PHC-3, and PHC-4 were 6.0 Ω, 6.2 Ω, 6.0 Ω, and 5.9 Ω, respectively. Taking 6.0 Ω as the natural resistance of the test piles in this environment, it can be calculated that with an applied voltage of 10 V, the corrosion of the piles corresponded to the corrosion for 30 years and 50 years in the natural environment after applying electricity for 96 d and 160 d.

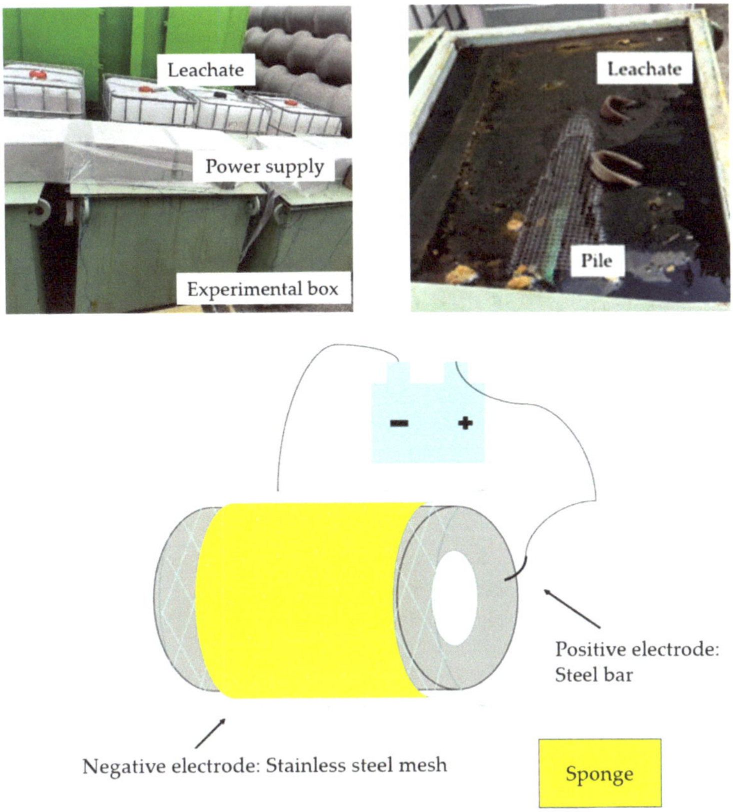

Figure 3. The electric acceleration device.

2.3. Test Methods

2.3.1. The Chloride Ion Concentration at Different Depths of Piles

The sampling procedure was performed with the aid of a core drilling machine. Then, the specimen was drilled to the location of the reinforcement and the pile was repaired after the coring procedure had been completed. The mortar in the concrete at the depths of 5 mm, 10 mm, 20 mm, 30 mm, and 40 mm from the surface of the test pile was ground into powder, respectively. According to JGJ/T 322-2013, *Technical specification for test of chloride ion content in concrete* [23], the mortar powder was mixed with 100 mL of distilled water, followed by stirring and boiling. Subsequently, the upper clear liquid was collected. The chloride ion content at different depths was calculated by measuring the Cl^- content in the clear liquid using $AgNO_3$.

Samples were taken after the production of the experimental piles was complete; the concentration of chloride ions in the concrete was 0.018%. In the experiment, the chloride ion concentration was measured every 16 days. The initial chloride ion concentration of PHC-1, PHC-2, and PHC-4 was taken when they had been soaked in the leachate for 10 days without electricity, and the subsequent chloride ion concentration was measured after the accelerated corrosion of the experimental pile had been energized.

2.3.2. Electric Current and Corrosion Potential

Faraday's first and second laws describe the relationship between electrons and matter in the electrochemical process. According to the laws, the mass of the substance precipitated (or dissolved) on the electrode (Δm) is proportional to the total amount (′) of electricity through the electrolyte, expressed by the formula

$$\Delta m = KQ \tag{1}$$

where Δm is the mass of precipitated (or dissolved) material on the electrode; K is the electrochemical equivalent, related to the precipitated (or dissolved) substance; the electrochemical equivalent is equal to the mass of the substance precipitated (or dissolved) by 1 coulomb of electricity; and Q is the amount of electricity passed through the electrolyte solution.

When the total amount (Q) of electricity through each electrolyte is the same, the mass of each substance precipitated (or dissolved) on the electric plate (Δm) is proportional to the chemical equivalent (C) of the substance, expressed by the formula

$$\Delta m \propto C \tag{2}$$

where C is the chemical equivalent, which is the ratio of the molar mass (M) of the substance to its valency (Z), which means that $C = M/Z$.

According to Formula (2), the following relation can be obtained by introducing Far-aday's constant F:

$$\Delta m = \frac{MIt}{ZF} \tag{3}$$

where M represents the molar mass of the substance (the molar mass of Fe is 56 g/mol); I represents the electrification current; t represents the electrification time; Z represents the valency of the substance (the valency of an iron atom is 2); and F represents Faraday's constant, which has a value of 96,484 C/mol.

The product of the electric current and time can be used to calculate the mass of steel corrosion in concrete, thus assessing the corrosion of the steel in the pile. The current change was recorded three times per day in the test and the current change curve was plotted. The corrosion potential was measured using TST-XS310, a steel corrosion meter from TAster (Beijing) Testing Technology Co., Ltd. (Beijing, China), with a measuring accuracy of ± 1 mV.

As shown in Figure 4, the test grid was first arranged on the pile surface during the corrosion potential test. The grid size was 20 × 20 cm (along the axis of the pile), with the grid intersection points serving as the test points. The DC power supply was disconnected one hour before data collection, and the wires connecting the main bar and the DC power supply were disconnected at the same time. The measured corrosion potential was compared with the corrosion potential evaluation standard to calculate the probability of steel bar corrosion.

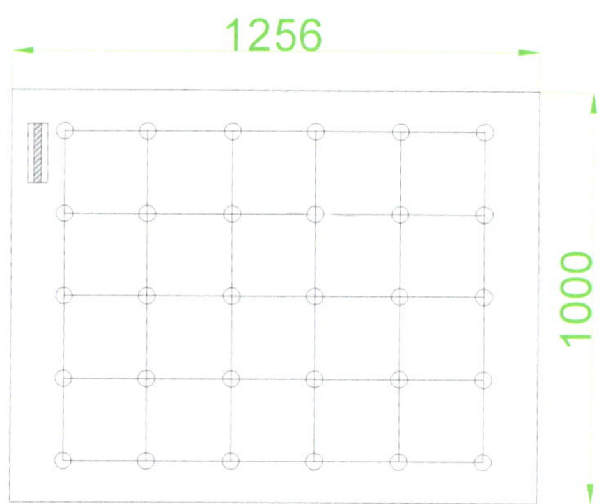

Figure 4. Test points of corrosion potential.

2.3.3. Changes in Weld Depths

Ten equidistant test points were marked at the welded joint of PHC-3. The rust pit depths were measured using a welding seam inspection ruler, and the changes in weld heights at different ages were recorded. The weld depth was measured by a weld gauge with an accuracy of 0.01 mm. Figure 5 illustrates test points 4, 5, and 6 at the welded joint.

Figure 5. Test points at the welded joint.

3. Results

3.1. The Chloride Ion Concentration at Different Depths of Prestressed Piles

Chloride ions are abundant in leachate, posing significant hazards to the safety of steel bars in precast piles. The samples were taken from the depths of 0–5 mm, 5–10 mm, 10–20 mm, 20–30 mm, and 30–40 mm from the surface of the test pile in mortar.

Figure 6 illustrates the chlorine ion concentration of the prestressed piles at different depths. In the early and middle stages, the chloride ion concentration on the surface increased faster than that in the interior. In the middle and later stages, the increase rate of the chloride ion concentration accelerated in the interior, eventually reaching a uniform concentration throughout the pile.

In the early stages of corrosion testing, the chloride ion concentration of PHC-1 and PHC-2 in the leachate at a depth of 5–10 mm started to increase after 0 d, while that of PHC-4 started to increase significantly after 16 d. The chloride ion concentration inside the prestressed pile was similar to that in the surface layer in the early stage. In the middle stage, the chloride ion concentration inside the pile began to increase continuously. The chloride ion concentration inside PHC-1 and PHC-2 gradually approached the surface chloride ion concentration. The increase rate of the chloride ion concentration at the depths of 0–5 mm, 5–10 mm, and 10–20 mm in PHC-4 decelerated after 96 d. However, the chloride concentration inside the prestressed pile increased faster, and its chloride concentration at the depths of 0–5 mm and 5–10 mm increased more slowly after 160 d. The chloride concentration in the prestressed pile was found to be higher than that at the depths of 0–5 mm and 5–10 mm.

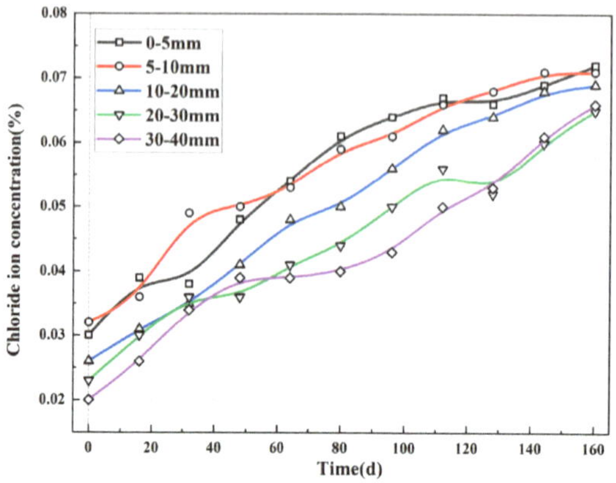

(**a**) PHC-1

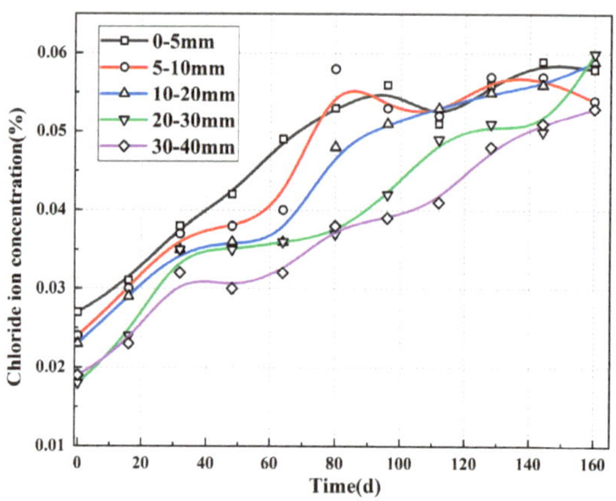

(**b**) PHC-2

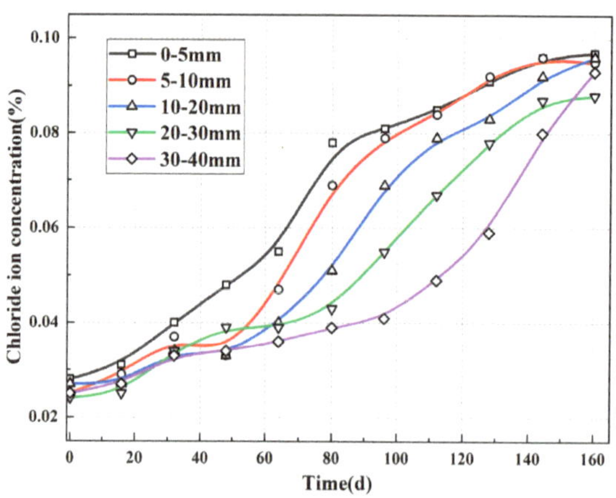

(**c**) PHC-4

Figure 6. The chloride concentration at different depths of the prestressed piles.

The epoxy resin protective layer inhibited leachate corrosion. From day 48, the chloride ion concentration in PHC-2 was lower than that in the other precast piles in the leachate environment. On day 0, the surface chloride ion concentration was 0.027%, while the chloride ion concentration at the depths of 5–10 mm, 10–20 mm, 20–30 mm, and 30–40 mm was 0.024%, 0.023%, 0.018%, and 0.19%, respectively, which was close to that of PHC-1 and PHC-4. On day 48, the gap between the chloride ion concentration in the surface layer of the pipe piles in PHC-2 and the rest of the leachate environment started to increase, until day 160, when the chloride ion concentration in the surface layer was only 0.059%, which was 0.010% less than PHC-1.

In terms of the chloride ion concentration, the epoxy applied to the surface of the prestressed pile inhibited 20%, 23%, 14%, 7%, and 19% of the chloride ions in the concrete 0–5 mm, 5–10 mm, 10–20 mm, 20–30 mm, and 30–40 mm from the surface, respectively, compared to PHC-1. The dry and wet cycling conditions accelerated the diffusion of the chloride ions into the concrete, increasing the chloride ions by 35%, 34%, 39%, 35%, and 41% in the concretes 0–5 mm, 5–10 mm, 10–20 mm, 20–30 mm, and 30–40 mm from the surface, respectively, compared to PHC-1.

Direct titration with $AgNO_3$ was performed on the core samples on day 160, as illustrated in Figure 7. The core samples of PHC-1 and PHC-4 had white precipitates at a depth of about 0–10 mm, while that of PHC-2 was protected by the epoxy resin, and the chloride ion concentration in the interior of the concrete was not enough to cause $AgNO_3$ to produce white precipitates.

(**a**) PHC-1 (**b**) PHC-2 (**c**) PHC-4

Figure 7. The staining of silver nitrate.

3.2. *Electric Current and Corrosion Potential*

The process of steel corrosion is accompanied by an increase in resistance, leading to a decrease in the current under a constant voltage. According to Faraday's law, the amount of steel corrosion is proportional to $\sum It$. Therefore, the changes in the image integral caused by the current changes during the pile corrosion testing can effectively reflect the law of steel corrosion in the pile.

As illustrated in Figure 8, the overall trend of the electric current in the pile foundations was decreasing and tended to level off eventually. The current integrals of PHC-1, PHC-2, and PHC-4 were 61.85 A·day, 47.45 A·day, and 65.37 A·day, respectively, from day 0 to 160, for values of 1.55 kg, 1.20 kg, and 1.64 kg after conversion into the corrosion amount.

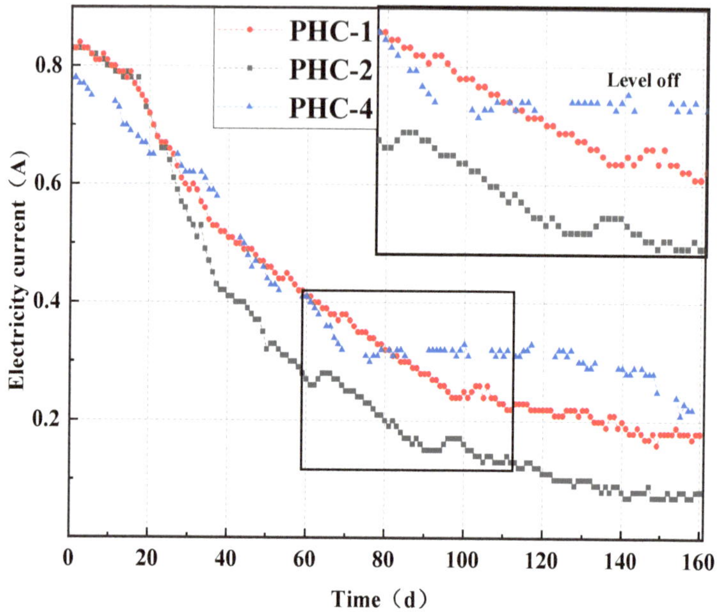

Figure 8. Current changes in prestressed piles.

The currents of PHC-1 and PHC-2 displayed the same trend in the early stage, and hazardous ions did not penetrate the interior completely at this stage. After 22 d, the electric current of PHC-1 was larger than that of PHC-2. Due to the protective layer, hazardous ions had difficulty entering PHC-2. The final corrosion amount of PHC-2 was 0.35 kg, which was smaller than that of PHC-1. The current of PHC-3 in the drying and watering cycle environment tended to level off on day 75 and decreased again after 140 d. The corrosion amount of PHC-3 was higher than that of PHC-1, but only by 0.09 kg. This indicated that although the drying–wetting cycle can accelerate precast pile corrosion, this effect was smaller compared with the effect of electrification.

When the experiment was over, PHC-2 with the epoxy resin avoided 22.6% of the amount of corrosion compared to PHC-1, while PHC-4 in a drying–wetting cycle leachate environment increased the amount of corrosion by 5.8%.

The corrosion potential of the test pile was measured every 16 d after energization, and energization was stopped 1 h before the test. The test piles were taken out, and the surface rust was cleaned. The grid area was divided according to the requirements. A steel bar was cut out and tested using a steel bar corrosion meter. Figure 9 illustrates the corrosion potential of the prestressed piles.

The change rule of the corrosion potential was similar to that of the current. Specifically, the three piles exhibited the same trend from day 0 to 32 but diverged from day 32 to 64. The corrosion potential of PHC-1 decreased faster in this stage and reached −341 mV on day 64. The corrosion potential of PHC-4 decreased rapidly after 48 d, approaching that of PHC-1 on day 80. The corrosion potential of these three piles tended to level off after 90 d, consistent with the changing trend of the electric current. On day 48, the corrosion potential of PHC-2 was −284 mV, 48 mV higher than that of PHC-1 and 5 mV higher than that of PHC-4. With the increase in time, there was a growing gap between PHC-2 and PHC-1 or PHC-4.

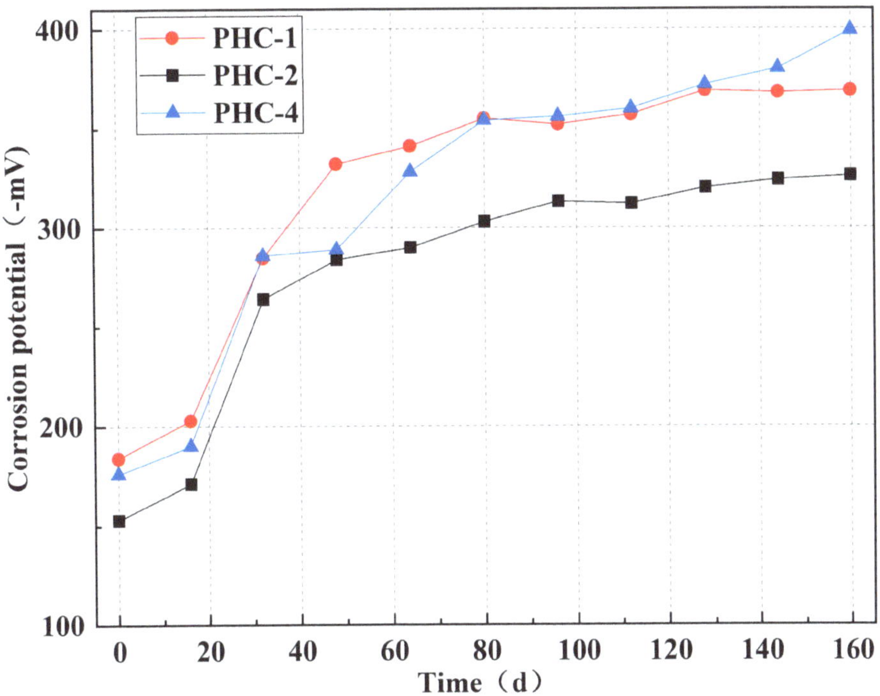

Figure 9. Corrosion potential changes in prestressed piles.

When the corrosion potentials of PHC-1, PHC-2, and PHC-3 were −369 mV, −326 mV, and −399 mV at the end of the experiment, there was an increase of 43 mV in the corrosion potential of PHC-2, which used the epoxy resin, and a decrease of 30 mV in the corrosion potential of PHC-4, which was subjected to the drying–wetting cycle.

The corrosion potential values of PHC-1 and PHC-4 were less than −350 mV on day 160, while those of PHC-2 coated with the epoxy resin protective layer were −324 mV and −326 mV on days 144 and 160, larger than −350 mV. According to the evaluation standard of the corrosion potential of steel reinforcement in structural concrete, PHC-1 and PHC-4 were at level 3 and PHC-2 was at level 2 [24].

3.3. Changes in Weld Depths

PHC piles are mainly connected by end plate welding to meet the design pile length requirements, and the durability of the welding in harsh environments needs to be emphasized.

The appearance of the welding seams of PHC-3 in harsh environments from day 0 to 160 is illustrated in Figure 10. The welding seam was a new one, with obvious metallic luster on day 0, and it was corroded by electricity by day 16. There was no metallic luster at the welding seam, and there were more rust deposits on the skirt plate and the welding seam. The thickness of the skirt plate was about 1.5 mm, and it had corroded completely within 50 years.

The depth change in the weld is illustrated in Figure 10. A large depth change was observed in the early stage and tended to level off in the later stage. The final corrosion depth was 3.2 mm.

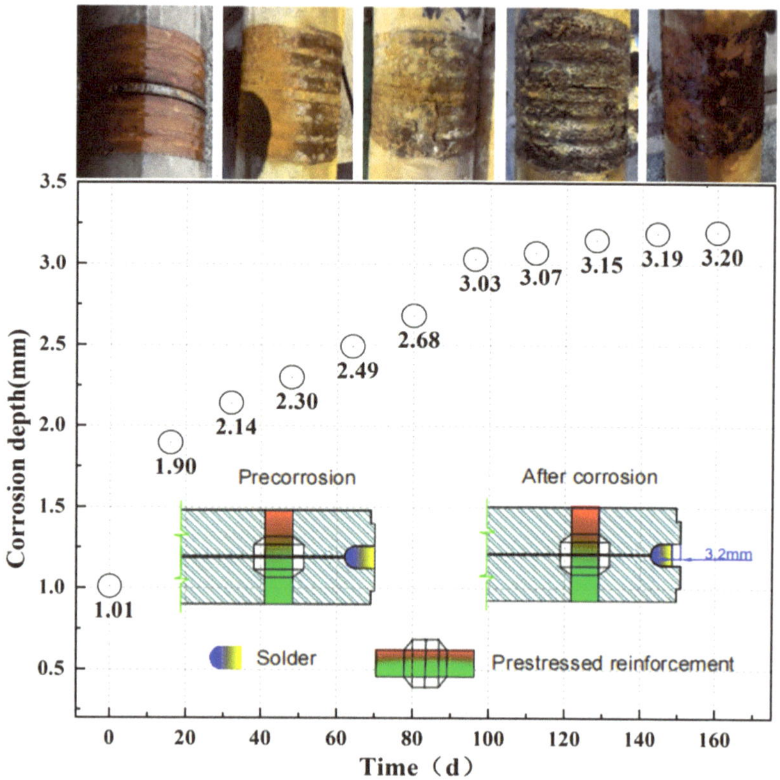

Figure 10. Changes in the corrosion depth of the welding seams.

The prestressed piles had a load value of 106 kN under the ultimate bending moment condition. When calculating the tensile bearing capacity of a single pile according to the connection strength of the prestressed concrete piles, the welded connection should be calculated according to the following formula:

$$N \leq \frac{1}{4}\left(d_1^2 - d_2^2\right) f^w \tag{4}$$

where N represents the single pile design value of the pullout force (kN); d_1 represents the weld outer diameter (mm), 398 mm; d_2 represents the weld inner diameter (mm), 376 mm; and f^w represents the weld tensile strength design value, 170 MPa.

When the weld depth decreased, the weld outer diameter d_1 and its pull-out bearing capacity decreased.

Table 3 shows the results corresponding to the corrosion allowance of the weld depth of pile joints under the simulated 50-year corrosive environment. This corresponds to the requirements of the pile cracking moment, and there was a considerable corrosion allowance for the welding seam. The accelerated test in the underground corrosive solution of the landfill indicated that the welded joints of prestressed piles had favorable durability after 50 years in the high Cl^- environment.

Table 3. The depth corrosion allowance of the welded joints of the prestressed piles under the simulated 50-year corrosive environment.

The Load Condition Corresponds to the Calculated Values of the Weld Depth (mm)			50-Year Equivalent Corrosion Depth (mm)	Corrosion Allowance (mm)
Loading Condition	Loading Intensity (kN)	Corresponding Weld Depth (mm)		
Cracking moment (kN·m)	64	4.36	3.20	4.44
Ultimate bending moment (kN·m)	106	6.53		2.27

3.4. Discussion

The external current method can be employed to accelerate the diffusion of chloride ions in pile concrete; in this case, it energized and accelerated the prestressed piles for 160 d, equivalent to 50 years of service in the landfill landscape. In general, the steel corrosion rate in the prestressed piles decelerated with an increase in age. This may be attributed to a decrease in the cross-sectional area of the steel during corrosion, which increasingly blocked the passage of the electric current. Therefore, the corrosion rate of the piles reached the maximum within 16–32 d, followed by a gradual decrease in the corrosion rate of the subsequent prestressed piles.

When the prestressed piles were in an immersion environment, the chloride ion concentration in the surface layer of the pile concrete started to rise first. The concentration of the chloride ions inside the concrete began to rise in the middle and late stages until the concentration of the chloride ions inside the concrete was equal to that on the surface. This conformed to the permeation law of chloride ions in concrete. From day 0 to 16, the chloride ion concentration of the prestressed piles in the drying–wetting cycle environment was smaller than that of the prestressed pile concrete in the immersion environment. As the drying–wetting cycle and energization exerted more significant effects on chloride ion diffusion, the chloride ion concentration of PHC-4 was higher than that of PHC-1. The surface layer of the core sample of PHC-1 and PHC-4 can be stained white by $AgNO_3$ on day 160. The drying–wetting cycle increased chloride penetration by 5.8%; a coating on the surface of pipe piles can effectively prevent the diffusion of chloride ions into the inside of the pipe piles, and the application of the epoxy resin reduced chloride penetration by 22.6%. This result was verified in the tests on electric current and half-cell corrosion potential, resulting in minimal rusting and maximum half-cell corrosion potential.

After the accelerated corrosion continued for 160 d, the skirt plate of the welded pile was corroded completely. The average change in the weld depth was 3.2 mm. Compared with the maximum corrosion depth of 6.53 m allowed by the standard, there was a large rust allowance to meet the requirements of the project.

The size, prestress, and reinforcement of prestressed piles will affect the corrosion. This paper only takes PHC 400(95) as the research object, which is the most used in a leachate environment, aiming to provide practical guidance for most projects.

The current acceleration experiment only accelerated the corrosion of chloride ions, and has little effect on sulfate and ammonia roots. How to couple the effects of these will be an interesting discussion point.

4. Conclusions

In the corrosion test of prestressed piles in a landfill leachate environment, the electrification-accelerated corrosion of prestressed piles has presented favorable durability. The implications of these findings for the design and maintenance of prestressed concrete structures in leachate environments are significant. In general, prestressed piles can meet the conditions of use in leachate-soaked environments with a building design life of 50 years. When in a drying–wetting cycle leachate environment, prestressed piles can also be used, but some protective measures are recommended. The specific conclusions are elucidated as follows:

(1) The chloride ion concentration in piles increased gradually with an increase in time and leveled off at the later stage of tests. The chloride ion concentration in the surface layer of the pile concrete began to rise first, while the internal chloride ion concentration started to rise in the middle and later stages. Finally, the internal chloride ion concentration was equal to the surface one.
(2) The corrosion rate of the pile in the drying–wetting cycle environment accelerated in the middle and late stages, and the drying–wetting cycle increased chloride penetration by 5.8% at 160 d. After 64 d, the corrosion potential of the pipe pile in the drying–wetting cycle decreased faster, and its resistance also began to increase faster. In addition, the current decreased faster, the chloride ion concentration increased rapidly, and the corrosion potential decreased faster.
(3) Applying a layer of epoxy resin protective layer on the pile can inhibit the chloride ion erosion and prevent the chloride ions from entering the pile concrete effectively, and the application of the epoxy resin reduced chloride penetration by 22.6% at 160 d. The chloride ion concentration was the lowest and the corrosion potential was the highest in the test piles coated with epoxy resin. In practical engineering applications, the test results showed that applying a protective layer on the surface of prestressed piles can improve the durability of precast piles.
(4) Under the accelerated corrosion for 160 d, the actual corrosion depth of the welding seam was 3.20 mm. Compared with the maximum corrosion depth of 6.53 mm permitted by the standard, there was a large corrosion allowance. Therefore, the prestressed piles can be welded in a leachate environment.

Author Contributions: Conceptualization, R.Z. and M.D.; methodology, M.D. and Y.W.; investigation, J.Z. and X.Y.; data curation, Y.W., X.Y. and J.X.; formal analysis, X.Y., R.Z. and J.X.; writing—original draft preparation, Y.W. and X.Y.; writing—review and editing, J.Z. and M.D. All authors have read and agreed to the published version of the manuscript.

Funding: This research was funded by the Science and Technology Innovation 2025 Major Special Project (Ningbo Major Science and Technology Task Tackling Project), 2022Z030, and Project: Research on the construction of underground total factor information model for large landfill sites, 20dz1203403.

Institutional Review Board Statement: Not applicable.

Informed Consent Statement: Not applicable.

Data Availability Statement: The raw data supporting the conclusions of this article will be made available by the authors on request.

Conflicts of Interest: Author Rihong Zhang was employed by the company Ningbo ZCONE High-tech Holdings Co., Ltd. Authors Xuming Yu and Junzhong Xue were employed by the company CGN Solar Energy (Jiaxing) Co., Ltd. Author Jing Zhang was employed by the company SGIDI Engineering Consulting (Group) Co., Ltd. The remaining authors declare that the research was conducted in the absence of any commercial or financial relationships that could be construed as a potential conflict of interest.

References

1. Feng, W. Evaluation of macro and meso-mechanical properties of concrete under the aggressiveness of landfill leachate. *Sci. Rep.* **2022**, *12*, 3976. [CrossRef]
2. Teng, C.; Zhou, K.; Peng, C.; Chen, W. Characterization and treatment of landfill leachate: A review. *Water Res.* **2021**, *203*, 117525. [CrossRef] [PubMed]
3. Tanveer, S.; Md, J.M.; Nehreen, M.; Mahmudul, H.; Tanbir, K. Pollutant removal from landfill leachate employing two-stage constructed wetland mesocosms: Co-treatment with municipal sewage. *Environ. Sci. Pol. Res.* **2020**, *27*, 13–32.
4. Feng, S.J.; Zhao, Y.; Zhang, X.L.; Zhen, B.B. Leachate leakage investigation, assessment and engineering countermeasures for tunneling underneath a MSW landfill. *Eng. Geol.* **2020**, *265*, 105447. [CrossRef]

5. Naveen, B.P.; Sumalatha, J.; Malik, R.K. A study on contamination of ground and surface water bodies by leachate leakage from a landfill in Bangalore, India. *Int. J. Geo-Eng.* **2018**, *9*, 27. [CrossRef]
6. Omeiza, J.O. Durability considerations for concrete containment liners use in engineered ladfills. *J. Civ. Environ. Syst. Eng.* 121–141. Available online: https://j-cese.com/storage/app/public/3fWLnAY29mHqHQSnAj9ZWVVjQr1xki9zsjXuNjVQ.pdf (accessed on 12 January 2019).
7. Han, X.F.; Du, Z.X.; Wang, P.G.; Zhang, R.H.; Gao, J.F.; Ling, Z.J.; Wei, D.X. Multi-scale microstructure quantitative characterization and anti-erosion performance of PHC pipe pile. *Constr. Build. Mater.* **2023**, *406*, 133464. [CrossRef]
8. Liu, F.; Feng, W.H.; Xiong, Z.; Li, L.J.; Yang, Y.B.; Lin, H.; Shen, Y.J. Impact performance of new prestressed high-performance concrete pipe piles manufactured with an environmentally friendly technique. *J. Clean. Prod.* **2019**, *231*, 683–697. [CrossRef]
9. Wang, H.; Chen, H.H.; Yang, C.Y.; Li, J.P. Durable life assessment of centrifugal stratified PHC piles in marine environment. *Eng. Struct.* **2022**, *263*, 114404. [CrossRef]
10. Liu, Z.Y.; Bu, L.K.; Wang, Z.X.; Hu, G.G. Durability and microstructure of steam cured and autoclaved PHC pipe piles. *Constr. Build. Mater.* **2019**, *209*, 679–689. [CrossRef]
11. Han, J.L.; Shui, Z.H.; Wang, G.M.; Sun, T.; Gao, X. Performance evaluation of steam cured HPC pipe piles produced with metakaolin based mineral additives. *Constr. Build. Mater.* **2018**, *189*, 719–727. [CrossRef]
12. Han, X.F.; Fu, H.; Li, G.G.; Tian, L.; Pan, C.G.; Chen, C.L.; Wang, P.G. Volume Deformation of Steam-Cured Concrete with Slag during and after Steam Curing. *Materials* **2021**, *14*, 1647. [CrossRef] [PubMed]
13. Jensen, U.G.; Hoang, L.C.; Joergensen, H.B.; Fabrin, L.S. Shear strength of heavily reinforced concrete members with circular cross section. *Eng. Struct.* **2010**, *32*, 617–626. [CrossRef]
14. Shao, W.; Li, J.P. Service life prediction of cracked RC pipe piles exposed to marine environments. *Constr. Build. Mater.* **2014**, *64*, 301–307. [CrossRef]
15. Watanabe, T.; Nagao, T.; Kobayashi, H.; Mashima, M. Durability study of existing concrete piles. *Geotech. Eng. Res.* **2003**, *38*, 1535–1536.
16. Yamashita, E.; Ikegami, Y.; Okada, E.; Oura, A. Durability high strength concrete pile after 16 years. *Proc. Jpn. Concr. Inst.* **1987**, *9*, 345–350.
17. Tatsuya, T.; Mizunuma, S.; Motoki, S.; Ikegami, Y. Durability high strength concrete pile after 32 years. *Proc. Jpn. Concr. Inst.* **2001**, *23*, 667–672.
18. Nakamura, K.; Kawakami, H.; Iwanaga, K.; Jitsufuji, S. Durability high strength concrete pile after 50 years. *Proc. Jpn. Concr. Inst.* **2020**, *42*, 449–454.
19. Zhan, Y.; Yang, Y.B.; Wu, X.H.; Xie, Y.H.; Luo, J.; Guo, W.J.; Hao, Y.H.; Wang, H.C. Experimental study on non-autoclaved PHC pile concrete applying polycarboxylate superplasticizer. *Key Eng. Mater.* **2014**, *629–630*, 345–350. [CrossRef]
20. Liu, W.Y.; Gong, S.F.; Xu, X.B.; Chen, G.; Liu, C.B.; Fan, H. Study on flexural performance of prestressed concrete square pile with resilient clamping connection joint after electricity accelerates corrosion. *China Concr. Cem. Prod.* **2021**, *5*, 33–38.
21. Li, L.; Gong, W.B.; Li, J.P. Service life of prestressed high-strength concrete pile in marine environment considering effects of concrete stratification and temperature. *Constr. Build. Mater.* **2020**, *253*, 119233. [CrossRef]
22. *2010-G22*; Pre-Tensioned Prestressed Concrete Pipe Pile. Beijing China Planning Publishing House: Zhejiang, China, 2010.
23. *JGJ/T 322-2013*; Technical Specification for Test of Chloride Ion Content in Concrete. China Building Industry Press: Beijing, China, 2013.
24. *ASTM C876-15*; Standard Test Method for Corrosion Potentials of Uncoated Reinforcing Steel in Concrete. ASTM International: West Conshohocken, PA, USA, 2015.

materials

MDPI

Article

Inhibition Studies of Expansion Damage in Medium–Low Reactivity Limestone by Fly Ash

Shaocong Dai [1], Xinyu Zhang [1], Wei Li [1], Zhongyang Mao [1], Xiaojun Huang [1], Min Deng [1,*] and Bi Chen [2,*]

[1] College of Material Science and Engineering, Nanjing Tech University, Nanjing 211816, China
[2] College of Emergency Management, Nanjing Tech University, Nanjing 211816, China
* Correspondence: dengmin@njtech.edu.cn (M.D.); chenbi@njtech.edu.cn (B.C.); Tel.: +86-136-0518-4865 (M.D.); +86-159-5195-0507 (B.C.)

Abstract: Expansion damage in medium–low reactivity dolomite limestone poses significant challenges in construction and engineering projects. This study investigates the potential of fly ash in inhibiting expansion damage in such limestone formations based on RILEM AAR-5 method. Through a series of laboratory experiments, various proportions of fly ash instead of cement, respectively, were prepared and subjected to varying alkali content conditions immersion tests to simulate expansion conditions. The expansion rates and extents were monitored and compared between pure limestone samples and those mixed with different proportions of fly ash. Additionally, scanning electron microscopy (SEM) analysis was employed to investigate the microstructure of the dolomite limestone–fly ash mixtures to understand the inhibition mechanisms. Results indicate that fly ash demonstrates promising inhibitory effects on expansion damage in medium–low reactivity dolomite limestone across the addition of 40% fly ash and alkali content of 0.70%. The reaction products are calcite, brucite, and a mixture of Mg-Si-Al phases and the reaction area is within 100 μm from the boundary when the cement alkali content is 1.50% without any fly ash. However, no reaction products were found at the boundary after adding 40% fly ash when lowering the cement alkali content to 0.70%. This research contributes to a better understanding of the interaction between fly ash and dolomite limestone in inhibiting expansion damage, providing valuable insights for engineering applications.

Keywords: alkali–carbonate reaction; dolomitic limestone; inhibition; cement alkali content; fly ash

Citation: Dai, S.; Zhang, X.; Li, W.; Mao, Z.; Huang, X.; Deng, M.; Chen, B. Inhibition Studies of Expansion Damage in Medium–Low Reactivity Limestone by Fly Ash. *Materials* **2024**, *17*, 2422. https://doi.org/10.3390/ma17102422

Academic Editor: Miguel Ángel Sanjuán

Received: 11 April 2024
Revised: 9 May 2024
Accepted: 13 May 2024
Published: 17 May 2024

1. Introduction

Alkali–aggregate reaction (AAR) is a phenomenon that may occur in concrete structures, typically referring to the reaction between the alkaline components in concrete and certain aggregates containing reactive minerals (such as siliceous or carbonate rocks) [1]. This reaction can result in expansion and cracking, ultimately leading to damage to the concrete structure. AAR typically falls into two types: alkali–silica reaction (ASR) and alkali–carbonate reaction (ACR). ASR refers to the reaction of alkalis with the siliceous minerals in the aggregate, producing a gel-like material that causes the expansion of concrete volume [2]. ACR involves the reaction of alkalis with carbonate minerals (such as dolomitic rocks), resulting in the formation of alkali metal carbonates, which also leads to volume expansion.

In the early 1950s, Canada and other countries successively discovered severe mesh cracking in concrete using carbonate aggregate. In 1963, Gillott [3] discovered and proposed the ACR, in which in the presence of alkali, dolomite crystals inside dolomite, dolomitic chert, and gray dolomite, which are used as concrete aggregates, react with alkali to form brucite, calcite, and CO_3^{2-}. According to thermodynamic calculations, this reaction process occurs spontaneously. In addition, the hydrated $Ca(OH)_2$ in the concrete slurry, which undergoes hydration, continues to react with the CO_3^{2-} ions in Equation (1) to produce hydroxide ions once again, with the reaction formula shown in Equations (1) and (2):

$$CaMg(CO_3)^2 + 2OH^- = Mg(OH)_2 + CaCO_3 + CO_3^{2-} \quad (1)$$

$$CO_3^{2-} + Ca(OH)_2 = CaCO_3 + 2OH^- \quad (2)$$

Cody and Deng [4,5] favored the dedolomitization reaction (ADR) to describe ACR. Chen et al. [6,7] illustrated the capability to differentiate between ASR and ACR within carbonate systems using Tetramethylammonium hydroxide (TMAH). They further asserted that the primary culprit behind ACR is the ADR. Li [8] discovered that the dolomitization reaction process primarily involves OH^- initially reacting on the surface of dolomite.

Dolomitic rocks are widely distributed in China (as seen in Figure 1) and are the most commonly used rock aggregate in concrete due to their ease of extraction and processing. Since Deng discovered severe cracking damage in concrete structures using carbonate rock aggregates in Shandong Province and Tianjin, China, some other locations have also experienced similar issues, as indicated in Figure 2. Analysis and research identified the cause as ACR. Due to the formation mechanism of carbonate rocks, the limitations of engineering conditions, and economic factors, it is impractical to use entirely non-alkaline reactive aggregates. Identifying aggregate activity and adopting corresponding inhibition measures against alkali–aggregate reaction types are the main methods to prevent damage from alkali–aggregate reactions. This study proposes corresponding inhibition measures and methods to avoid significant economic losses caused by alkali–aggregate reactions in concrete structural engineering, which is of great guiding significance.

To mitigate ACR, numerous researchers both domestically and internationally have conducted extensive investigations into preventive measures. A plethora of experimental studies and engineering practices have demonstrated that the inclusion of fly ash in concrete mixtures not only retards or suppresses ACR but also enhances other concrete properties, conserves resources, and safeguards the environment [9–11]. Mineral admixtures such as silica fume, pulverized fly ash and granulated blast furnace slag are widely used to inhibit ACR expansion [12–14]. The effectiveness of mineral admixtures in controlling ACR is generally derived from the decreasing of content of cementitious materials, lowering OH^- concentrations in pore solutions by the adsorption of low Ca/Si C-S-H gels and resisting the migration of ions in pore solutions of concrete by fining pores [15]. Nevertheless, there has been limited research into ACR inhibition. Deng [5] conducted an assessment of the inhibitory effects of mineral admixtures, sulfoaluminate cement, and persulfate cement on rocks in the Kingston, Canada, utilizing the concrete microbar method. Sulfoaluminate cement, persulfate cement, and low-alkali Portland cement, blended with high dosages of PFA (pulverized fuel ash), BFS (blast furnace slag), or SF (silica fume), effectively prevented concrete expansion and cracking induced by alkali–dolomite reaction. In cases where Portland cement with an equivalent alkali content of 0.43% and reactive dolomitic limestone from Kingston, Canada, was employed, it became imperative to substitute the cement with 70% PFA, 90% BFS, or 30% SF. The essence of the issue lies in significantly lowering the pH of the concrete pore solution. Shehata [16] assessed the prolonged impacts of supplementary cementitious materials (SCMs) on ACR over a span of up to 10 years utilizing a concrete prism test. While certain types of SCMs exhibited greater effectiveness in diminishing expansion, none proved efficacious in curbing it over the long haul. Blending 10% reactive aggregate with 90% non-reactive aggregate proved effective in attaining the threshold of achieving expansion below 0.040% at the end of the first year. However, the expansion of the specimen escalated to 0.074% after a decade of placement at room temperature. Joshaghani [17] evaluated the effectiveness of ACR inhibition in both short and long terms using two types of admixtures, namely trass and fly ash. The experiments revealed that trass was not highly effective in inhibiting ACR compared to fly ash. Ren [18] conducted studies using the concrete microbar method and concrete prism method to investigate the impact of mineral admixtures on ACR expansion in dolomitic rocks. They found that 30% fly ash could temporarily inhibit alkali carbonate reaction, with the primary mechanism being the refinement of pore structure and reduction in the migration rate of Na^+ and K^+ ions into the pore solution.

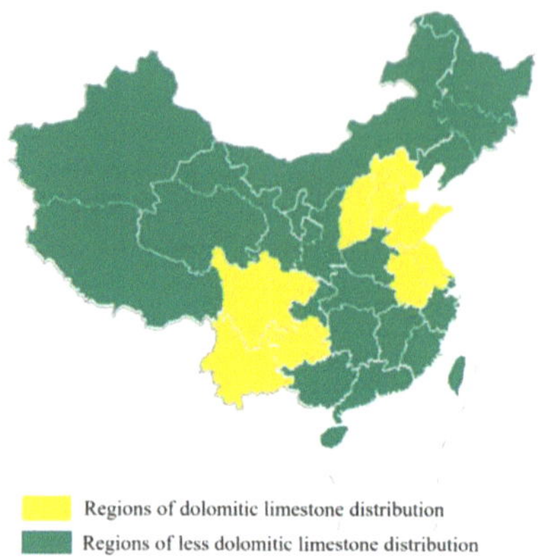

Figure 1. Regions of dolomitic limestone distribution, including Beijing, Tianjin, Hebei, Shanxi, Sichuan, Shandong, Guizhou, Anhui, Jiangsu, Yunnan, and Guizhou.

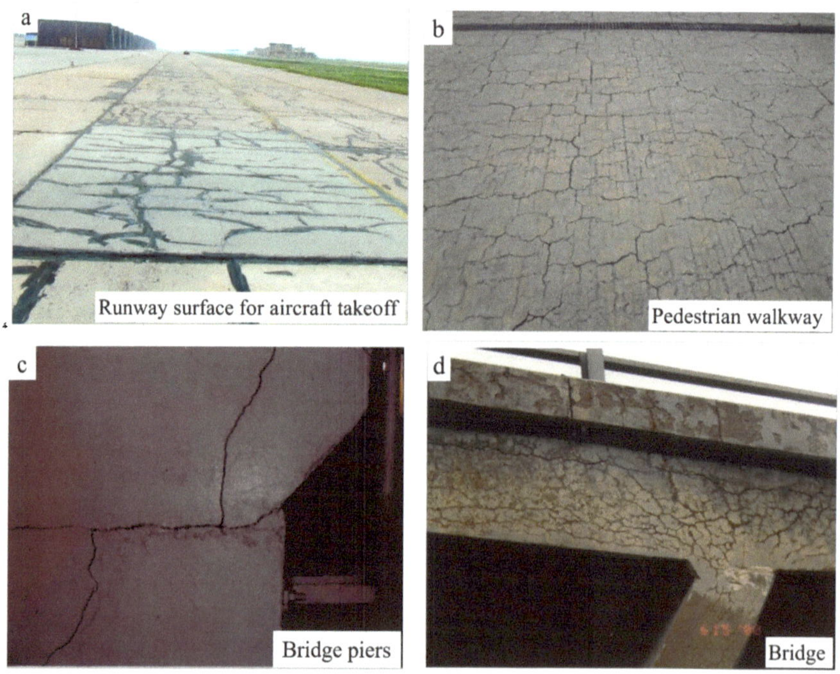

Figure 2. Cases of ACR damage. (**a**) Runway surface for aircraft takeoff in Shandong, China (**b**) Pedestrain walkway in Shandong, China (**c**) Bridge piers in Tianjin, China (**d**) Bridge in Tianjin, China.

In published papers, although there are many reports related to AAR inhibition by fly ash, this is mostly for ASR, with very few reports for ACR. There is still no definitive answer as to whether the use of fly ash has an inhibitory effect on ACR, and many of the inhibition methods are currently based on a modification of the alkali activity test method (RILEM AAR-5; RILEM AAR-2), which is based on a conditioning environment of 80 °C, 1 mol/L NaOH solution with a cement alkali content of 1.50%. This will have an accelerating effect in the experiment, but due to the harshness of the curing environment in the past, it could not be suppressed even with the use of fly ash, and cements with such a high alkali content would not be used in practical engineering applications.

The objective of this study was to investigate the effect of fly ash on the inhibition of alkali carbonate reaction (ACR) in low and moderately reactive dolomitic limestones and to assess the suitable conditions for the application of fly ash in these two types of dolomitic limestones by using different alkaline environments. In this study, the efficacy of fly ash in inhibiting the alkali carbonate reaction (ACR) swelling of dolomitic limestone in different alkaline environments was evaluated by adjusting the admixture ratio of fly ash using the concrete micro-column test method based on the RILEM AAR-5 standard. Scanning electron microscopy (SEM) was utilized to observe the morphology of dolomite limestone after the reaction in order to observe the inhibition effect of fly ash on dolomite limestone, and then application guidelines are proposed for the specific conditions of dolomite limestone.

2. Materials and Methods

2.1. Materials

2.1.1. Cement

The cement used in the experiment was P•II 52.5 cement produced by Nanjing Jiangnan Onoda Co., Ltd., Nanjing, China with an alkali content of ($Na_2O_{eq} = Na_2O + 0.658K_2O$) of 0.54%. The chemical composition is shown in Table 1, and the XRD is shown in Figure 3.

Table 1. Chemical compositions of cement from the Jiangnan Onoda Cement Company/wt%.

Component	LOI.	CaO	MgO	SiO_2	Al_2O_3	Fe_2O_3	K_2O	Na_2O	SO_3	Total
wt	2.81	64.00	2.35	19.43	4.73	2.96	0.47	0.26	2.58	99.59

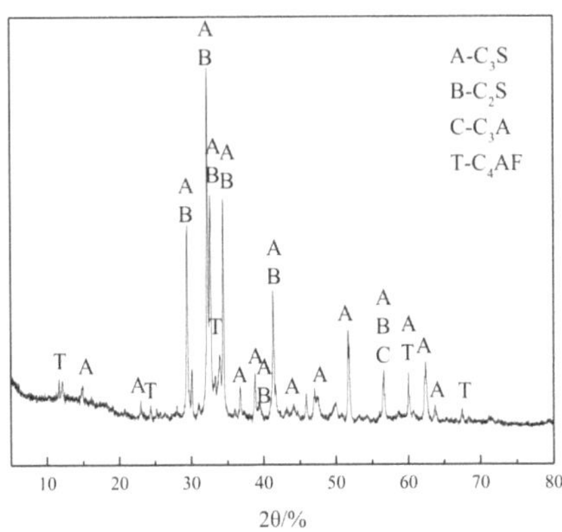

Figure 3. XRD pattern of P•II 52.5 cement.

2.1.2. Fly Ash

The fly ash (FA) used in the experiment is Class II, which complies with the GB/T 1596-2017 standard [19], produced by Pudi Mixing Station in Nanjing, China. Its density is 2.24 g/cm^3. Its chemical composition is shown in Table 2. Figure 4 shows the XRD pattern of FA, wherein the primary components identified are mullite and quartz.

Table 2. Chemical composition of FA/wt%.

Component	LOI.	CaO	MgO	SiO_2	Al_2O_3	Fe_2O_3	K_2O	Na_2O	SO_3	Total
wt	1.81	4.40	1.11	50.10	29.77	8.95	0.89	0.39	1.15	98.57

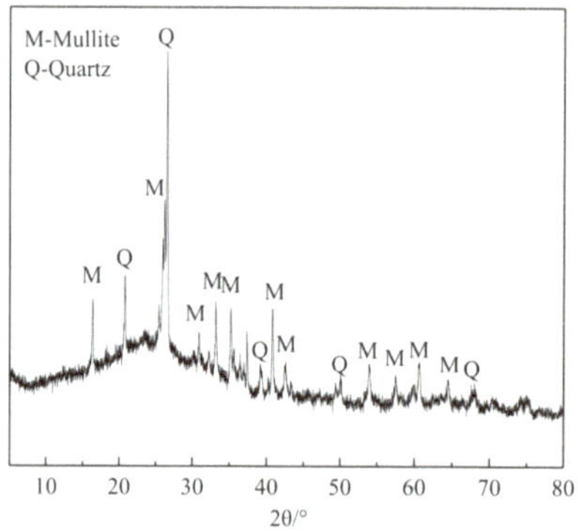

Figure 4. XRD pattern of FA.

2.1.3. Aggregates

The dolomitic rocks selected for this experiment were sourced from the dolomitic limestone designated as SJW and YM, located in the quarry of Sansui County, Guizhou. The experiment used aggregate with a particle size of 5–10 mm, with a single gradation. Figure 5 shows the petrographic map of SJW and YM, revealing a mosaic-type structure in the dolomite of these rocks. From the chemical composition table provided in Table 3 and the XRD pattern depicted in Figure 6, it is evident that the primary components of both rocks are calcite and dolomite.

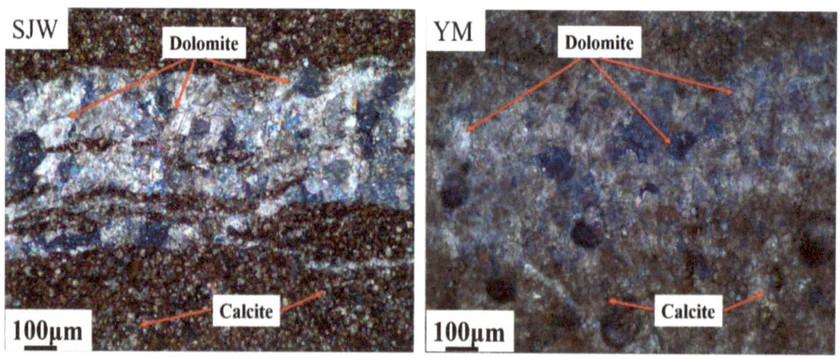

Figure 5. Petrographic micrographs of dolomitic rocks SJW and YM.

Table 3. Chemical compositions of rocks/wt%.

Simple	LOI.	SiO_2	Fe_2O_3	Al_2O_3	CaO	MgO	Total
SJW	40.58	5.99	0.26	1.49	45.68	2.55	96.55
YM	37.19	11.18	1.03	2.80	42.26	2.48	96.94

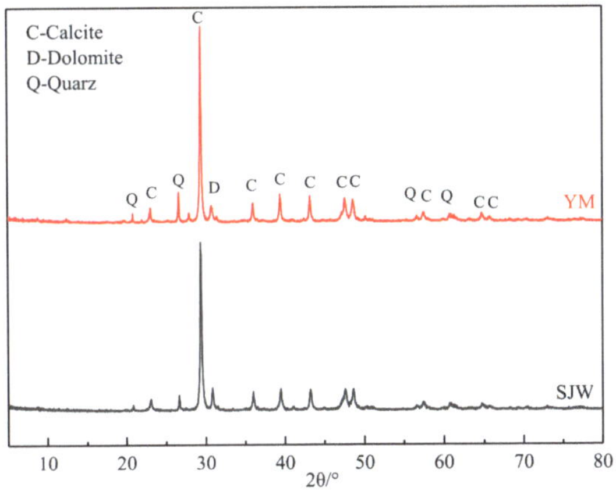

Figure 6. XRD patterns of rocks SJW and YM.

2.2. Experimental Methods

2.2.1. Inhibition Experiments with Simulated Pore Solutions

In this study, the ultra-accelerated mortar bar method (RILEM AAR-2) [20] was employed to investigate the potential alkali–silica reactivity of SJW and YM rock samples, while rapid preliminary screening test for the carbonate aggregate method (RILEM AAR-5) [21] was utilized to assess their alkali–carbonate reactivity. Subsequently, based on the RILEM AAR-5 concrete microbar method, concrete microbars with dimensions of 40 mm × 40 mm × 160 mm were prepared. The aggregate mass was 900 g, and the cement-to-aggregate mass ratio was 1:1, with an aggregate particle size ranging from 5 mm to 10 mm. P•II 52.5 silicate cement with an alkali content of 0.54% was used. The alkali content of the cement was adjusted to 1.50 wt% and 0.70 wt% using analytically pure NaOH reagent. Additionally, 0, 20%, 30%, and 40% of fly ash were substituted for an equal mass of cement, with a water-to-binder ratio of 0.32. After molding for 24 h and demolding, the initial length L_0 of the specimens was measured using a length comparator. Subsequently, the concrete microbars with different alkali contents were placed in 80 °C solutions of 60 mL 1.5 mol/L and 60 mL 0.7 mol/L NaOH for curing. To minimize the influence of external curing solutions on the microbars, they were placed in boxes measuring 45 mm × 45 mm × 180 mm and filled with two 5 mm × 40 mm × 160 mm acrylic plates to fill the gaps, the model is as shown in Figure 7. After a specific curing period, the concrete microbar specimens were retrieved and their lengths, L_D, were accurately measured after cooling. Through this series of experiments, the aim was to elucidate the reaction activity of SJW and YM rock samples under different alkali concentration environments.

The formula for calculation is as follows:

$$Pt = (L_0 - 2b)/(L_t - L_0) \times 100\%$$

where:

Pt is the expansion rate after t days of curing, in %;

L_t is the test piece length after t days of curing, in mm;

L_0 is the initial length of the test piece, in mm;

b is the length of the nail embedded in the concrete, in mm.

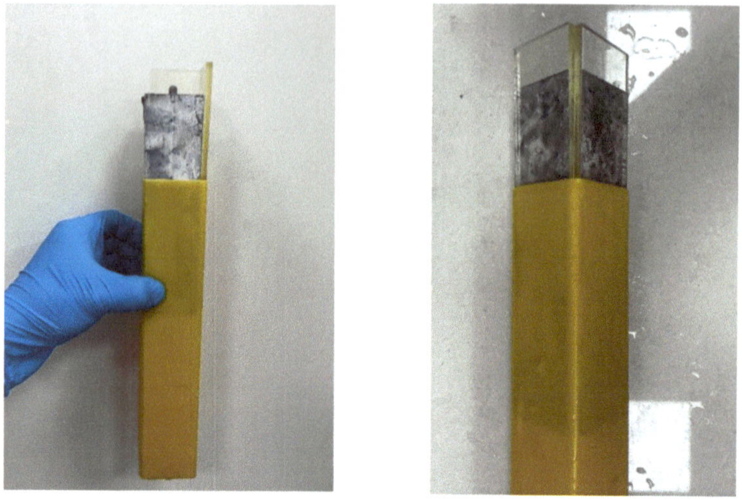

Figure 7. Curing of concrete microbar specimens.

2.2.2. Micro-Structure Analysis

To investigate the microstructural changes in concrete microbars after 120 days of reaction, SEM samples need to be prepared. Selected samples of SJW with a cement alkali content of 1.50%, maintained at 80 °C, SJW with a cement alkali content of 1.50% doped with 40% FA, and SJW with a cement alkali content of 0.70% doped with 40% FA were selected for the study. Firstly, the concrete microbars were gently crushed, and the rock particles were carefully extracted. Next, the reacted rock particles were placed into a mold and filled with epoxy resin for curing. Then, precise cutting was performed until the rock surface was fully exposed. Subsequently, the surfaces of the rock samples were finely polished and made into slides using the AutoMet 250 grinder–polisher produced by Buehler, Lake Bluff, IL, USA. After ensuring a smooth and traceless surface, the samples were polished. This preparation work aimed to enable the better observation and analyzation of the microstructures of the rock samples. Figure 8 shows a photomicrograph of SJW sample production. Finally, the rock samples were observed using the Ultra55 field emission scanning electron microscope from Carl Zeiss, Oberkochen, Germany. Additionally, the composition of the samples was analyzed using energy dispersive X-ray spectroscopy (EDS) to understand their chemical compositions.

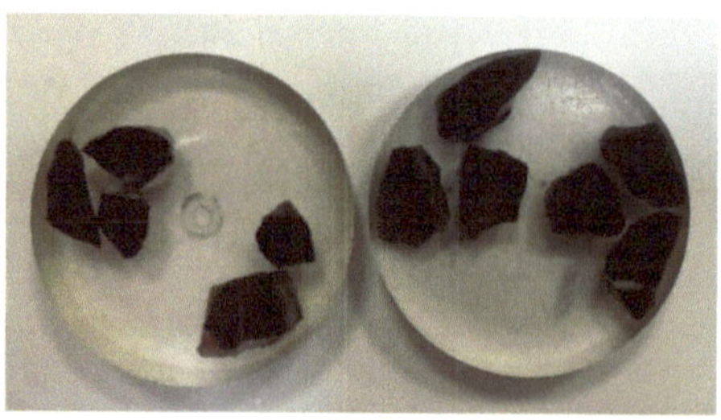

Figure 8. Photomicrograph of SJW sample production.

Through these meticulous operations and the use of advanced equipment, a deeper understanding of the microstructural changes in concrete microbars after 120 days of reaction can be achieved, providing valuable data support for research.

3. Results

3.1. Concrete Microbar Experiment

3.1.1. Aggregate Alkali Activity

In Figure 9, the expansion rates of SJW and YM at 28 days are 0.128% and 0.227%, respectively. According to RILEM AAR-5 standard, their expansion levels have exceeded the threshold of 0.1%, indicating the alkali–carbonate reaction activity of both SJW and YM aggregates. On the other hand, in Figure 10, the expansion rates of SJW and YM at 14 days are 0.062% and 0.682% (could be rounded down to 0.1%), respectively. According to RILEM AAR-2 standard, they do not exceed the critical value of 0.1%, indicating that SJW aggregates do not possess alkali–silica reaction activity. Therefore, combining the data from Figures 8 and 9, it can be inferred that SJW and YM aggregates exhibit alkali–carbonate reaction activity but do not possess alkali–silica reaction activity.

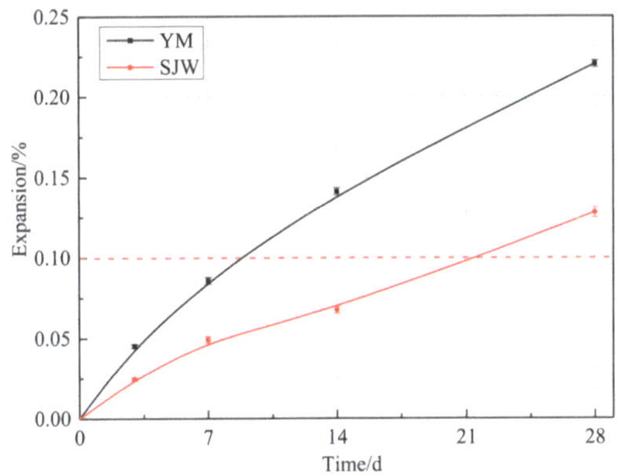

Figure 9. Expansion of the concrete microbars prepared according to RILEM AAR-5.

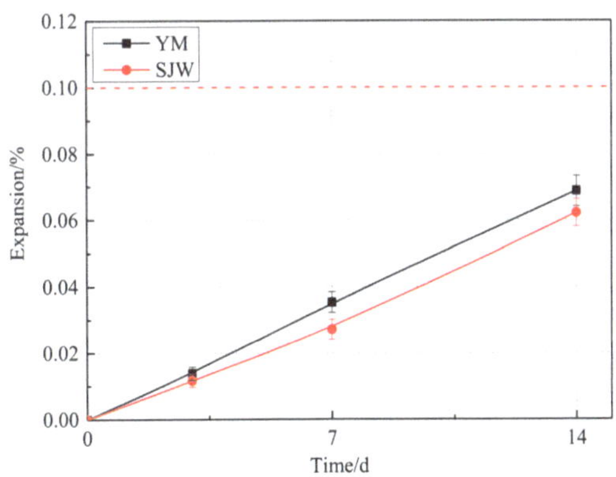

Figure 10. Expansion of the mortar bars prepared according to RILEM AAR-2.

3.1.2. Effect of Cement Alkali Content on Concrete Microbar

Figure 11 illustrates that when the cement alkali content is 1.50%, the expansion rates of concrete microbars made with SJW and YM rapidly increase, reaching 0.406% and 0.416%, respectively, after 120 days of reaction. As the cement alkali content decreases to 0.70%, the expansion rates of both types of rocks continue to show an increasing trend, with expansion rates reaching 0.22% and 0.32%, respectively. This result indicates that reducing the cement alkali content does not completely eliminate the expansion tendency of concrete microbars but rather exerts an inhibitory effect, slowing down the rate of expansion. It suggests that changes in cement alkali content have a direct and significant impact on the expansion performance of concrete microbars. Excessive or insufficient cement alkali content may lead to an increase in the expansion rate of concrete microbars, thereby affecting their long-term performance and durability. Therefore, when designing concrete mixtures, careful consideration should be given to the cement alkali content to optimize both the microstructure and macroscopic performance of concrete.

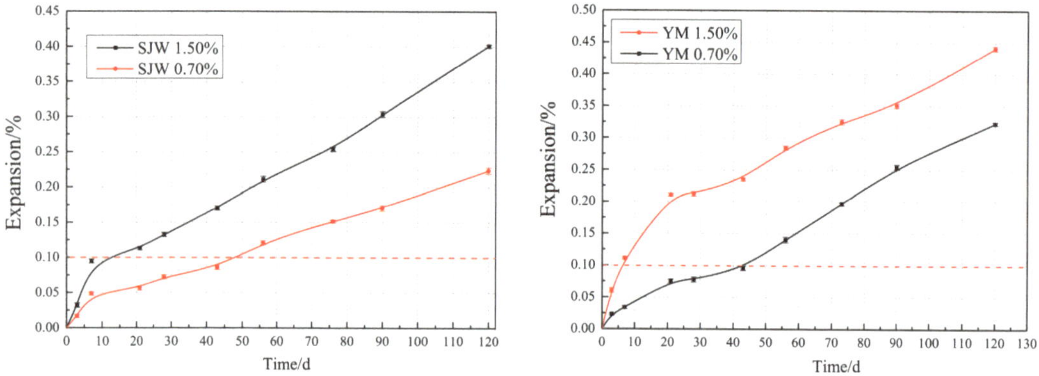

Figure 11. Expansion of SJW and YM when cement alkali content of 1.50% and 0.70%.

3.1.3. Effect of Fly Ash on Concrete Microbars

In Figure 12, the comparative curve chart illustrates the influence of fly ash content at 0%, 20%, 30%, and 40% on the AAR expansion of concrete microbars made with SJW and YM, with a cement alkali content of 1.50%. It can be observed from the graph that without the addition of fly ash, the expansion rates of concrete microbars made with SJW and YM are relatively high. However, as the content of fly ash increases, there is a decreasing trend in the expansion rates. This result suggests that the addition of fly ash appears to suppress the alkali–aggregate reaction expansion of concrete microbars made with SJW and YM.

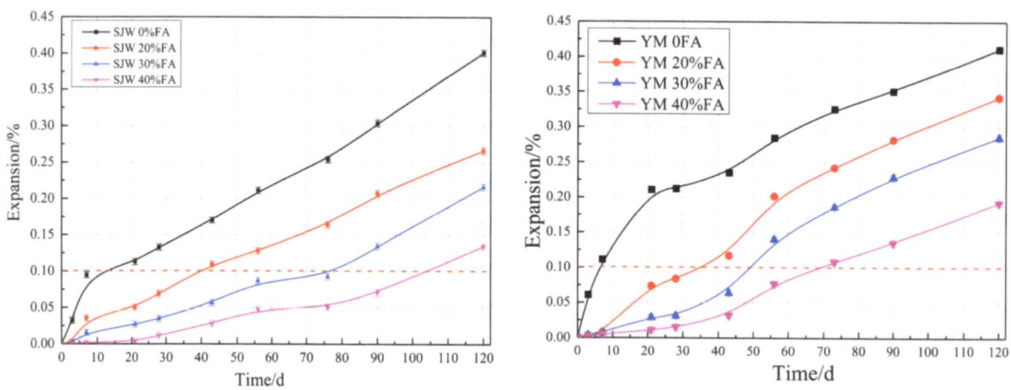

Figure 12. Expansion curves of concrete microbars with different fly ash contents under the same curing conditions and alkali content of 1.50% for SJW and YM.

Based on the experimental results, the following conclusion can be drawn: With the increase in fly ash content, the AAR expansion of concrete microbars made from both types of rocks gradually decreases. This indicates that the addition of fly ash helps to mitigate the expansion of concrete microbars.

At 28 days, the expansion rates of concrete microbars without fly ash exceeded the failure criterion of 0.10%, reaching 0.13% and 0.21%. In contrast, the expansion rates of concrete microbars containing 20%, 30%, and 40% fly ash were 0.06%, 0.03%, and 0.02%, respectively, significantly lower than those without fly ash.

After 120 days of curing, the expansion rates of concrete microbars containing 20%, 30%, and 40% fly ash for both SJW and YM were all below the failure criterion of 0.10%, with values of 0.08%, 0.05%, and 0.03% for SJW and 0.04%, 0.06%, and 0.04% for YM.

Comparing the two types of rocks, although concrete microbars made with YM exhibited higher expansion rates at the same fly ash content, this difference may be attributed to the inherent properties of YM rock and requires further investigation.

Overall, the addition of fly ash significantly reduced the expansion rate of concrete microbars. The experimental data also reveals a notable trend: as the fly ash content increases, the extent of AAR expansion in concrete microbars made from SJW and YM significantly decreases. When the fly ash content is increased to 20%, 30%, and 40%, the expansion rate of SJW decreases by 48%, 74%, and 91%, respectively, while the expansion rate of YM decreases by 65%, 86%, and 95%, respectively. This series of data indicates that fly ash addition can significantly suppress AAR-induced expansion in the short term (28 days) and remains effective in the long term (120 days). After 120 days of curing, the expansion rates of SJW concrete microbars containing 20%, 30%, and 40% fly ash decreased by 34%, 47%, and 67%, respectively, while those of YM concrete microbars decreased by 8%, 18%, and 53%, respectively. This suggests that with increasing fly ash content, the long-term expansion performance of concrete microbars is further improved.

These results demonstrate that the addition of fly ash can effectively suppress AAR expansion in concrete microbars, enhancing their durability.

Based on the experimental data in Figure 13, the following analysis can be conducted: When the cement alkali content is 0.70%, the expansion rates of SJW and YM concrete microbars without fly ash at 56 days are 0.12% and 0.14%, respectively, exceeding the failure criterion of 0.10%. This indicates that without fly ash, the concrete microbars exhibit some AAR expansion. After adding 30% and 40% fly ash, the expansion rates of SJW and YM concrete microbars at 56 days significantly decreased. The expansion rates of SJW decreased by 83% and 88%, respectively, while those of YM decreased by 72% and 78%, respectively. This suggests that the addition of fly ash effectively suppresses AAR expansion in concrete microbars. After 120 days of curing, although the expansion rates of SJW and YM concrete microbars with 30% and 40% fly ash increased slightly, they still remained below the 0.10% failure criterion. The expansion rates of SJW decreased by 48% and 76%, respectively, while those of YM decreased by 61% and 79%, respectively. This further confirms the effectiveness of fly ash addition in the long-term suppression of AAR expansion.

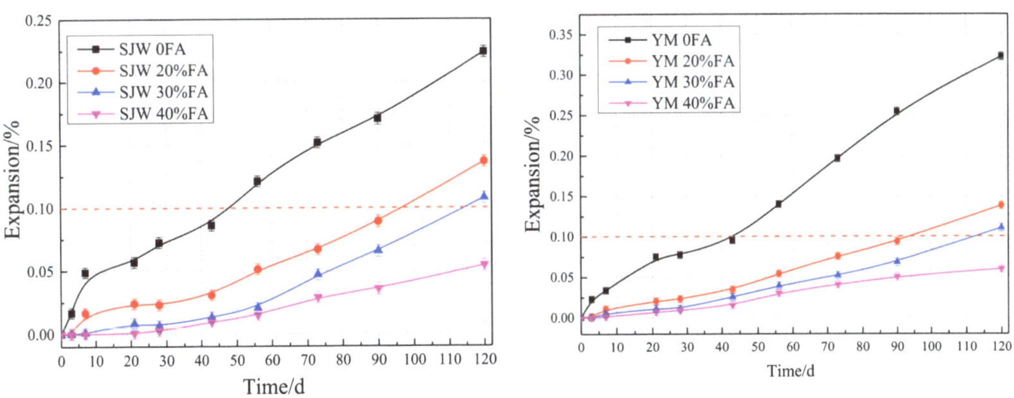

Figure 13. Expansion curves of concrete microbars with different fly ash contents under the same curing conditions and alkali content of 0.70% for SJW and YM.

3.2. Micro-Structure Analysis

Concrete microbars with a cement alkali content of 1.50%, without fly ash, derived from SJW samples, were subjected to 120 days of curing at 80 °C. Detailed observations were conducted using BSEM. Figure 14 presents the BSEM images of SJW rock after reacting, revealing evident dedolomitization reactions within the region approximately 100 μm from the boundary, while no reactions were observed in the central area.

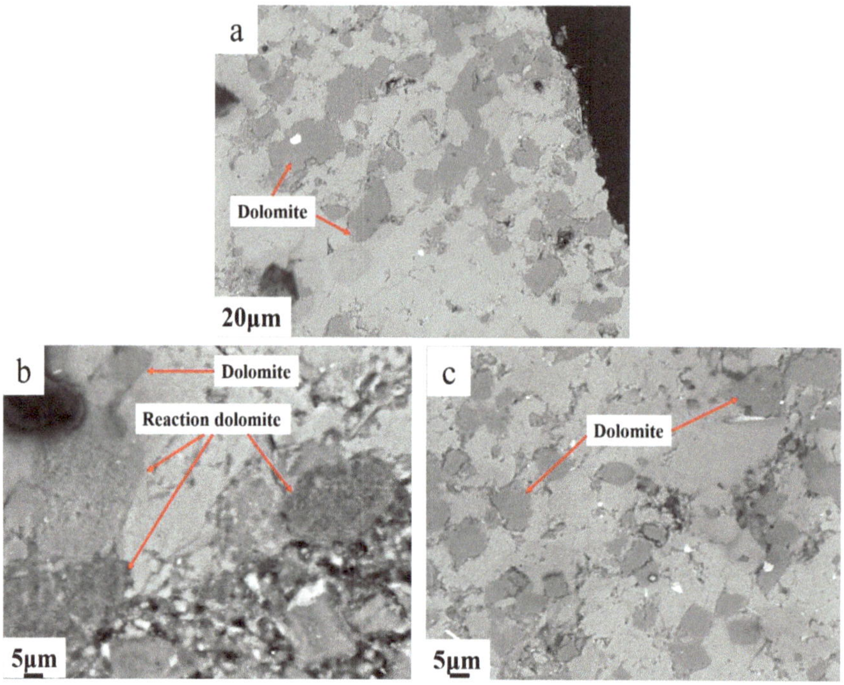

Figure 14. (**a**) BSEM images of dolomitic limestone SJW rock after reacting; (**b**)dedolomitization reactions within the region approximately 100 μm from the boundary; (**c**) no reactions were observed in the central area.

Through BSEM examination, the reactions were predominantly manifested in three forms: in Figure 15b, the dolomite region was magnified from Figure 15a, revealing distinct reaction rims around the dolomite grains. According to the EDS analysis results in Figure 15c–f, the surroundings of dolomite exhibited significant enrichment of clay minerals (primarily consisting of silicon and aluminum elements), with the presence of magnesium–silicon–aluminum phases detected within the reaction rims. Based on Li's research findings [22], this indicates that upon dissolution of dolomite, magnesium ions react with clay minerals to form magnesium–silicon–aluminum phases. The formation of such phases impedes further erosion by alkaline substances, thereby resulting in relatively lower levels of erosion experienced by dolomite.

In Figure 16a, we observed the image of another dolomite, finding two distinct reaction regions. In reaction region 1, the surface of dolomite underwent a reaction, displaying a rough granular texture. Through EDS analysis, seen in Figure 16b,c, we identified the reaction products such as calcite and brucite, likely occurring in the initial stages of the reaction, forming a structure where reacted and unreacted dolomite are intergrown [23]. In contrast, reaction region 2 represents a fully reacted area, where the predominant reaction product is brucite. From these observations, we can conclude that the reaction of dolomite in cement is a complex process involving various forms of reactions and products.

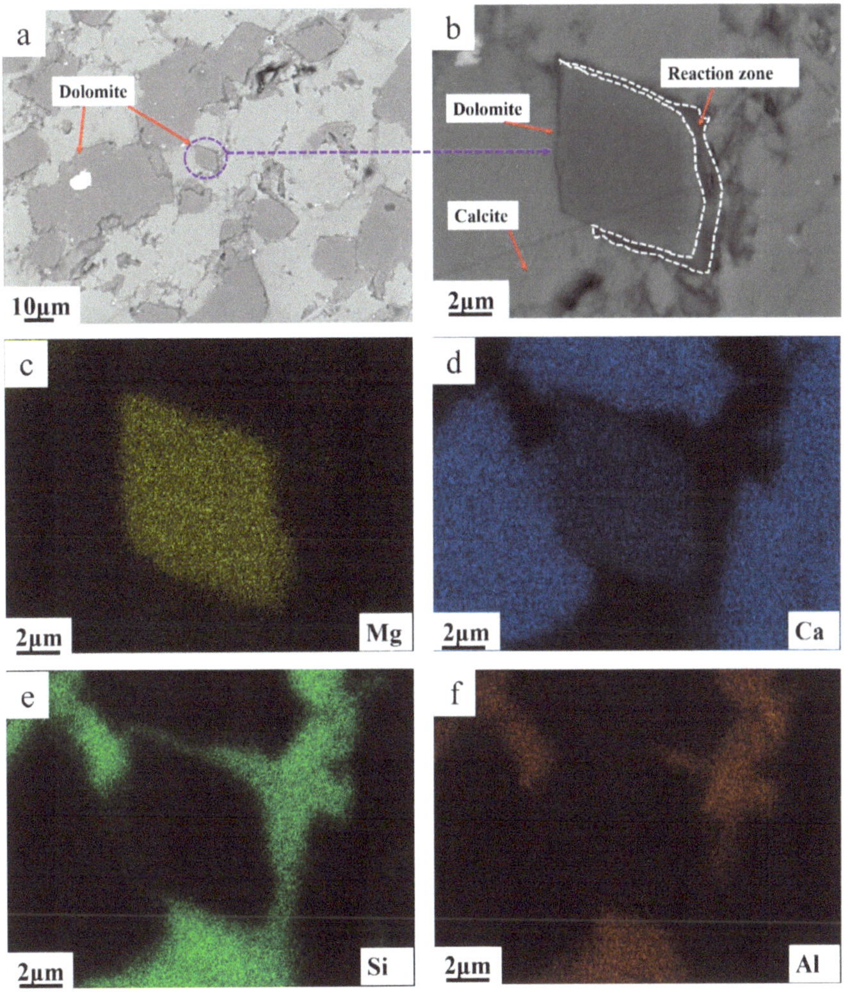

Figure 15. (**a**) BSEM images of SJW; (**b**) magnification of zone dolomite; (**c**) Mg map of magnified zone dolomite; (**d**) Ca map of magnified zone dolomite; (**e**) Si map of magnified zone dolomite; (**f**) Al map of magnified zone dolomite.

In Figure 17, the BSEM image after adding 40% fly ash is depicted. Figure 16b clearly reveals subtle variations at the edges of rock particles. It is noteworthy that in these edge regions, only a small amount of dolomite has undergone dedolomitization reaction. Upon further observation, it is found that in the areas closer to the interior, the effects of alkali attack are no longer present, and dolomite retains its original morphology. This finding is thought-provoking, as it suggests that the addition of fly ash not only inhibits the diffusion of alkali–aggregate reactions but also limits the extent of their impact.

In Figure 18, BSEM images after reducing the cement alkali content to 0.70% are presented. In these images, dolomite near the edges exhibits intact and distinct boundaries, contrasting sharply with previous observations. In this region, no generation of black-and-white intergrown structures or black reaction rims is observed.

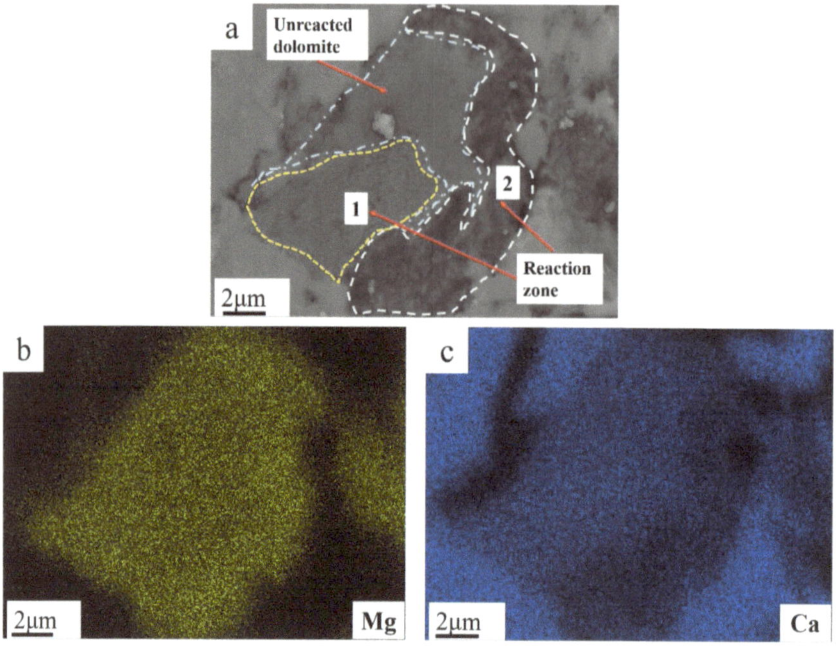

Figure 16. (**a**) BSEM images of SJW; (**b**) Mg map of magnified dolomite; (**c**) Ca map of magnified dolomite.

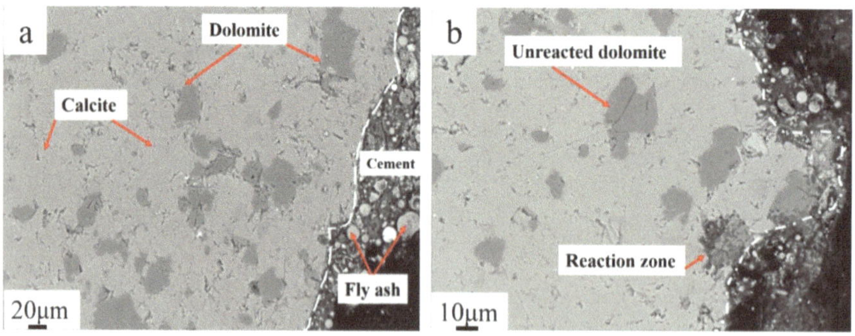

Figure 17. (**a**) BSEM image of SJW; (**b**) magnification of (**a**).

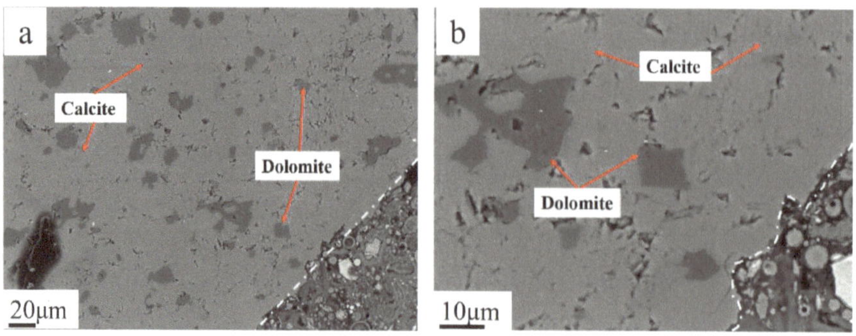

Figure 18. (**a**) BSEM image of SJW; (**b**) magnification of dolomite zone.

4. Discussion

A comparison of the swelling curves of the two rocks at different cement alkali contents revealed a sharp increase in the swelling rate at a cement alkali content of 1.50%. However, by reducing the alkali content to 0.70%, a significant reduction in the expansion rate was observed, although expansion was still observed. This suggests that simply lowering the cement alkali content is not an effective strategy for inhibiting the alkali carbonate reaction for aggregates with moderate to low reactivity. In practice, although cement alkali content is typically maintained below 1.50%, a cement alkali content of 0.70% is considered reasonable according to national standards. The causes of AAR include water, sufficient alkali, and active aggregate [2]. To inhibit ACR, the alkali content can be reduced. However, previous studies have shown that while reducing the cement alkali content can alleviate ACR, it cannot fully inhibit it [24]. There are two main reasons for this. Firstly, reducing the cement alkali content does not alter the pore structure of the slurry. Excessive harmful pores [25] in the slurry still facilitate the migration of OH^- into the aggregate. Secondly, cement hydration continues to produce $Ca(OH)_2$, which also provides OH^- for ACR [26]. Further experiments revealed that even at a cement alkali content of 1.50% and a fly ash content of up to 40%, the alkali–carbonate reaction of medium- and low-reactive dolomitic limestone aggregates (SJW and YM) was not significantly suppressed. According to the alkali activity test criteria, when the cement alkali content was reduced to 0.70%, the late expansion exceeded 0.10%, which still indicated that the aggregates were potentially reactive and might pose a threat to the concrete structure. Data for fly ash contents between 20% and 40% showed that 20% and 30% fly ash contents failed to control expansion effectively. However, when the fly ash content was raised to 40%, the expansion curves leveled off and no upward trend was observed, implying the effectiveness of 40% fly ash content in suppressing the alkali–carbonate reaction expansion in both types of rocks. This is because the addition of fly ash refines the porosity, and at the same time, fly ash reacts with cement hydration $Ca(OH)_2$ to generate C-S-H, which lowers the pH value inside the slurry and reduces the migration of OH^- to the inside of the aggregate, thus inhibiting the ACR [18,25].

In Qian's study [27], it was found that the pH of the aggregate surface is higher than that of the interior, indicating a significant presence of OH^- on the aggregate surface, initiating the reaction from the surface. Li's study [22] further demonstrated that the reaction products are dolomite and calcite. When the clay content is excessively high, Mg-Si-Al rings are formed, and the reaction gradually migrates towards the interior of the aggregate. This aligns with the experimental results. Using this criterion, the observation of the reaction area can reveal the effectiveness of adding fly ash. BSEM images further revealed the specific products of dedolomitization reactions. The study indicated that the main cause of alkali carbonate reaction is the dedolomitization reaction, with its products mainly including brucite and calcite. In cases where the clay content around dolomite is high, a reaction ring of Mg-Si-Al phase can form. In rock samples without fly ash, reaction products were mainly concentrated within 100 μm of the rock edge. However, with a 40% fly ash content, the reaction was significantly inhibited, with only a small amount of reaction found at the edge, possibly due to excessive external alkali content. When the cement alkali content decreased to 0.70%, no significant reaction was observed in the edge area. These results corroborate the concrete microbar expansion results, providing solid evidence for the study.

In summary, the addition of fly ash is an effective measure to improve the durability and long-term performance of concrete microbars, especially for concrete structures used in alkaline environments. However, it should be noted that the amount of fly ash added should be appropriate to ensure that the mechanical properties of concrete are not affected.

5. Conclusions

The aim of this study is to evaluate the ACR inhibition effect of medium- and low-activity dolomite limestone by observing the effect of fly ash. The concrete microcolumn method (RILEM AAR-5) was used as an evaluation criterion to experimentally seek the optimal combination of conditions for ACR inhibition in medium- and low-activity dolomite limestone. FESEM was utilized in the study to evaluate the ACR inhibition effect. Through the experimental results, this paper draws the following conclusions:

(1) Experimental results indicate that under a cement alkali content of 1.50%, it is challenging to accurately assess the inhibitory effect of fly ash on alkali carbonate reaction. When the alkali content is reduced to 0.70%, although the expansion rate decreases compared to the 1.50% alkali content condition, it still exceeds the threshold of 0.10% in the short term, indicating that reducing the alkali content may not effectively inhibit alkali carbonate reaction.

(2) In concrete microbars made with two different cement alkali contents and incorporating 20–40% fly ash, when the cement alkali content is 1.50%, none of the different proportions can inhibit alkali carbonate reaction. When the alkali content is reduced to below 0.70%, incorporating 20% and 30% fly ash still fails to suppress the reaction, while 40% fly ash achieves inhibitory effects.

(3) Through BSEM images, the products of dedolomitization reactions can be clearly observed. Analysis and observation of the reaction areas reveal that dedolomitization reactions mainly produce brucite, calcite, and Mg-Si-Al phases, with reactions primarily concentrated at the boundaries of the rocks. With the addition of fly ash, the reaction area significantly decreases, and when the cement alkali content decreases to 0.70%, successful control of the alkali carbonate reaction is achieved.

The results of these studies highlight that the incorporation of 40% fly ash in concrete structures containing moderately low reactive dolomitic limestone aggregates is effective in mitigating the alkali carbonate reaction, especially when the cement alkali content is controlled at 0.70%. Based on the experimental results, future work can include large-scale concrete exposure experiments for long-term observation. These experiments should include testing for flexural and compressive strength, as well as density, to verify the applicability of the working conditions.

Author Contributions: Conceptualization, M.D., W.L. and Z.M.; methodology, W.L. and Z.M.; data curation, S.D. and X.H.; writing—original draft, S.D.; investigation, S.D. and X.Z., formal analysis, X.Z. and B.C., resources X.H.; writing—review and editing, M.D. and B.C.; supervision M.D. All authors have read and agreed to the published version of the manuscript.

Funding: The work was funded by the Jiangsu Funding Program for Excellent Postdoctoral Talent (2022ZB362, 2023ZB286), and funded by the Open Project Program of Anhui Institute of Strategic Study on Carbon Dioxide Emissions Peak and Carbon Neutrality in Urban-Rural Development (Grant No. STY-023-01), the Graduate Student Scientific Research Innovation Projects in Jiangsu province, and the Priority Academic Program Development (PAPD) of Jiangsu Higher Education Institutions.

Institutional Review Board Statement: Not applicable.

Informed Consent Statement: Not applicable.

Data Availability Statement: The raw data supporting the conclusions of this article will be made available by the authors on request.

Acknowledgments: The authors would like to sincerely thank Zhongyang Mao, Wei Li, and Bi Chen of NJTECH and the staff from the State Key Laboratory of Materials-Oriented Chemical Engineering.

Conflicts of Interest: The authors declare no conflicts of interest.

References

1. Tang, M.S. Some Theoretical problems of alkali-aggregate reaction. *J. Chin. Ceram. Soc.* **1990**, *4*, 365–373.
2. Stanton, T.E. Expansion of concrete through reaction between cement and aggregate. *Trans. Am. Soc. Civ. Eng.* **1942**, *107*, 54–84. [CrossRef]
3. Gillott, J.E. Mechanism and kinetics of expansion in the alkali-carbonate rock reaction. *Can. J. Earth Sci.* **1964**, *1*, 121–145. [CrossRef]
4. Pagano, M.A.; Cody, P.D. A chemical approach to the problem of alkali-reactive carbonate aggregates. *Cem. Concr. Res.* **1982**, *12*, 1–12. [CrossRef]

5. Deng, M.; Tang, M.S. Measures to inhibit alkali-dolomite reaction. *Cem. Concr. Res.* **1993**, *23*, 1115–1120.
6. Chen, B.; Deng, M.; Huang, X.; Mo, L.; Huang, B.; Lan, X. Siliceous and dolomitic-bearing aggregates reaction in Tetramethylammonium hydroxide. *Constr. Build. Mater.* **2021**, *299*, 123948. [CrossRef]
7. Shikida, M.; Sato, K.; Tokoro, K.; Uchikawa, D. Differences in anisotropic etching properties of KOH and TMAH solutions. *Sens. Actuators A Phys.* **2000**, *80*, 179–188. [CrossRef]
8. Li, W.; Deng, M.; Mo, L.; Panesar, D.K.; Mao, Z. Alkali carbonate reaction (ACR): Investigations on mechanism of dedolomitization of dolomite in dolostones. *Constr. Build. Mater.* **2022**, *351*, 128942. [CrossRef]
9. Tapas, M.J.; Thomas, P.; Vessalas, K.; Nsiah-Baafi, E.; Martin, L.; Sirivivatnanon, V. Comparative study of the efficacy of fly ash and reactive aggregate powders in mitigating alkali-silica reaction. *J. Build. Eng.* **2023**, *63*, 105571. [CrossRef]
10. Krüger, M.E.; Heisig, A.; Hilbig, H.; Eickhoff, H.; Heinz, D.; Machner, A. Effect of aluminum on the structure of synthetic alkali-silica gels. *Cem. Concr. Res.* **2023**, *166*, 107088. [CrossRef]
11. Mccoy, W.J.; Caldwell, A.G. New approach to inhibiting alkali-aggregate expansion. *J. ACI* **1951**, *22*, 693–706.
12. Vrenderburgh, L.D.; Lemish, J. Evaluation of autoclave induced expansion of some Iowa carbonate rocks. *Lowa Acad. Sci.* **1964**, *71*, 335–341.
13. Diamond, S. A review of alkali-silica reaction and expansion mechanisms 1. Alkalis in cements and in concrete pore solutions. *Cem. Concr. Res.* **1975**, *5*, 329–345. [CrossRef]
14. Abd-Elssamd, A.; Ma, Z.J.; Hou, X.; Le Pape, Y. Influence of mineralogical and chemical compositions on alkali-silica-reaction of Tennessee limestones. *Constr. Build. Mater.* **2020**, *261*, 119916. [CrossRef]
15. Bakhoum, E.S.; Mater, Y.M. Decision analysis for the influence of incorporating waste materials on green concrete properties. *Int. J. Concr. Struct. Mater.* **2022**, *16*, 63. [CrossRef]
16. Shehata, M.H.; Jagdat, S.; Rogers, C.; Lachemi, M. Long-term effects of different cementing blends on alkali-carbonate reaction. *ACI Mater. J.* **2017**, *114*, 661–672. [CrossRef]
17. Joshaghani, A. The effect of trass and fly ash in minimizing alkali-carbonate reaction in concrete. *Constr. Build. Mater.* **2017**, *150*, 583–590. [CrossRef]
18. Ren, X.; Li, W.; Mao, Z.; Deng, M. Inhibition of the alkali-carbonate reaction using fly ash and the underlying mechanism. *Crystals* **2020**, *10*, 484. [CrossRef]
19. *GB/T 1596-2017*; Standardization Administration of China. Fly Ash for Cement and Concrete. Standards Press of China: Beijing, China, 2017.
20. Nixon, P.; Sims, I. RILEM Recommended Test Method: AAR-2-Detection of potential alkali-reactivity -Accelerated mortar-bar test method for aggregates. *Mater. Sci. Eng.* **2016**, *33*, 61–77. [CrossRef]
21. Sommer, H.; Nixon, P.J.; Sims, I. AAR-5: Rapid preliminary screening test for carbonate aggregates. *Mater. Struct.* **2005**, *38*, 787–792. [CrossRef]
22. Li, W. Mechanism for Dedolomitization of Dolomite in Dolomitic Rocks Soaked in NaOH Solutions. PhD Thesis, Nanjing Tech University, Nanjing, China, 2022.
23. Milanesi, C.A.; Marfil, S.A.; Locati, F. Expansive behavior of an alkali-carbonate reactive dolostone from Argentina: Proposal of an osmotic theory-based model to explain the expansion caused by the alkali attack. *Cem. Concr. Res.* **2020**, *138*, 106239. [CrossRef]
24. Wang, H.; Wu, D.; Zhen, M. Effect of fly ash and limestone powder on inhibiting alkali aggregate reaction of concrete. *Constr. Build. Mater.* **2019**, *210*, 620–626. [CrossRef]
25. Mehta, P.K. Study on blended portland cements containing santirin earth. *Cem. Concr. Res.* **1981**, *11*, 507–518. [CrossRef]
26. Cao, H.; Mao, Z.; Huang, X.; Deng, M. Inhibition of alkali-carbonate reaction by fly ash and metakaolin on dolomitic limestones. *Materials* **2022**, *15*, 3538. [CrossRef] [PubMed]
27. Qian, G. Textural and Geological Characteristics of Carbonate Rock and Alkali Carbonate Reaction. Ph.D. Thesis, Nanjing University of Chemical Technology, Nanjing, China, 1999.

materials

MDPI

Review

Early-Age Cracking of Fly Ash and GGBFS Concrete Due to Shrinkage, Creep, and Thermal Effects: A Review

Yingda Zhang [1], Xinyue Liu [1], Ziyi Xu [2], Weiguang Yuan [3], Yong Xu [1], Zuobang Yao [4], Zihao Liu [5] and Ruizhe Si [2,*]

1 School of Architecture and Civil Engineering, Xihua University, Chengdu 610039, China; yingda.zhang@mail.xhu.edu.cn (Y.Z.); liuxinyue1@stu.xhu.edu.cn (X.L.); 0120180009@mail.xhu.edu.cn (Y.X.)
2 Institute of Civil Engineering Materials, School of Civil Engineering, Southwest Jiaotong University, Chengdu 610031, China; xzy2020@my.swjtu.edu.cn
3 School of Environment and Civil Engineering, Chengdu University of Technology, Chengdu 610059, China; yuanweiguang19@cdut.edu.cn
4 Centre for Infrastructure Engineering and Safety, School of Civil and Environmental Engineering, The University of New South Wales, Sydney, NSW 2052, Australia; joe.zb.yao@outlook.com
5 Department of Architecture, Faculty of Environmental Engineering, The University of Kitakyushu, 1-1 Hibikino Wakamatsu, Kitakyushu 8080135, Fukuoka, Japan; z-liu@kitakyu-u.ac.jp
* Correspondence: ruizhesi@swjtu.edu.cn

Abstract: Supplementary cementitious materials (SCMs) are eco-friendly cementitious materials that can partially replace ordinary Portland cement (OPC). The occurrence of early-age cracking in OPC-SCM blended cement is a significant factor impacting the mechanical properties and durability of the concrete. This article presents a comprehensive review of the existing research on cracking in OPC-SCM concrete mix at early ages. To assess the effects of SCMs on the early-age cracking of concrete, the properties of blended cement-based concrete, in terms of its viscoelastic behavior, evolution of mechanical performance, and factors that affect the risk of cracking in concrete at early ages, are reviewed. The use of SCMs in OPC-SCM concrete mix can be an effective method for mitigating early-age cracking while improving the properties and durability of concrete structures. Previous research showed that the shrinkage and creep of OPC-SCM concrete mix are lower than those of conventional concrete. Moreover, the lower cement content of OPC-SCM concrete mix resulted in a better resistance to thermal cracking. Proper selection, proportioning, and implementation of SCMs in concrete can help to optimize the performance and reduce the environmental impact of OPC-SCM concrete mix.

Keywords: restrained shrinkage; early-age cracking; SCMs; degree of restraint; thermal cracking; model prediction

Citation: Zhang, Y.; Liu, X.; Xu, Z.; Yuan, W.; Xu, Y.; Yao, Z.; Liu, Z.; Si, R. Early-Age Cracking of Fly Ash and GGBFS Concrete Due to Shrinkage, Creep, and Thermal Effects: A Review. *Materials* **2024**, *17*, 2288. https://doi.org/10.3390/ma17102288

Academic Editor: Weiting Xu

Received: 15 April 2024
Revised: 4 May 2024
Accepted: 9 May 2024
Published: 12 May 2024

1. Introduction

In the field of civil engineering, the shrinkage, creep, and temperature cracking of concrete are crucial issues that have a profound impact on the structural stability and durability of buildings. The causes and effects of these problems require in-depth understanding and research to effectively control and prevent them during the design and construction process.

Concrete shrinkage refers to the decrease in volume during the drying or hardening process. When concrete dries, the internal moisture gradually evaporates, resulting in a decrease in the concrete volume. This shrinkage is particularly evident during the early stages of concrete hardening and may result in surface cracks.

Creep refers to the deformation of concrete under long-term loading. Unlike elastic deformation, creep deformation is slow and increases with time. Creep can reduce the load-bearing capabilities of structures and may lead to a decrease in structural safety and stability.

Temperature cracks are caused by temperature changes in concrete. Due to poor thermal conductivity, when the temperature changes, there is an uneven temperature distribution inside and outside the concrete, leading to thermal expansion and contraction and ultimately the formation of temperature cracks. These cracks may affect the durability and waterproofness of the structure.

With the ongoing progress in science and technology, researchers have been searching for more environmentally friendly and efficient building materials. One increasingly popular option is the application of fly ash (FA) and slag to partially replace cement in concrete [1]. This new type of concrete not only has excellent engineering properties but also significantly reduces carbon emissions, contributing to the green development of the construction industry.

Fly ash and slag are both industrial waste products, but they can play unique roles in concrete. Fly ash, a fine ash collected from the smoke of coal combustion, has high volcanic activity and can partially replace cement while enhancing the density of concrete. Slag, on the other hand, is a waste product produced during manufacturing that can be used as a binder in concrete after proper treatment. This not only reduces the production cost of concrete but also achieves the resource utilization of waste products.

The application of fly ash and slag to replace some of the cement in concrete has significant advantages. First, this type of concrete has higher strength and durability [2,3]. Due to the characteristics of fly ash and slag, the internal structure of concrete is denser than that of the other materials, decreasing susceptibility to external environmental factors and improving durability. Second, the production process of this type of concrete has lower carbon emissions. Compared to traditional cement production processes involving significant carbon emissions, using fly ash and slag significantly reduces carbon emissions during concrete production, which aligns with the green building development trend. Finally, this type of concrete has better economic benefits. The production cost of concrete can be reduced, and the utilization rate of waste products can be improved by using industrial waste products to partially replace cement, achieving a win–win situation for both economic and environmental benefits.

However, using fly ash and slag to replace cement in concrete also comes with some challenges. For example, ensuring the quality and stability of concrete is a crucial concern. Additionally, the rational utilization of waste products to avoid secondary pollution to the environment is also a consideration. Therefore, when using this new type of concrete, it is necessary to combine specific project needs with actual circumstances and implement reasonable blending and quality control measures.

Fly ash and GGBFS can lead to different cracking risks compared to conventional concrete. For instance, Gao et al. [4] conducted a self-designed elliptical ring test and the results showed that introduction of fly ash can effectively improve the crack resistance of concrete. Similar results were also found by Wang et al. [5], who demonstrated that the effect of the pozzolanic reaction of fly ash was of benefit to the thermal cracking resistance of mass concrete. For GGBFS concrete, Liu and Ma [6] illustrated that the cracking risks of GGBFS concrete were lower than those of OPC concrete because GGBFS can fill the space among the cement particles, leading to a reduced surface area of the interfacial transition zone (ITZ).

Moreover, researchers have explored the impact of supplementary cementitious materials (SCMs) on various concrete properties, including mechanical strength, shrinkage, and creep characteristics. However, there is limited documentation on the evaluation of cracks in concrete incorporating SCMs at early ages. The occurrence of cracking in an OPC-SCM concrete mix at early ages is a multifaceted phenomenon influenced by factors like the mechanical properties, thermal gradient, restraint, shrinkage, and creep characteristics of the concrete [7]. The existing review articles mainly focus on discussing the behavior of early-age cracking of OPC concrete [8–10]. As such, a comprehensive review of cracking of SCMs blended with cementitious materials at early ages is lacking. This paper presents relevant research studies on the cracking of cementitious materials at early ages and identifies the gaps in the existing literature, as well as the proposed methodology for further research.

2. Supplementary Cementitious Materials (SCMs)

2.1. Fly Ash

Fly ash (FA) is an industrial waste material with volcanic ash activity that can be used as an admixture in concrete to improve its performance. When fly ash is added to concrete, its performance undergoes a series of changes. First, the flexural strength and compressive strength of concrete mixtures can be enhanced by fly ash. This is due to the small particles of fly ash filling the gaps between cement particles, which densifies the concrete and thereby improves its mechanical properties. Second, fly ash can reduce the drying shrinkage of concrete, improving its durability and crack resistance. This is because fly ash can absorb excess moisture in concrete and form a dense film that effectively prevents moisture evaporation and the erosion of harmful substances, thereby enhancing the durability of the concrete [11–13]. As such, fly ash can increase the autogenous shrinkage [14], but the total shrinkage of fly ash concrete is comparable to that of OPC concrete due to the compensation of low drying shrinkage [15].

Furthermore, fly ash enhances concrete durability by bolstering resistance against sulfate attack and inhibiting chloride ion penetration. This is because certain chemical components in fly ash can react with sulfates and chloride ions, reducing their erosive effect on concrete and improving the density of the microstructure [16].

Fly ash can notably diminish the early-age heat of hydration, attributed to a dilution effect, resulting in a reduced risk of thermal cracking [17,18]. Dockter [19] suggested that FA concrete could mitigate the volume changes due to the alkali–silica reaction (ASR). The ASR mitigation of fly ash in concrete is influenced by alkali binding, diffusion control, and reduction in available $Ca(OH)_2$ [20]. Ling et al. [21] observed a slower strength development in FA concrete compared to normal concrete. Additionally, Naik et al. [22] reported that the permeability and porosity of concrete could be reduced by incorporation of FA in the mixtures.

Fly ash alters the early-age behavior of concrete at the microscopic level. Fládr et al. [23] reported that the increase in the level of fly ash replacement led to an increase in micro-porosity, and fly ash reduced the presence of low-density C-S-H but increased that of high-density C-S-H. When fly ash was incorporated, the concrete pore structure was influenced due to the pore-filler effect and densification due to the pozzolanic reaction. The cumulative pore volume in fly ash concrete notably rises within the medium capillary pore zone [24].

As such, when using fly ash in concrete, issues such as activity requirements, particle size distribution, dosage control, curing conditions, and storage and transportation need to be considered. Selecting fly ash that meets the requirements and comprehensively considering various performance indicators of the concrete can improve the performance and stability of the concrete [25].

2.2. Slag

Slag is a common industrial waste that can be used as an admixture in concrete to improve its performance. When slag is added to concrete, its performance undergoes a series of changes [26]. First, slag contributes substantially to the enhancement of both flexural and compressive strength in concrete. This is due to the small particles of slag filling the gaps between cement particles, which densifies the concrete and thereby improves its mechanical properties. In addition, certain chemical components in slag can react with certain components in cement to form a stronger crystal structure. Second, slag can enhance the toughness and crack resistance of concrete [27,28]. This is because slag can absorb excess moisture in the concrete and form a dense film that effectively prevents moisture evaporation and the erosion of harmful substances, thereby enhancing the durability of the concrete. In addition, the small particles of slag can also form small voids in the concrete, which can relieve stress concentration and thereby improve the toughness of the concrete [29].

Regarding the microscopic behavior of GGBFS-based concrete, Zhang et al. [30] illustrated that the addition of GGBFS can decrease the content of portlandite and improve the compactness of the ITZ, as well as refine the pore structure of concrete. Yalçınkaya and Çopuroğlu [31] reported that GGBFS can lead to elephant skin formation of UHPC because most of the air bubbles of UHPC cannot escape from the fresh cement matrix.

Furthermore, slag can also improve concrete durability by bolstering resistance against sulfate attack and inhibiting chloride ion penetration. This is because certain chemical components in slag can react with sulfates and chloride ions, reducing their erosive effect on the concrete. According to the work of Li et al. [32], the hydration reaction of ground granulated blast furnace slag (GGBFS) occurs at a slower rate compared to that of cement, resulting in a prolonged setting time.

3. Time-Dependent Behaviors

3.1. Autogenous Shrinkage

Concrete autogenous shrinkage refers to the decrease in volume during the hardening process due to chemical reactions and capillary action. This type of shrinkage typically begins immediately after concrete placement and may continue over a period. The autogenous shrinkage of ordinary concrete is affected by different factors, such as the aggregate content and type, water-to-cement (w/c) ratio, and type of cement. This process is influenced by many factors, including environmental humidity, temperature, and concrete curing conditions [33]. Autogenous shrinkage is regarded as an inherent aspect of shrinkage as the reduction in volume takes place without the concrete losing moisture to the surrounding environment [34].

However, when certain additives are added to concrete, such as FA and GGBFS, the nature and degree of autogenous shrinkage may change. The effect of fly ash on autogenous shrinkage is still being debated. First, it was reported that incorporation of FA can mitigate autogenous shrinkage in high-performance concrete (HPC) [35]. Yan and Chen [36] stated that autogenous shrinkage of fly ash concrete quasi-linearly decreased with the increase in the addition of fly ash. The results showed that the greater the FA substitution rate, the less autogenous shrinkage occurred. This is mainly because fly ash can fill the capillaries in concrete, reducing moisture evaporation and thus reducing autogenous shrinkage. Some studies reported that fly ash can increase autogenous shrinkage. For example, according to the study of Termkhajornkit et al. [14], FA exerted a notable impact on autogenous shrinkage during the early ages but had only a minimal effect at later ages.

On the contrary, GGBFS yielded contrasting outcomes regarding autogenous shrinkage compared to FA. As illustrated in Figure 1, GGBFS concrete exhibited a late reaction and contributed to higher autogenous shrinkage than the reference concrete [37]. The HPC containing GGBFS tended to enhance autogenous shrinkage [38]. Lim et al. [39] observed accelerated compressive strength development and heightened autogenous shrinkage in concrete mixtures containing GGBFS. Wei et al. [40] asserted that GGBFS has the potential to decrease early-age shrinkage due to the delayed pozzolanic reactions compared with the cement hydration; however, in the long-term, autogenous shrinkage surpassed that of the plain concrete due to sustained reactions involving GGBFS.

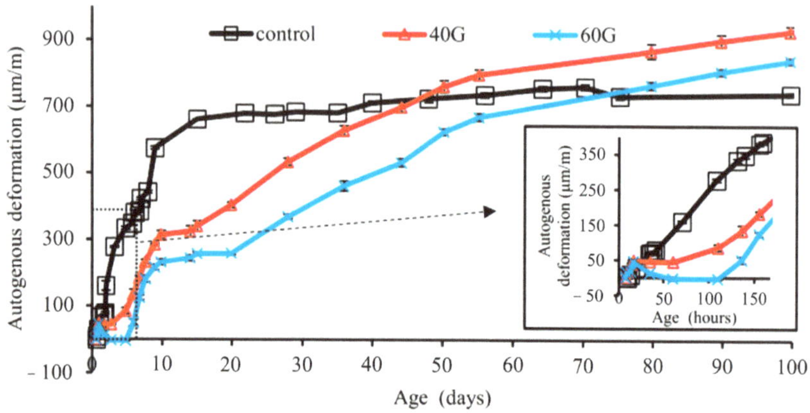

Figure 1. Influence of GGBFS on autogenous shrinkage [37].

3.2. *Drying Shrinkage*

Concrete drying shrinkage refers to the decrease in volume of concrete after hardening caused by moisture evaporation. This type of shrinkage typically begins within a period after concrete placement and may continue for a period of months or even years [41]. The drying shrinkage of ordinary concrete is affected by different factors, including the water-to-cement ratio, type of cement, and aggregate type and content. This process is influenced by many factors, including environmental humidity, temperature, and concrete curing conditions.

Extensive research has been conducted on the drying shrinkage of concrete mixes containing FA and GGBFS. Drying shrinkage of the concrete can be effectively reduced with the increased FA replacement ratio in the mixture [41]. It was reported that incorporation of 40% of FA in concrete can reduce drying shrinkage by up to 20% after 180 days of drying [41]. Comparable findings were presented by Seo et al. [42], who emphasized the advantageous reduction in drying shrinkage for mitigating the risk of concrete cracking under restrained conditions. The drying shrinkage reduction effect of FA is mainly because FA can fill the capillaries in concrete, reducing moisture evaporation and thus reducing drying shrinkage.

In terms of GGBFS, Yuan et al. [43] concluded that GGBFS-containing concrete tends to have lower drying shrinkage than a pure cement mixture. The decline in drying shrinkage became more pronounced with an increase in the level of cement replacement with GGBFS. This phenomenon is attributed to the slag particles, which can effectively occupy the pores in the concrete, mitigating capillary strength and consequently decreasing drying shrinkage. In the study by Saluja et al. [44], the investigation focused on the drying shrinkage behavior of GGBFS-containing concrete exposed to varying temperature conditions. The findings revealed a direct correlation with drying shrinkage temperature, as temperature significantly influences water loss.

For exposed concrete surfaces without sealing, the speed of moisture dissipation relies significantly on external conditions like airflow, humidity, sunlight exposure, and surrounding temperature [45]. In the case of substantial concrete structures prone to early-age thermal and autogenous distortions, the impact of drying shrinkage strains is typically less pronounced in the initial days post-concrete placement. Ensuring appropriate curing methods and adhering to sound construction practices becomes crucial in curtailing the repercussions of confined drying shrinkage [46,47]. Moreover, incorporating substances like fly ash or slag can contribute to alleviating these consequences [48].

3.3. Creep

The phenomenon where the deformation of concrete progressively intensifies over time under a sustained, long-term load is known as concrete creep. This deformation initiates promptly after the casting of concrete and can endure for numerous months or potentially extend over several years. The creep behavior of concrete is primarily associated with time, typically exhibiting a rapid increase in the early stage followed by a gradual decline, reaching stability within 2 to 5 years [49,50]. In situations where concrete shrinkage encounters restriction, tensile creep emerges, resulting in tensile stress relaxation. This aspect requires consideration while the tensile stress needs to be calculated.

Extensive research has been conducted on the creep performance of concrete incorporating FA and GGBFS. In terms of the difference of the creep behavior of concrete with the added FA and GGBFS, broadly speaking, the inclusion of fly ash tends to diminish concrete creep, whereas the incorporation of slag may increase the creep. Zhao et al. [51] investigated the creep properties of the concrete with FA under varying curing temperatures. Their findings revealed that the microstructure of C-S-H gel and reaction degree of the raw materials played a critical role in influencing the creep behavior of FA concrete. This is primarily because FA can fill the capillaries in concrete, mitigating moisture evaporation and thereby decreasing creep.

Wei et al. [52] reported that the evolution of tensile strain and stress of the concrete containing GGBFS was more rapid and substantial under higher curing temperatures. The influence of GGBFS was not pronounced within a 12-day testing period [52]. Khan et al. [53] found that the tensile creep strength of GGBFS-containing concrete exceeded that observed in cement-only samples at early ages. Similar findings were reported by Shariq et al. [54]. The presence of micro-spherical slag particles, which can fill the pores in concrete and reduce capillary action, might contribute to an escalation in concrete creep, as shown in Figure 2.

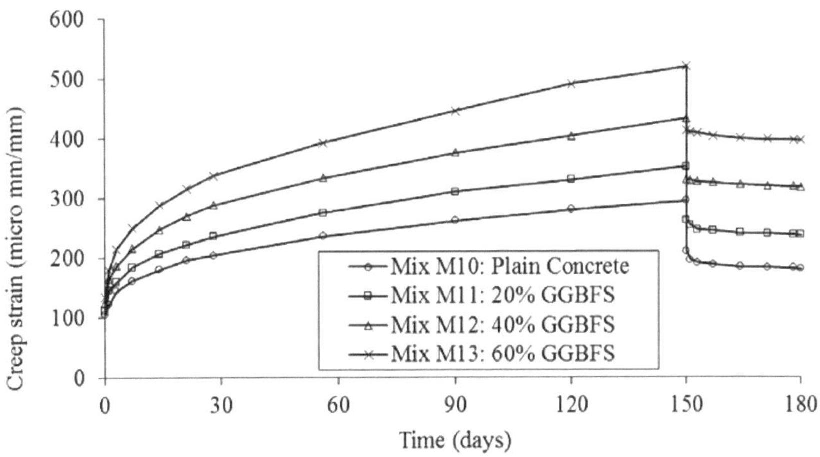

Figure 2. Effect of GGBFS on creep [54].

3.4. Thermal Effects

Temperature-induced cracks are the cracks that are presented both internally and on the concrete surface as a consequence of fluctuations in temperature. These cracks typically emerge during the hardening process of concrete and are primarily caused by factors such as internal and external temperature differences and cement hydration heat [55]. The shape and size of temperature cracks vary, potentially impacting the structural strength and durability of concrete [56].

Since the thermal conductivity of the concrete is low, the enhanced internal temperature of the concrete caused by the hydration of cement can result in the thermal gradients between the surface and interior of concrete [55]. The tensile stress can be introduced by the thermal gradients, potentially causing surface cracks, while the element's interior experiences compressive stress. When the environment temperature is lower than that of the surface of the concrete, the interior encounters tensile stress due to the external constraints and thermal gradient. This scenario may result in the development of penetrating cracks. Figure 3 illustrates the mechanisms behind cracking induced by temperature variations in concrete.

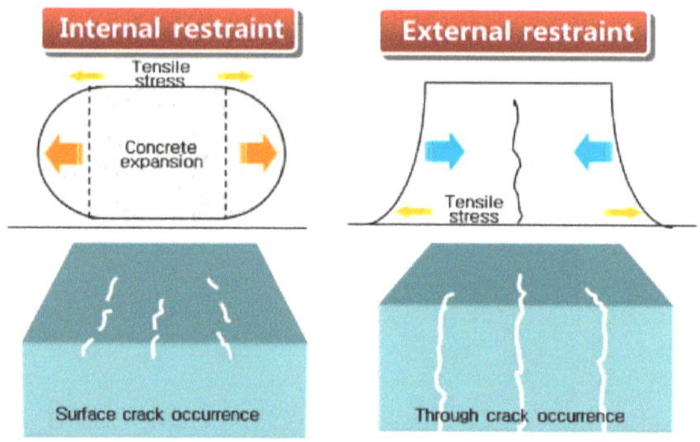

Figure 3. Schematic of the thermal crack generation in concrete [57].

The formation of cracks in concrete due to temperature is closely connected to the deformations induced by temperature variations in the concrete. The water-binder ratio, and content and type of cementitious material, affect the quantity of generated heat of the mixture. The thermal strain resulting from temperature changes is determined by the concrete's coefficient of thermal expansion (CTE). Therefore, the CTE serves as a crucial parameter in concrete, offering a method for calculating thermal strain and facilitating precise determination of thermal stress. The CTE refers to the rate of change in the concrete length or volume per unit temperature change [58]. In the work of Kada et al. [59], the significant variations in the CTE of curing concrete within initial hours after the casting of the concrete were observed. Consequently, when assessing the behavior of thermal stress of concrete in its early stages, depending on an unchanged CTE value calculated for fully cured concrete is unreliable. Neglecting to consider changes in CTE during the early stages may lead to an undervaluation of temperature-induced stress evolution.

Regarding the changes in concrete temperature cracks after adding fly ash and slag, generally speaking, fly ash can reduce concrete temperature cracks, while slag may increase concrete temperature cracks. This is mainly because fly ash can lower the hydration heat of concrete, thereby reducing the generation of temperature cracks. Slag, due to its micro-spherical particles, can fill the pores in concrete, reducing capillary action and potentially increasing temperature cracks. Moreover, it has been reported that GGBFS particle hydration is highly related to the ettringite formation because of aluminate phases for ettringite formation, leading to the long setting time and low early-age mechanical properties and hydration heat flow [60].

3.5. Mechanical Properties

The emergence of tensile stress from restricted shrinkage is influenced by both deformation and evolution of mechanical properties, such as modulus of elasticity, characteristics of tensile relaxation, the extent of external constraint, and tensile strength. Moreover, in the case of mass concrete experiencing a substantial temperature rise from cement hydration, especially within 24 to 36 h post-placement, the mechanical properties of concrete, particularly the modulus of elasticity, exhibit rapid development [61]. Research findings from ref. [62,63] indicate that the increase in the elastic modulus surpasses that of tensile strength during this period, suggesting a notable escalation in tensile stress. Nevertheless, investigations into stress progression in concrete at early ages have highlighted the importance of concrete's viscoelastic behavior, which can lead to substantial stress relaxation. Previous studies suggested that up to 60% of the stress generated from shrinkage can be alleviated by tensile relaxation [63,64]. As shown in Figure 4, it has been reported that the residual stress of concrete reduced with increasing GGBFS replacement [65]. The above discussion emphasizes the necessity of comprehensively examining the behavior of concrete while investigating the stresses induced by restrained shrinkage, including factors like tensile creep, free shrinkage, susceptibility to cracking, and temperature elevation. Specifically, the evolution of mechanical properties has to be performed with a temperature history akin to that of the core section of the concrete element.

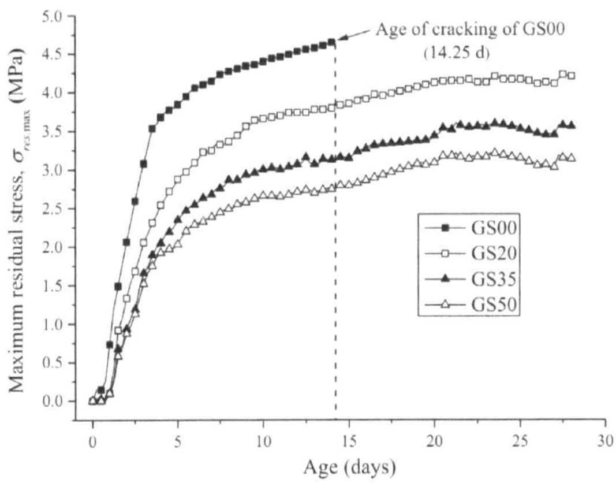

Figure 4. Residual stress development of OPC concrete and GGBFS concrete [65].

As previously mentioned, comprehensively examining concrete behavior is necessary when investigating the evolution of tensile stresses induced by restrained shrinkage. This examination should consider temperature elevation, tensile creep, shrinkage, and sensitivity of cracking. In particular, the assessment of mechanical properties has to be performed under a temperature history akin to that experienced by the core section of the concrete.

4. Factors Influencing Cracking in Concrete

4.1. Degree of Restraint

In concrete restrained cracking, the degree of constraint refers to the degree to which external factors restrict the free deformation of the concrete. These constraints can be caused by factors such as internal shrinkage, external loads, and temperature changes.

This constraint is closely related to the cracking behavior of concrete [56,66,67]. When the constraint on concrete is low, the free deformation is large, and the concrete is less likely to crack. As the constraint increases, the free deformation of the concrete is restricted, increasing the susceptibility to cracking. Furthermore, the generated stress is a combination of distinct stress components arising from external and internal constraints. The stress from internal constraints originates from the uneven distribution of temperature inside the concrete elements. The stress from external restraints can be generated as the contraction or expansion of a structural element encountering impediments from foundations, surrounding structures, and the underlying subsoil. The magnitude of external restraint is mainly contingent upon the comparative modulus of elasticity and dimensions of both the restraining elements and concrete.

Kawabata et al. [68] examined the impact of the restraint degree on the patterns of cracking and expansive pressure in concrete. Their observations indicated a significant correlation between expansion behavior and the restraint degree, as shown in Figure 5. Zych [69] investigated the external constraint of wall elements in reinforced concrete. In engineering practice, the degree of constraint can be adjusted by controlling factors, such as the mix ratio of concrete, curing conditions, and load conditions, to avoid or mitigate concrete cracking. The cracking of concrete is related not only to the degree of restraint, but also to other factors, such as the material properties of the concrete and the environmental conditions. Therefore, in practical engineering, various factors need to be comprehensively considered based on specific situations to prevent or mitigate concrete cracking.

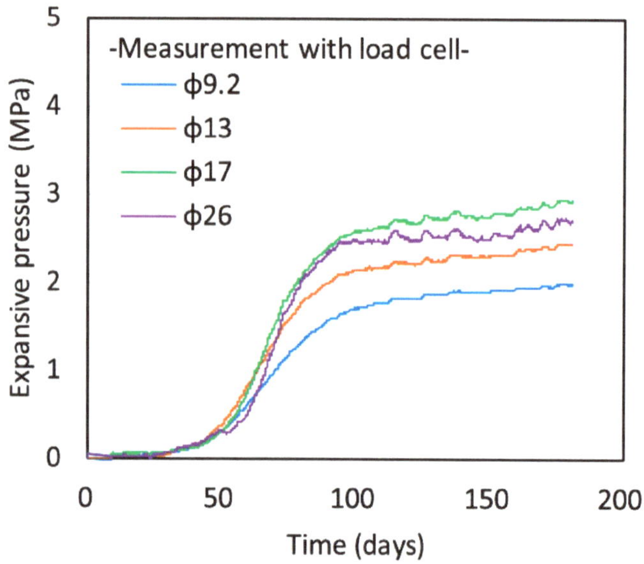

Figure 5. Influence of expansive behavior and degree of restraint [68].

4.2. Effect of Concrete Constituents

The mechanical properties of concrete can be negatively affected by the added SCMs (such as FA and GGBFS) at early ages. Consequently, the occurrence of cracking in concrete at early ages is notably influenced by the type of cementitious materials [70]. Altoubat et al. [71] investigated the influence of fly ash on early-age cracking by performing a restrained shrinkage test. The findings indicated that the relaxation behavior and cracking resistance of the mixture can be enhanced by introduction of FA compared to a cement-only mixture. Furthermore, the curing conditions, including air curing, 3-day water curing, and 7-day water curing, also play a role in determining the potential for cracking. The application of water curing can notably reduce cracking at early ages compared to air curing alone. Regarding concrete with GGBFS, Shen et al. [72] asserted that the addition of 50% slag can mitigate the restrained strain, resulting in improved cracking resistance compared to that observed in the cement-only mixture.

The selection of the binding materials determined the temperature evolution of the mass concrete. The previous literature [73–77] has indicated that the use of ASTM Type II or Type IV cement, as well as partial replacement of OPC with SCMs, tends to markedly diminish the temperature elevation in concrete compared with the concrete made by Type I cement. Furthermore, elevating the w/b ratio results in a decrease in the hydration heat and decelerates the release rate of the heat during the early-age hydration.

Batog and Giergiczny [76] investigated the impact of incorporating FA and GGBFS into cement on the hydration heat and hardening temperature of concrete. The outcomes of their investigation are depicted in Figure 6. As the proportions of FA or GGBFS in the cement mixtures increased, there was a notable reduction in both the quantity and rate of heat generated. This effect was particularly obvious in mixtures containing FA.

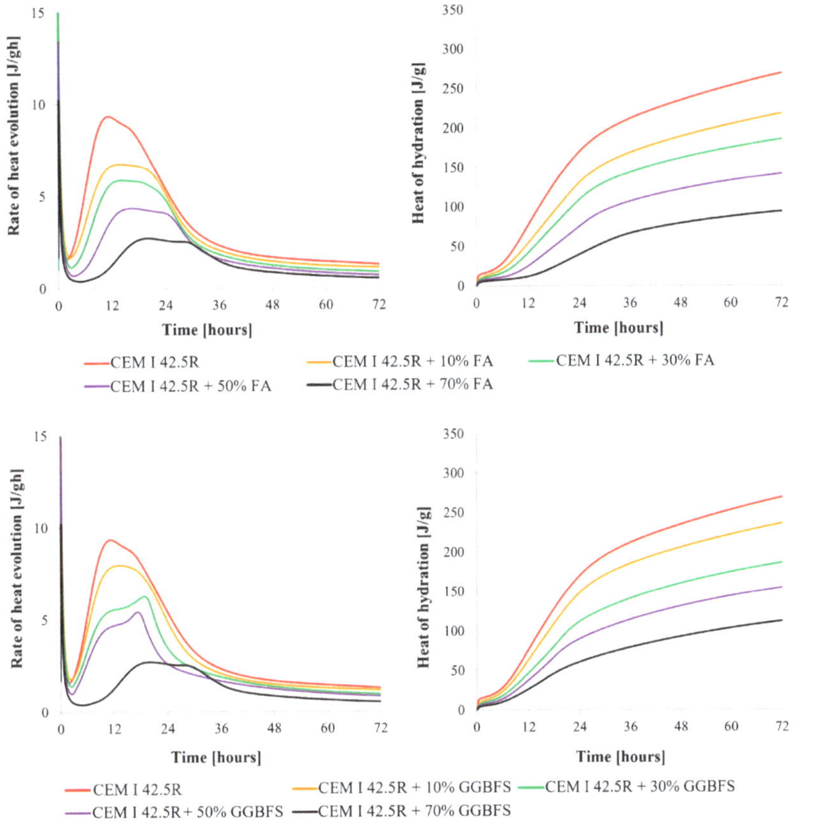

Figure 6. Influence of siliceous FA and GGBFS on the hydration heat and heat release rate under isothermal conditions (20 °C, w/c = 0.5 [76].

Despite the potential for SCMs to mitigate the rise of temperature in concrete structures, the occurrence of cracking remains a possibility. As noted by Springenschmid et al. [61], this is attributed to the slow development of the elastic modulus in the initial hardening stage, specifically within the first 12 h. Consequently, there is insufficient formation of compressive stresses within this timeframe to counteract the tensile stresses arising at the subsequent cooling stage. Furthermore, the decreased hydration rate leads to a lower tensile strength, which hinders the impact of thermal stresses.

As previously mentioned, further investigation is needed to explore the impact of SCMs on both total shrinkage and the evolution of mechanical properties, including strength, modulus of elasticity, heat evolution, and creep coefficient in mass concrete.

5. Analytical Models That Predict Factors Affecting Early-Age Concrete Cracking

5.1. Analytical Models for Predicting Shrinkage

The shrinkage behavior is a highly intricate physical and chemical process, influenced by numerous factors. Material-related aspects, including the quantity and type of cement, content of aggregate, and w/c ratio, as well as the type and quantity of SCMs and admixtures, exert a substantial impact on shrinkage. Construction and environmental considerations, such as the relative temperature surrounding the concrete structure, play a pivotal role in affecting shrinkage. Structurally, factors like nominal section size and volume-to-surface ratio, among others, bear a significant influence on drying shrinkage. Therefore, to quantitatively estimate concrete shrinkage, it is also necessary to comprehensively consider the above three aspects.

The ACI209-92 model proposed by The American Concrete Institute (ACI) in 1992 was subsequently reconfirmed and adopted in 2008 [78]. This model considers numerous factors, including the content of sand, type and content of cement, porosity, slump, duration and temperature of curing, curing humidity, temperature of concrete, content of water, and shape and size. The European Concrete Commission and the International Prestressed Concrete Association proposed the fib Model Code 2010 model [79]. This model accounts for the age of the concrete element at the beginning of drying, duration of drying, concrete age, 28-day compressive strength, type of cement, notional size, relative humidity, etc. Bazant et al. derived statistics and conducted analysis based on the B3 model and proposed a more applicable B4 model in 2015 [80]. In the B4 model, the autogenous and drying shrinkage models are split. The B4 model considers the aggregate type, effective thickness, cement type, elastic modulus, relative humidity, shape, age of concrete at the beginning of drying, etc. Note that this model also applies to concrete-containing admixtures. In 1993, the GZ 1993 model was introduced by Gardner and Zhao, formulated through an analysis of numerous results from long-term shrinkage tests; subsequently, Gardner and Lockman proposed the GL 2000 model by improving the GZ 1993 model [81]. The GL 2000 model requires the 28-day compressive strength, element size, and relative humidity as inputs. This model can be used regardless of the type of admixture, temperature, or curing regime.

In 2018, a shrinkage prediction model was introduced by the Australian Concrete Association AS3600 [82]. This model requires compressive strength, time, location factor, notional size, etc., as inputs. The shrinkage prediction model introduced by the European Concrete Association in the Eurocode 2 model [83] considers the type of cement, relative humidity, notional size, compressive strength, etc. Table 1 summarizes the parameters needed for different prediction models.

Table 1. Required parameters for different shrinkage prediction models.

Parameters	ACI209-R92	fib 2010	GL 2000	B4	AS3600-2018	Eurocode 2
Compressive strength at 28 days		Yes	Yes		Yes	Yes
Elastic modulus of concrete at loading	Yes					
Type of cement	Yes	Yes		Yes		Yes
Curing method	Yes		Yes	Yes		
Age of concrete at loading	Yes	Yes	Yes	Yes	Yes	Yes
Relative humidity	Yes	Yes	Yes	Yes		Yes
Volume to surface ratio	Yes	Yes	Yes	Yes		
Slump	Yes					
Fine aggregate content	Yes					
Air content	Yes					
Shape of section				Yes		
w/c ratio				Yes		
a/c ratio				Yes		
Admixtures				Yes		
Hypothetical thickness					Yes	Yes
Location					Yes	

Goel et al. [84] compared shrinkage test results and prediction models for various grades of concrete. They found that the shrinkage prediction by the GL 2000 model was the closest to the corresponding experimental results. Al-Manaseer and Prado [85] summarized and analyzed the RILEM and NU-ITI concrete databases and concluded that the ACI209R model was the closest to the test results, followed by the B3 and fib 2010 models. Shariq et al. [54] compared various models and shrinkage test results for slag concrete and found that the ACI209-R92 model had the lowest prediction results, while the GL2000 model had the highest prediction results. Kataoka et al. [86] also conducted a shrinkage test and reported that the ACI209R and GL 2000 models provided acceptable predictions. However, these prediction models are calibrated for OPC-based concrete instead of FA- and GGBFS-based concrete. As a result, additional shrinkage tests should be performed on the basis of the SCM-based concrete.

5.2. Analytical Models for Predicting Creep

The choice of a creep prediction model for concrete is related to the accuracy of the creep prediction values. Researchers worldwide have proposed different creep prediction models, but these models are all empirical formulas based on many experiments.

The ACI 209R model [78] can estimate the ultimate creep value by employing a hyperbolic function and expressing the computed value as a function of time. This model can calculate the creep of Type I and Type III cement under the relative humidity range from 40% to 100%, considering both steam and standard curing conditions. The concrete age at loading, sample size, surrounding temperature, relative humidity, conditions of curing, and the portion of the fine aggregate are the main factors determining the shape and limit of the curve.

The fib Model Code 2010 model [79] emerged from a multitude of experimental tests. It can be applied to concrete under ambient temperature (5–30 °C) and relative humidity levels from 40% to 100%, with a strength within the range of 12 to 80 MPa. The applied maximum strength has to be lower than 40% of the compressive strength. This model is divided into two classifications, namely basic creep and drying creep. The compressive strength, specimen size, ambient relative humidity over time, etc., are considered. The model also innovatively proposes that the temperature effect of concrete is equivalent to the adjusted age of concrete, and the adjusted age is used to predict the creep of the concrete.

In 1995, Bazant et al. [87] established the B3 model and divided creep into two classifications: drying creep and basic creep. They deduced the calculation expressions of drying creep and basic creep through consolidation theory and diffusion theory. Thus, this model is a semi-empirical and semi-theoretical formula according to theoretical and experimental data. The B3 model considers the influence of the w/c ratio, cement type, cement content, coarse and fine aggregate contents, concrete drying period, ambient relative humidity, concrete density, age of concrete at loading, etc. The B3 model also calculates the concrete creep strain through the compliance function, which is divided into the elastic compliance function, basic creep compliance, and drying creep compliance. Nevertheless, this model is usually not suitable for early-age creep. Østergaard et al. [88] introduced the parameter q5 to adjust viscoelastic parameter 2 of the B3 model to obtain the modified early-age creep model. D'Ambrosia [89] also affirmed the suitability of the modified B3 model for early-age concrete through performing the early-age creep tests.

On the other hand, the B4 model can be applied to estimate the creep of more types of concrete, such as concrete mixes with FA, silica fume, superplasticizer, air-entraining agent, and other admixtures, while the effect of admixtures on creep has not been considered by other models [80]. Moreover, a simplified model was also proposed, and concrete creep can be predicted accurately when only the 28-day compressive strength of the concrete is known.

The GL2000 model [81] takes into account various factors, including relative humidity, volume/surface ratio, duration of loading, initial loading, and drying period of concrete. This model offers enhanced convenience in calculation as it does not differentiate between basic and drying creep, ensuring accurate calculations.

The AS3600-2018 model [82] takes into consideration the total creep without separate calculations for basic and drying creep. The strength factor, creep factor, age of loading thickness, location, temporal development of creep are considered in this model.

For the Eurocode 2 model [83], the relative humidity, the time of loading, nonlinear creep factor, compressive strength, and temporal development of creep are considered. This model can be employed for predicting the creep of concrete under ambient temperature from −40 to 40 °C and relative humidity from 40% to 100%. Table 2 summarizes the parameters needed for the different prediction models.

Table 2. Required parameters for different creep prediction models.

Parameters	ACI209-R92	fib 2010	GL 2000	B4	AS3600-2018	Eurocode 2
Compressive strength at 28 days	Yes	Yes	Yes		Yes	Yes
Elastic modulus of concrete at loading	Yes	Yes	Yes	Yes	Yes	Yes
Type of cement	Yes	Yes		Yes		Yes
Curing method	Yes		Yes	Yes		
Age of concrete at loading	Yes	Yes	Yes	Yes	Yes	Yes
Relative humidity	Yes	Yes	Yes	Yes		Yes
Volume to surface ratio	Yes	Yes	Yes	Yes		
Slump	Yes					
Fine aggregate content	Yes					
Air content	Yes					
Shape of section				Yes		
w/c ratio				Yes		
a/c ratio				Yes		
Admixtures				Yes		
Hypothetical thickness					Yes	Yes
Location					Yes	
Nonlinear creep		Yes			Yes	Yes

Goel et al. [84] compared creep tests and prediction models for various grades of concrete. They found that the compressive creep prediction by the GL 2000 model was the closest to its experimental results. Al-Manaseer and Prado [85] summarized and analyzed the RILEM and NU-ITI concrete databases and concluded that the ACI209-R92 model was the closest to the test results, followed by the B3 and GL2000 models. Shariq et al. [54] compared various models and compressive creep test results for slag concrete and found that the ACI209-R92 model had the lowest prediction results, while the GL2000 model had the highest prediction results.

The above models do not specify whether they can be used to predict compressive or tensile creep. However, these models were deduced based on compressive creep test results. According to the previous discussion, the mechanisms and characteristics of tensile and compressive creep are different under a certain stress level. As such, if the above models are used to predict tensile creep directly, considerable deviations and variations between the prediction and test results will be observed.

Dabarera et al. [90] measured the tensile creep of HPC employing TSTM and dog-bone samples and compared the test results with the basic creep model of the fib Model Code 2010. The findings showed that the fib Model Code 2010 model significantly underestimates the tensile basic creep of HPC at early ages. Thus, they introduced a modified basic tensile creep model derived from the basic compressive creep model of the fib 2010 model. The validity of the proposed model was affirmed through validation against experimental results conducted by Ji et al. [91] and Atrushi [92]. Zhang et al. [93] proposed a tensile creep model for FA and GGBFS concrete based on specific test conditions (23 °C and 50%). They successfully verified the applicability of the proposed model for estimating tensile stress during the restrained shrinkage ring test.

There are still many concerns to be addressed in the study of tensile creep models: (1) The establishment of a compressive creep model is based on experimental statistical data, the statistical data of tensile creep are insufficient, and the uncertainty of the model is difficult to measure quantitatively. (2) The linear creep coefficient model constitutes a binary function characterized by two parameters: the initial loading age and the load holding time. However, the comprehensive consideration of the influence of the initial loading age on the creep model remains incomplete. (3) The size effect of tensile creep and the influence of temperature and humidity parameters still lack systematic research.

5.3. Analytical Models for Predicting Evolution of Tensile Stress during the Restrained Ring Test

Normally, the cracking of concrete due to the constrained shrinkage at early ages can be tested by using three test methods: the restrained ring test (RRT) [94,95], the restrained shrinkage test in one dimension [96,97], and the restrained shrinkage test in two dimensions [98,99]. Among these approaches, the RRT is the most frequently utilized experimental technique to assess the susceptibility of concrete to cracking under constrained shrinkage. This preference arises from its practical implementation compared to the other two test methods [100–102].

Hossain and Weiss [100] introduced a method to assess both the evolution of stress relaxation and residual stress through the RRT. Both residual stress and theoretical elastic stress can be quantitatively estimated using their approach. To study the influence of tensile creep on cracking performance of concrete at early ages, Khan et al. [66] used both the tensile creep test and RRT. Their findings revealed that the coefficients of tensile creep obtained from tensile creep tests can effectively be applied to assess the tensile creep-induced stress relaxation during the RRT.

The method of computing tensile stress evolution in concrete, based on testing the strain of the steel of the steel ring, is not suitable for predicting such development and can only be used to analyze the results of ring tests [100]. Moreover, the theoretical elastic stress model proposed by Hossain et al. [100] enables the estimation of concrete tensile stress based on the restraint degree, modulus of elasticity, and free shrinkage, but does not consider the effect of tensile creep. To address this limitation, Zhang et al. [67] introduced a model to analyze the RRT that captures the effects of tensile creep and restrained shrinkage using the method of the age-adjusted effective modulus, as shown in Equations (1) and (2). According to experimental data capturing the time-dependent evolution of the tensile creep, total free shrinkage, and modulus of elasticity, the analytical model effectively forecasts the tensile stress in restrained concrete rings. The model's accuracy and reliability were confirmed through successful validation using numerical finite element simulation.

$$\sigma_{cs} = D_R \, \varepsilon_{sh} \, \overline{E}_e \tag{1}$$

where σ_{cs} is the tensile stress experienced by the restrained concrete rings, D_R is the restraint degree and can be estimated using Equation (2), and $\overline{E}_e$ is the age-adjusted effective modulus.

$$D_R = \frac{\frac{R_{OC}^2 + R_{OS}^2}{R_{OC}^2 - R_{OS}^2}}{\frac{\overline{E}_e}{E_s} \cdot \frac{\left[(1+\nu_s)R_{IS}^2 + (1-\nu_s)R_{OS}^2\right]}{\left(R_{OS}^2 - R_{IS}^2\right)} + \frac{\left[(1+\nu_c)R_{OC}^2 + (1-\nu_c)R_{OS}^2\right]}{\left(R_{OC}^2 - R_{OS}^2\right)}} \tag{2}$$

where R_{OS} is outer radius of the steel ring and R_{IS} is inner radius of the steel ring; R_{OC} is the exterior circumference of the concrete ring; and ν_c and ν_s are the Poisson's ratios of concrete and steel, respectively.

5.4. Analytical Models That Predict the Risk of Cracking at Early Ages

The risk coefficient, determined by the RRT, is employed to evaluate the potential for early-age concrete cracking. It can be expressed as the ratio of time-dependent tensile stress (σ_{act}) to the time-dependent tensile strength (f_t), as shown in Equation (3).

$$R(t) = \frac{\sigma_{act}(t)}{f_t(t)} \tag{3}$$

A value of 0 for the risk coefficient indicates no evolution of tensile stress in the mixture, preventing the occurrence of cracking. As the risk coefficient increases with the gradual development tensile stress, the risk of cracking also rises. In the work by Khan et al. [53], for GGBFS concretes with strengths greater than 40 MPa, the cracking time was slightly greater than that of the reference concretes. Moreover, the 60% GGBFS concrete mix cracked earlier than the 40% GGBFS mixture because the tensile strength of the concrete with 60% GGBFS was less than that of the concrete with 40% GGBFS. Among the tested mixes, the FA concretes exhibited the best performance. This is because the tensile creep and tensile strength of FA concrete were similar to those of reference concretes, but the lower free shrinkage development of FA concretes led to less tensile stress being generated, and the cracking time was subsequently delayed. Thus, the risk coefficient is a frequently employed parameter to evaluate the potential of concrete cracking at early ages.

6. Conclusions and Recommendations

This paper conducts a thorough review of a diverse range of literature associated with cracking in OPC-SCM cement-based concrete at early ages. The experimental investigations into early-age cracking of normal concreate have predominantly centered on time-dependent behaviors, such as crack patterns, creep, and shrinkage. However, there has been a limited comprehensive examination of the combined effects of these factors. Additionally, the environmental advantages of SCMs have garnered research attention, but their specific impacts on concrete cracking at early ages remain inadequately explored. Furthermore, there is a noticeable gap in the study of modeling concrete cracking at early ages, particularly for the prediction of time-dependent behaviors of concrete. Consequently, future research exploration of factors influencing the cracking of FA and GGBFS concrete at early ages can be conducted, with a focus on temperature, tensile creep, shrinkage, tensile stress, and temperature evolution of concrete, as follows:

1. In terms of shrinkage, past research has shown that OPC-SCM concrete mixes have a lower shrinkage rate than ordinary concrete, primarily due to the optimized combination of fine aggregates and high-efficiency water-reducing agents. However, the shrinkage rate of these materials is still relatively high, increasing susceptibility to cracking. Future research should focus on further reducing the shrinkage rate and improving the anti-cracking performance. This can be achieved through optimized selection of raw materials, fine-tuned mix design, and improved construction techniques.

2. In terms of creep, OPC-SCM concrete mixes exhibit significantly better creep performance than ordinary concrete. However, as time passes and as environmental conditions change, creep performance may vary, potentially impacting structural performance. Therefore, future research should focus on the long-term creep performance of OPC-SCM blended cement-based concrete to better assess its impact on structural safety.

3. In terms of thermal cracking, OPC-SCM concrete mixes have a lower cement content and higher early strength, resulting in good resistance to thermal cracking. However, at high temperatures, this material has a larger coefficient of thermal expansion, which can lead to cracking. Future research should explore methods to reduce the coefficient of thermal expansion and enhance the thermal cracking resistance through mix ratio optimization and the addition of external admixtures.

4. Future research should also focus on studying the durability, environmental performance, and compatibility of OPC-SCM concrete mixed with other materials in Portland cement-based concrete. Additionally, establishing and refining relevant design specifications and construction guidelines will facilitate the practical application of this material in engineering projects.

Author Contributions: Y.Z.: Conceptualization, Funding acquisition, Project administration, Writing—original draft. X.L.: Data curation, Formal analysis, Investigation, Validation. Z.X.: Methodology, Supervision. W.Y.: Data curation, Investigation. Y.X.: Data curation, Validation. Z.Y.: Methodology, Investigation. Z.L.: Methodology, Writing—review and editing. R.S.: Funding acquisition, Project administration, Resources, Writing—review and editing. All authors have read and agreed to the published version of the manuscript.

Funding: This paper was funded by the Xihua University Project (Z231001) and Jinhua Key Science and Technology Plan Project (2023-1-128). This study was partially supported by the China Postdoctoral Science Foundation (2022M712639) and the Sichuan Science and Technology Program (2023NSFSC0897).

Institutional Review Board Statement: Not applicable.

Informed Consent Statement: Not applicable.

Data Availability Statement: Data are contained within the article.

Conflicts of Interest: The authors declare no conflict of interest.

References

1. von Greve-Dierfeld, S.; Lothenbach, B.; Vollpracht, A.; Wu, B.; Huet, B.; Andrade, C.; Medina, C.; Thiel, C.; Gruyaert, E.; Vanoutrive, H.; et al. Understanding the carbonation of concrete with supplementary cementitious materials: A critical review by RILEM TC 281-CCC. *Mater. Struct.* **2020**, *53*, 136. [CrossRef]
2. Elahi, M.; Shearer, C.; Reza, A.; Saha, A.; Khan, M.; Hossain, M.; Sarker, P. Improving the sulfate attack resistance of concrete by using supplementary cementitious materials (SCMs): A review. *Constr. Build. Mater.* **2021**, *281*, 1226280. [CrossRef]
3. Beglarigale, A.; Yazici, H. Mitigation of detrimental effects of alkali-silica reaction in cement-based composites by combina-tion of steel microfibers and ground-granulated blast-furnace slag. *J. Mater. Civ. Eng.* **2014**, *26*, 04014091. [CrossRef]
4. Gao, Y.; Zhang, H.; Tang, S.; Liu, H. Study on early autogenous shrinkage and crack resistance of fly ash high-strength lightweight aggregate concrete. *Mag. Concr. Res.* **2013**, *65*, 906–913. [CrossRef]
5. Wang, J.; Liu, S.; Yang, W. Influence of fly ash on crack resistance of mass concrete. *Adv. Mater. Res.* **2011**, *194–196*, 825–828. [CrossRef]
6. Liu, Y.; Ma, D. The research of GGBS′ influence on the cracking resistance of concrete. *Adv. Mater. Res.* **2014**, *919–921*, 1864–1867. [CrossRef]
7. Qu, Z.; Si, R.; Jia, P.; Zhang, Y. Creep and relaxation responses of fly ash concrete: Linear and nonlinear cases. *Case Stud. Constr. Mater.* **2022**, *17*, e01562. [CrossRef]
8. Bentz, D. A review of early-age properties of cement-based materials. *Cem. Concr. Res.* **2008**, *38*, 196–204. [CrossRef]
9. Maruyama, I.; Lura, P. Properties of early-age concrete relevant to cracking in massive concrete. *Cem. Concr. Res.* **2019**, *123*, 105770. [CrossRef]
10. Safiuddin, M.; Kaish, A.; Woon, C.; Raman, S. Early-age cracking in concrete: Causes, consequences, remedial measures, and recommendations. *Appl. Sci.* **2018**, *8*, 1730. [CrossRef]
11. Yu, Z.; Ye, G. New perspective of service life prediction of fly ash concrete. *Constr. Build. Mater.* **2013**, *48*, 764–771. [CrossRef]
12. Khaliq, W.; Kodur, V. Behavior of high strength fly ash concrete columns under fire conditions. *Mater. Struct.* **2013**, *46*, 857–867. [CrossRef]
13. Wang, S.; Baxter, L.E.; Fonseca, F. Durability of biomass fly ash concrete: Freezing and thawing and rapid chloride permeability tests. *Fuel* **2008**, *87*, 359–364. [CrossRef]
14. Termkhajornkit, P.; Nawa, T.; Nakai, M.; Saito, T. Effect of fly ash on autogenous shrinkage. *Cem. Concr. Res.* **2005**, *35*, 473–482. [CrossRef]
15. Afroz, S.; Zhang, Y.; Nguyen, Q.; Kim, T.; Castel, A. Shrinkage of blended cement concrete with fly ash or limestone calcined clay. *Mater. Struct.* **2023**, *56*, 15. [CrossRef]
16. Zhang, J.; Zhang, S.; Liu, B. Degradation technologies and mechanisms of dioxins in municipal solid waste incineration fly ash: A review. *J. Clean. Prod.* **2020**, *250*, 119507. [CrossRef]
17. De la Varga, I.; Castro, J.; Bentz, D.; Zunino, F.; Weiss, J. Evaluating the hydration of high volume fly ash mixtures using chemically inert fillers. *Constr. Build. Mater.* **2018**, *161*, 221–228. [CrossRef]
18. Wongkeo, W.; Thongsanitgarn, P.; Poon, C.; Chaipanich, A. Heat of hydration of cement pastes containing high-volume fly ash and silica fume. *J. Therm. Anal. Calorim.* **2019**, *138*, 2065–2075. [CrossRef]
19. Dockter, B. Using class C fly ash to mitigate alkali-silica reactions in concrete. In Proceedings of the 3rd World of Coal Ash, WOCA Conference Proceedings, Lexington, KY, USA, 4–7 May 2009.
20. Hay, R.; Ostertag, C. New insights into the role of fly ash in mitigating alkali-silica reaction (ASR) in concrete. *Cem. Concr. Res.* **2021**, *144*, 106440. [CrossRef]
21. Ling, X.; Setunge, S.; Patnaikuni, I. Effect of different concentrations of lime water on mechanical properties of high volume fly ash concrete. In *From Materials to Structures: Advancement Through Innovation, Proceedings of the 22nd Australasian Conference on the Mechanics of Structures and Materials, ACMSM 2012, Sydney, Australia, 11–14 December 2012*; CRC Press: Leiden, The Netherlands, 2013.
22. Naik, T.; Singh, S.; Hossain, M. Permeability of concrete containing large amounts of fly ash. *Cem. Concr. Res.* **1994**, *24*, 913–922. [CrossRef]
23. Fládr, J.; Bílý, P.; Chylík, R.; Prošek, Z. Macroscopic and microscopic properties of high performance concrete with partial replacement of cement by fly ash. *Solid State Phenom.* **2019**, *292*, 108–113. [CrossRef]
24. Liu, J.; Ou, G.; Qiu, Q.; Chen, X.; Hong, J.; Xing, F. Chloride transport and microstructure of concrete with/without fly ash under atmospheric chloride condition. *Constr. Build. Mater.* **2017**, *146*, 493–501. [CrossRef]
25. Shen, W.; Zhang, Z.; Li, J.; Li, Z.; Wang, Z.; Cao, L.; Rong, G.; Wu, M.; Zhao, D.; Zhao, Z. Experimental investigation on the high-volume fly ash ecological self-compacting concrete. *J. Build. Eng.* **2022**, *60*, 105163. [CrossRef]

26. Niu, Q.; Feng, N.; Yang, J.; Zheng, X. Effect of superfine slag powder on cement properties. *Cem. Concr. Res.* **2002**, *32*, 615–621. [CrossRef]
27. Liu, S.; Li, L. Influence of fineness on the cementitious properties of steel slag. *J. Therm. Anal. Calorim.* **2014**, *117*, 629–634. [CrossRef]
28. Tavasoli, S.; Nili, M.; Serpoosh, B. Effect of GGBS on the frost resistance of self-consolidating concrete. *Constr. Build. Mater.* **2018**, *165*, 717–722. [CrossRef]
29. Gao, J.; Qian, C.; Liu, H.; Wang, B.; Li, L. ITZ microstructure of concrete containing GGBS. *Cem. Concr. Res.* **2005**, *35*, 1299–1304. [CrossRef]
30. Zhang, S.; Niu, D.; Wang, Y.; Tian, W.; Luo, D.; Zhang, L. Insight into mechanical properties and microstructure of concrete containing steel slag and ground-granulated blast-furnace slag. *J. Sustain. Cem.-Based Mater.* **2023**, *12*, 1169–1180. [CrossRef]
31. Yalçınkaya, Ç.; Çopuroğlu, O. Elephant skin formation on UHPC surface: Effects of climatic condition and blast furnace slag content. *Constr. Build. Mater.* **2021**, *268*, 121126. [CrossRef]
32. Li, Z.; Ding, Z. Property improvement of Portland cement by incorporating with metakaolin and slag. *Cem. Concr. Res.* **2003**, *33*, 579–584. [CrossRef]
33. Kovler, K.; Zhutovsky, S. Overview and future trends of shrinkage research. *Mater. Struct.* **2006**, *39*, 827–847. [CrossRef]
34. Zhang, Y.; Afroz, S.; Nguyen, Q.; Kim, T.; Nguyen, D.; Castel, A.; Nairn, J.; Gilbert, R. Autogenous Shrinkage of Fly Ash and Ground Granulated Blast Furnace Slag Concrete. *Mag. Concr. Res.* **2023**, *75*, 283–295. [CrossRef]
35. Lee, H.; Lee, K.; Kim, B. Autogenous shrinkage of high-performance concrete containing fly ash. *Mag. Concr. Res.* **2003**, *55*, 507–515. [CrossRef]
36. Yan, P.; Chen, Z. Autogenous Shrinkage of Fly Ash Concrete with Different Water Binder Ratio. In Proceedings of the 8th International Symposium on Cement & Concrete, Nanjing, China, 20–23 September 2013.
37. Afroz, S.; Zhang, Y.; Nguyen, Q.; Kim, T.; Castel, A. Effect of limestone in General Purpose cement on autogenous shrinkage of high strength GGBFS concrete and pastes. *Constr. Build. Mater.* **2022**, *327*, 126949. [CrossRef]
38. Zhao, Y.; Gong, J.; Zhao, S. Experimental study on shrinkage of HPC containing fly ash and ground granulated blast-furnace slag. *Constr. Build. Mater.* **2017**, *155*, 145–153. [CrossRef]
39. Lim, S.; Wee, T. Autogenous shrinkage of ground-granulated B last-furnace slag concrete. *ACI Struct. J.* **2000**, *97*, 587–593.
40. Wei, Y.; Hansen, W.; Biernacki, J.; Schlangen, E. Unified Shrinkage Model for Concrete from Autogenous Shrinkage Test on Paste with and without Ground-Granulated Blast-Furnace Slag. *ACI Mater. J.* **2011**, *108*, 13–20.
41. Nath, P.; Sarker, P. Effect of Mixture Proportions on the Drying Shrinkage and Permeation Properties of High Strength Concrete Containing Class F Fly Ash. *KSCE J. Civ. Eng.* **2013**, *17*, 1437–1445. [CrossRef]
42. Seo, T.; Lee, M.; Choi, C.; Ohno, Y. Properties of drying shrinkage cracking of concrete containing fly ash as partial replacement of fine aggregate. *Mag. Concr. Res.* **2010**, *62*, 427–433. [CrossRef]
43. Yuan, J.; Lindquist, W.; Darwin, D.; Browning, J. Effect of slag cement on drying shrinkage of concrete. *ACI Mater. J.* **2015**, *112*, 267–276. [CrossRef]
44. Saluja, S.; Kaur, K.; Goyal, S.; Bhattacharjee, B. Long-Term Drying Shrinkage of GGBFS-Incorporated RCC under Various Temperature Exposures. *J. Mater. Civ. Eng.* **2021**, *33*, 04021122. [CrossRef]
45. Liu, J.; Yu, J.; Hu, S.; Hu, C.; Wang, J.; Yu, Z.; Huang, K. Analysis of crack evolution of expansive soil embankment under extreme arid climate. *Case Stud. Constr. Mater.* **2024**, *20*, e02902. [CrossRef]
46. Wei, J.; Farzadnia, N.; Khayat, K. Synergistic effect of macro synthetic fiber and shrinkage-reducing admixture on engineering properties of fiber-reinforced super-workable concrete. *Constr. Build. Mater.* **2024**, *414*, 134566. [CrossRef]
47. Zhou, F.; Huang, J.; Jian, S.; Tan, H.; Lv, Y.; Hu, H.; Wang, W.; Yang, R.; Manuka, M.; Yin, Y.; et al. Photocurable resin as rapid in-situ protective coating for slag concrete against dry shrinkage. *Constr. Build. Mater.* **2023**, *396*, 132171. [CrossRef]
48. Dong, D.; Huang, Y.; Pei, Y.; Zhang, X.; Cui, N.; Zhao, P.; Hou, P.; Lu, L. Effect of spherical silica fume and fly ash on the rheological property, fluidity, setting time, compressive strength, water resistance and drying shrinkage of magnesium ammonium phosphate cement. *J. Build. Eng.* **2023**, *63*, 105484. [CrossRef]
49. Neville, A.; Dilger, W.; Brooks, J. *Creep of Plain and Structural Concrete*; Construction Press: London, UK; New York, NY, USA, 1983.
50. Han, B.; Xie, H.; Zhu, L.; Jiang, P. Nonlinear model for early age creep of concrete under compression strains. *Constr. Build. Mater.* **2017**, *147*, 203–211. [CrossRef]
51. Zhao, Q.; Liu, X.; Jiang, J. Effect of curing temperature on creep behavior of fly ash concrete. *Constr. Build. Mater.* **2015**, *96*, 326–333. [CrossRef]
52. Wei, Y.; Hansen, W. Tensile Creep Behavior of Concrete Subject to Constant Restraint at Very Early Ages. *J. Mater. Civ. Eng.* **2013**, *25*, 1277–1284. [CrossRef]
53. Khan, I.; Xu, T.; Khan, M.; Castel, A.; Gilbert, R. Effect of Various Supplementary Cementitious Materials on Early-Age Concrete Cracking. *J. Mater. Civ. Eng.* **2020**, *32*, 04020049. [CrossRef]
54. Shariq, M.; Prasad, J.; Abbas, H. Creep and drying shrinkage of concrete containing GGBFS. *Cem. Concr. Compos.* **2016**, *68*, 35–45. [CrossRef]
55. Zhu, B. *Thermal Stresses & Temperature Control of Mass Concrete*; Tsinghua University Press: Beijing, China, 2014.
56. ACI Committee 207. *ACI207.1R-05: Guide to Mass Concrete*; American Concrete Institute: Farmington Hills, MI, USA, 2012.

57. Ha, J.; Jung, Y.; Cho, Y. Thermal crack control in mass concrete structure using an automated curing system. *Autom. Constr.* **2014**, *45*, 16–24. [CrossRef]
58. Bamforth, P.; Oenton, S. Early-age Thermal Crack Control in Concrete—Changes in UK design resulting from the introduction of EN1992. In Proceedings of the 11th Annual International Fib Symposium—Concrete: 21st Century Superhero: Building a Sustainable Future, London, UK, 22–24 June 2009.
59. Kada, H.; Lachemi, M.; Petrov, N.; Bonneau, O.; Aitcin, P.-C. Determination of the coefficient of thermal expansion of high performance concrete from initial setting. *Mater. Struct.* **2002**, *35*, 35–41. [CrossRef]
60. Tang, P.; Wen, J.; Fu, Y.; Liu, X.; Chen, W. Improving the early-age properties of eco-binder with high volume waste gypsum: Hydration process and ettringite formation. *J. Build. Eng.* **2024**, *86*, 108988. [CrossRef]
61. Springenschmid, R.; Breitenbücher, R. Influence of Constituents, Mix Proportions and Temperature on Cracking Sensitivity of Concrete. In *Prevention of Thermal Cracking in Concrete*; RILEM Report 15; CRC Press: London, UK, 1998.
62. De Schutter, G.; Taerwe, L. Degree of Hydration-Based Description of Mechanical Properties of Early Age Concrete. *Mater. Struct.* **1996**, *29*, 335–344. [CrossRef]
63. Gutsch, A. Properties of early age concrete—Experiments and modelling. *Mater. Struct.* **2002**, *34*, 76–79.
64. Igarashi, S.; Bentur, A.; Kovler, K. Autogenous shrinkage and induced restraining stresses in high-strength concretes. *Cem. Concr. Res.* **2000**, *30*, 1701–1707. [CrossRef]
65. Shen, D.; Liu, K.; Wen, C.; Shen, Y.; Jiang, G. Early-age cracking resistance of ground granulated blast furnace slag concrete. *Constr. Build. Mater.* **2019**, *222*, 278–287. [CrossRef]
66. Khan, I.; Castel, A.; Gilbert, R. Tensile creep and early-age concrete cracking due to restrained shrinkage. *Constr. Build. Mater.* **2017**, *149*, 705–715. [CrossRef]
67. Zhang, Y.; Afroz, S.; Nguyen, Q.; Kim, T.; Eisentrager, J.; Castel, A.; Xu, T. Analytical model predicting the concrete tensile stress development in the restrained shrinkage ring test. *Constr. Build. Mater.* **2021**, *307*, 124930. [CrossRef]
68. Kawabata, Y.; Ueda, N.; Miura, T.; Multon, S. The influence of restraint on the expansion of concrete due to delayed ettringite formation. *Cem. Concr. Compos.* **2021**, *121*, 104062. [CrossRef]
69. Zych, M. Degree of external restraint of wall segments in semi-massive reinforced concrete tanks: Part I rectangular segments. *Struct. Concr.* **2018**, *19*, 820–828. [CrossRef]
70. Qu, Z.; Liu, Z.; Si, R.; Zhang, Y. Effect of Various Fly Ash and Ground Granulated Blast Furnace Slag Content on Concrete Properties: Experiments and Modelling. *Materials* **2022**, *15*, 3016. [CrossRef] [PubMed]
71. Altoubat, S.; Junaid, M.; Leblouba, M.; Badran, D. Effectiveness of fly ash on the restrained shrinkage cracking resistance of self-compacting concrete. *Cem. Concr. Compos.* **2017**, *79*, 9–20. [CrossRef]
72. Shen, D.; Jiao, Y.; Kang, J.; Feng, Z.; Shen, Y. Influence of ground granulated blast furnace slag on early-age cracking potential of internally cured high performance concrete. *Constr. Build. Mater.* **2020**, *233*, 117083. [CrossRef]
73. Mehta, P.; Monteiro, J. *Concrete: Microstructure, Properties and Materials*; McGraw-Hill: New York, NY, USA, 2014.
74. Ballim, Y.; Graham, P. The effects of supplementary cementing materials in modifying the heat of hydration of concrete. *Mater. Struct.* **2009**, *42*, 803–811. [CrossRef]
75. Kim, G.; Lee, E.; Nam, J.; Koo, K. Analysis of hydration heat and autogenous shrinkage of high-strength mass concrete. *Mag. Concr. Res.* **2011**, *63*, 377–389. [CrossRef]
76. Batog, M.; Giergiczny, Z. Influence of mass concrete constituents on its properties. *Constr. Build. Mater.* **2017**, *146*, 221–230. [CrossRef]
77. Markandeya, A.; Shanahan, N.; Gunatilake, D.; Riding, K.; Zayed, A. Influence of slag composition on cracking potential of slag-portland cement concrete. *Constr. Build. Mater.* **2018**, *164*, 820–829. [CrossRef]
78. ACI Committee 209. *ACI209R: Prediction of Creep, Shrinkage, and Temperature Effects in Concrete Structures*; American Concrete Institute Committee 209: Farmington Hills, MI, USA, 2008.
79. FIB. *fib Model Code for Concrete Structures*; Wiley-VCH Verlag GmbH & Co. KGaA: Weinheim, Germany, 2010.
80. Bazant, Z. RILEM draft recommendation: TC-242-MDC multi-decade creep and shrinkage of concrete: Material model and structural analysis* Model B4 for creep, drying shrinkage and autogenous shrinkage of normal and high-strength concretes with multi-decade applicability. *Mater. Struct.* **2015**, *48*, 753–770.
81. Gardner, N.; Lockman, M. Design provisions for drying shrinkage and creep of normal strength concrete. *ACI Mater. J.* **2001**, *98*, 159–167.
82. *AS3600-2018*; Australian Standard for Concrete Structures. Standards Australia: Sydney, Australia, 2018.
83. *Eurocode 2 and EN 19920101*; Design of Concrete Structures. General Rules and Rules for Buildings. European Committee for Standardization: Brussels, Belgium, 2004.
84. Goel, R.; Kumar, R.; Paul, D. Comparative study of various creep and shrinkage prediction models for concrete. *J. Mater. Civ. Eng.* **2007**, *19*, 249–260. [CrossRef]
85. Al-Manaseer, A.; Prado, A. Statistical comparisons of creep and shrinkage prediction models using RILEM and NU-ITI databases. *ACI Mater. J.* **2015**, *112*, 125–136. [CrossRef]
86. Kataoka, L.; MacHado, M.; Bittencourt, T. Short-term experimental data of drying shrinkage of ground granulated blast-furnace slag cement concrete. *Mater. Struct.* **2011**, *44*, 671–679. [CrossRef]

87. Bazant, Z.; Murphy, W. Creep and shrinkage prediction model for analysis and design of concrete structures—Model B3. *Mater. Struct.* **1995**, *28*, 357–365.
88. Østergaard, L.; Lange, D.; Altoubat, S.; Stang, H. Tensile basic creep of early-age concrete under constant load. *Cem. Concr. Res.* **2001**, *31*, 1895–1899. [CrossRef]
89. D'Ambrosia, M.; Lange, D.A.; Grasley, Z.C. *Measurement and Modeling of Concrete Tensile Creep and Shrinkage at Early Age*; ACI Special Publication; American Concrete Institute: Farmington Hills, MI, USA, 2004; Volume SP-220, pp. 99–112.
90. Dabarera, A.; Li, L.; Dao, V. Experimental evaluation and modelling of early-age basic tensile creep in high-performance concrete. *Mater. Struct.* **2021**, *54*, 130. [CrossRef]
91. Ji, G.; Kanstad, T.; Bjøntegaard, O.; Sellevold, E. Tensile and compressive creep deformations of hardening concrete containing mineral additives. *Mater. Struct.* **2013**, *46*, 1167–1182. [CrossRef]
92. Atrushi, D. Tensile and Compressive Creep of Early Age Concrete: Testing and Modelling. Ph.D. Thesis, The Norwegian University of Science and Technology, Trondheim, Norway, 2003.
93. Zhang, Y.; Afroz, S.; Nguyen, Q.; Kim, T.; Castel, A.; Xu, T. Modeling blended cement concrete tensile creep for standard ring test application. *Struct. Concr.* **2022**, *24*, 2170–2188. [CrossRef]
94. Carlson, R.; Reading, T. Model of studying shrinkage cracking in concrete building walls. *ACI Struct. J.* **1988**, *85*, 395–404.
95. Swamy, R.; Stavrides, H. Influence of fiber reinforcement on restraint shrinkage and cracking. *ACI Mater. J.* **1979**, *76*, 443–460.
96. Bloom, R.; Bentur, A. Free and restrained shrinkage of normal and high-strength concretes. *ACI Mater. J.* **1995**, *92*, 211–217.
97. Altoubat, S.; Lange, D. A New Look at Tensile Creep of Fiber Reinforced Concrete. In *ACI Special Publication on Fiber Reinforced Concrete*; American Concrete Institute: Farmington Hills, MI, USA, 2003.
98. Banthia, N.; Yan, C.; Mindess, S. Restrained shrinkage cracking in fiber reinforced concrete—A novel test technique. *Cem. Concr. Res.* **1996**, *26*, 9–14. [CrossRef]
99. Weiss, J.; Yang, W.; Shah, S. Shrinkage cracking of restrained concrete slabs. *J. Eng. Mech.* **1998**, *124*, 765–774. [CrossRef]
100. Hossain, A.; Weiss, J. Assessing residual stress development and stress relaxation in restrained concrete ring specimens. *Cem. Concr. Compos.* **2004**, *26*, 531–540. [CrossRef]
101. Shen, D.; Shi, H.; Tang, X.; Ji, Y.; Jiang, G. Effect of internal curing with super absorbent polymers on residual stress development and stress relaxation in restrained concrete ring specimens. *Constr. Build. Mater.* **2016**, *120*, 309–320. [CrossRef]
102. See, H.; Attiogbe, E.; Miltenberger, M. Shrinkage cracking characteristics of concrete using ring specimens. *ACI Mater. J.* **2003**, *100*, 239–245.

MDPI AG
Grosspeteranlage 5
4052 Basel
Switzerland
Tel.: +41 61 683 77 34

Materials Editorial Office
E-mail: materials@mdpi.com
www.mdpi.com/journal/materials

www.ingramcontent.com/pod-product-compliance
Lightning Source LLC
Chambersburg PA
CBHW041350210926
51850CB00013B/3041
* 9 7 8 3 7 2 5 8 2 1 3 8 9 *